# Catalytic Methods for the Synthesis of Carbon Nanodots and Their Applications

# Catalytic Methods for the Synthesis of Carbon Nanodots and Their Applications

Editors

Indra Neel Pulidindi
Archana Deokar
Aharon Gedanken

Basel • Beijing • Wuhan • Barcelona • Belgrade • Novi Sad • Cluj • Manchester

*Editors*

Indra Neel Pulidindi
Department of Chemical Sciences
GSFC Univeristy
Vadodara
India

Archana Deokar
Department of Chemical Sciences
GSFC Univeristy
Vadodara
India

Aharon Gedanken
Chemistry
Bar Ilan University
Ramat Gan
Israel

*Editorial Office*
MDPI
St. Alban-Anlage 66
4052 Basel, Switzerland

This is a reprint of articles from the Special Issue published online in the open access journal *Catalysts* (ISSN 2073-4344) (available at: www.mdpi.com/journal/catalysts/special_issues/Q281281448).

For citation purposes, cite each article independently as indicated on the article page online and as indicated below:

Lastname, A.A.; Lastname, B.B. Article Title. *Journal Name* **Year**, *Volume Number*, Page Range.

**ISBN 978-3-0365-9225-1 (Hbk)**
**ISBN 978-3-0365-9224-4 (PDF)**
doi.org/10.3390/books978-3-0365-9224-4

© 2023 by the authors. Articles in this book are Open Access and distributed under the Creative Commons Attribution (CC BY) license. The book as a whole is distributed by MDPI under the terms and conditions of the Creative Commons Attribution-NonCommercial-NoDerivs (CC BY-NC-ND) license.

# Contents

**About the Editors** . . . . . . . . . . . . . . . . . . . . . . . . . . . . . . . . . . . . . . . . . . . . . . . . . . . . . . . . . . . . . . . vii

**Preface** . . . . . . . . . . . . . . . . . . . . . . . . . . . . . . . . . . . . . . . . . . . . . . . . . . . . . . . . . . . . . . . . . . . . . . . ix

**Anjali Banger, Sakshi Gautam, Sapana Jadoun, Nirmala Kumari Jangid, Anamika Srivastava, Indra Neel Pulidindi, et al.**
Synthetic Methods and Applications of Carbon Nanodots
Reprinted from: *Catalysts* 2023, *13*, 858, doi:10.3390/catal13050858 . . . . . . . . . . . . . . . . . . . . 1

**Pradeep Kumar Yadav, Subhash Chandra, Vivek Kumar, Deepak Kumar and Syed Hadi Hasan**
Carbon Quantum Dots: Synthesis, Structure, Properties, and Catalytic Applications for Organic Synthesis
Reprinted from: *Catalysts* 2023, *13*, 422, doi:10.3390/catal13020422 . . . . . . . . . . . . . . . . . . . . 27

**Lerato L. Mokoloko, Roy P. Forbes and Neil J. Coville**
The Behavior of Carbon Dots in Catalytic Reactions
Reprinted from: *Catalysts* 2023, *13*, 1201, doi:10.3390/catal13081201 . . . . . . . . . . . . . . . . . . . 49

**Siti Hasanah Osman, Siti Kartom Kamarudin, Sahriah Basri and Nabila A. Karim**
Anodic Catalyst Support via Titanium Dioxide-Graphene Aerogel ($TiO_2$-GA) for A Direct Methanol Fuel Cell: Response Surface Approach
Reprinted from: *Catalysts* 2023, *13*, 1001, doi:10.3390/catal13061001 . . . . . . . . . . . . . . . . . . . 73

**Siti Hasanah Osman, Siti Kartom Kamarudin, Sahriah Basri and Nabilah A. Karim**
Three-Dimensional Graphene Aerogel Supported on Efficient Anode Electrocatalyst for Methanol Electrooxidation in Acid Media
Reprinted from: *Catalysts* 2023, *13*, 879, doi:10.3390/catal13050879 . . . . . . . . . . . . . . . . . . . 97

**Ravichandran Manjupriya and Selvaraj Mohana Roopan**
Unveiling the Photocatalytic Activity of Carbon Dots/g-$C_3N_4$ Nanocomposite for the O-Arylation of 2-Chloroquinoline-3-carbaldehydes
Reprinted from: *Catalysts* 2023, *13*, 308, doi:10.3390/catal13020308 . . . . . . . . . . . . . . . . . . . 113

**Ruchi Singh, Rajesh K. Yadav, Ravindra K. Shukla, Satyam Singh, Atul P. Singh, Dilip K. Dwivedi, et al.**
Highly Selective Nitrogen-Doped Graphene Quantum Dots/Eriochrome Cyanine Composite Photocatalyst for NADH Regeneration and Coupling of Benzylamine in Aerobic Condition under Solar Light
Reprinted from: *Catalysts* 2023, *13*, 199, doi:10.3390/catal13010199 . . . . . . . . . . . . . . . . . . . 131

**Vaibhav Gupta, Rajesh K. Yadav, Ahmad Umar, Ahmed A. Ibrahim, Satyam Singh, Rehana Shahin, et al.**
Highly Efficient Self-Assembled Activated Carbon Cloth-Templated Photocatalyst for NADH Regeneration and Photocatalytic Reduction of 4-Nitro Benzyl Alcohol
Reprinted from: *Catalysts* 2023, *13*, 666, doi:10.3390/catal13040666 . . . . . . . . . . . . . . . . . . . 147

**Gabriela Rodríguez-Carballo, Cristina García-Sancho, Manuel Algarra, Eulogio Castro and Ramón Moreno-Tost**
One-Pot Synthesis of Green-Emitting Nitrogen-Doped Carbon Dots from Xylose
Reprinted from: *Catalysts* 2023, *13*, 1358, doi:10.3390/catal13101358 . . . . . . . . . . . . . . . . . . . 165

**Chau Thi Thanh Thuy, Gyuho Shin, Lee Jieun, Hyung Do Kim, Ganesh Koyyada and Jae Hong Kim**
Self-Doped Carbon Dots Decorated TiO$_2$ Nanorods: A Novel Synthesis Route for Enhanced Photoelectrochemical Water Splitting
Reprinted from: *Catalysts* **2022**, *12*, 1281, doi:10.3390/catal12101281 . . . . . . . . . . . . . . . . . **181**

**Somasundaram Chandra Kishore, Suguna Perumal, Raji Atchudan, Thomas Nesakumar Jebakumar Immanuel Edison, Ashok K. Sundramoorthy, Muthulakshmi Alagan, et al.**
Eco-Friendly Synthesis of Functionalized Carbon Nanodots from Cashew Nut Skin Waste for Bioimaging
Reprinted from: *Catalysts* **2023**, *13*, 547, doi:10.3390/catal13030547 . . . . . . . . . . . . . . . . . **195**

**Madushmita Hatimuria, Plabana Phukan, Soumabha Bag, Jyotirmoy Ghosh, Krishna Gavvala, Ashok Pabbathi, et al.**
Green Carbon Dots: Applications in Development of Electrochemical Sensors, Assessment of Toxicity as Well as Anticancer Properties
Reprinted from: *Catalysts* **2023**, *13*, 537, doi:10.3390/catal13030537 . . . . . . . . . . . . . . . . . **209**

# About the Editors

**Indra Neel Pulidindi**

Dr. Indra Neel Pulidindi received his PhD degree from the Indian Institute of Technology Madras in 2010 under the supervision of Professor T K Varadarajan and Professor (Em) B Viswanathan. He worked in the laboratory of Professor Aharon Gedanken from 2010 to 2016 in Israel on research related to biomass conversion of biofuels and biochemicals. From 2016 to 2017, he worked in the laboratory of Professor Tae Hyun Kim at Hanyang University and conducted systematic studies on biomass composition analysis and conversion of biomass to biochemicals. Subsequently, he worked in the laboratory of Professor Xinling Wang at Shanghai Jiao Tong University on carbon fiber-reinforced materials. Dr. Neel has 49 research papers, 1 patent, 4 patent applications, 1 book, 1 ebook and 6 book chapters to his credit. Dr. Neel has supervised several PhD, master, and undergraduate students in their academic research curriculum and helped them earn their degrees. Dr. Neel is currently working as an assistant professor at GSFC University, Vadodara.

**Archana Deokar**

Dr. Archana Deokar is currently working as an Assistant Professor of Chemistry at GSFC University, Vadodara. Her research interests include synthesis of magnetic nanoparticles and metal or metal oxide nanoparticles anchored on graphene and carbon nanotubes for biomedical applications. She earned her PhD degree from National Tsing Hua University in 2013. She gained postdoctoral experience by working with Professor Aharon Gedanken from 2013 to 2015 at Bar-Ilan University, Israel. She has published eight papers in peer-reviewed scientific journals of international repute, one US patent and two book chapters. She is a recipient of a Taiwan Government fellowship.

**Aharon Gedanken**

Professor Aharon Gedanken is an outstanding and well-known scientist with over 900 papers published in peer-reviewed journals of international repute and with a high impact factor. He has 37 patent applications, 1 book on biofuels, and over 10 book chapters to his credit. Professor Gedanken has made remarkable contributions to the fields of sonochemistry and microwave technology and their application to nanomaterials, biomaterials and biofuels. He served as a faculty member in the Department of Chemistry at Bar-Ilan University for over 34 years (1975–2009) and has been an Emeritus Professor at the same institute for over 13 years (2009–present). Professor Gedanken's research metrics, namely, H-index and citations, are 117 and 51510, respectively. His research interests include solid-state chemistry, catalysis, energy, materials science and biochemistry.

# Preface

Carbon quantum dots can be designated as a new allotropic form of carbon materials. Carbon materials are ever-interesting arousing materials, and it is no wonder that the works related to two of the previously known allotropic forms of carbon, namely, fullerenes and graphene, have won two Nobel prizes. In 1996, Robert F. Curl Jr., Sir Harold Kroto and Richard E. Smalley were jointly awarded the Nobel prize in Chemistry for their discovery of fullerenes. Likewise, in 2010, Andre Geim and Konstantin Novoselov were jointly awarded the Nobel prize in Physics for their groundbreaking work on the two-dimensional carbon material called graphene. However, much remains to be explored in carbon materials research. The discovery of carbon dots, the newest allotrope of carbon, is no ordinary discovery. Unlike their predecessors, carbon dots are hydrophilic, water-soluble, nano-sized and spherical. They are also highly polar and functionalized. Carbon dots exhibit peculiar light absorption and emission properties. They are susceptible to tuning in terms of size, composition, surface functionality, light absorption and emission properties. They have found applications in almost all spheres of human activity. The applications of carbon dots in the fields of catalysis, electrocatalysis, photocatalysis, photoelectrocatalysis, medicine and materials science are well known. Many new applications are being unravelled. The Editors would like to express their gratitude to the research groups that have contributed their papers that formed the 12 chapters in this edited reprint. Grateful thanks are due to Mrs. Cathy Yang, the Editorial Manager, for her steadfast support to the Editors. Finally, we bow down before our Lord and Savior Jesus Christ for His all-sufficient grace that has enabled the successful completion of this endeavor.

Dedicated to "My LORD and my God, Jesus Christ. John 20:28" "My grace is sufficient for three. 2 Corinthians 9:12"

**Indra Neel Pulidindi, Archana Deokar, and Aharon Gedanken**
*Editors*

*Review*

# Synthetic Methods and Applications of Carbon Nanodots

Anjali Banger [1], Sakshi Gautam [1], Sapana Jadoun [2], Nirmala Kumari Jangid [1,*], Anamika Srivastava [1], Indra Neel Pulidindi [3], Jaya Dwivedi [1] and Manish Srivastava [4,*]

[1] Department of Chemistry, Banasthali Vidyapith, Banasthali 304022, India
[2] Facultaed de Ciencias Quimicas, Departamento de Quimica Analitica e Inorganica, Universidad de Concepcion, Edmundo Larenas 129, Concepcion 4070371, Chile
[3] School of Science, GSFC University, Vadodara 391750, India
[4] Department of Chemistry, Central University of Allahabad, Allahabad 211002, India
* Correspondence: nirmalajangid.111@gmail.com (N.K.J.); msrivastava@allduniv.ac.in (M.S.)

**Abstract:** In the recent decade, carbon dots have drawn immense attention and prompted intense investigation. The latest form of nanocarbon, the carbon nanodot, is attracting intensive research efforts, similar to its earlier analogues, namely, fullerene, carbon nanotube, and graphene. One outstanding feature that distinguishes carbon nanodots from other known forms of carbon materials is its water solubility owing to extensive surface functionalization (the presence of polar surface functional groups). These carbonaceous quantum dots, or carbon nanodots, have several advantages over traditional semiconductor-based quantum dots. They possess outstanding photoluminescence, fluorescence, biocompatibility, biosensing and bioimaging, photostability, feedstock sustainability, extensive surface functionalization and bio-conjugation, excellent colloidal stability, eco-friendly synthesis (from organic matter such as glucose, coffee, tea, and grass to biomass waste-derived sources), low toxicity, and cost-effectiveness. Recent advances in the synthesis and characterization of carbon dots have been received and new insight is provided. Presently known applications of carbon dots in the fields of bioimaging, drug delivery, sensing, and diagnosis were highlighted and future applications of these astounding materials are speculated.

**Keywords:** carbon nanodots; synthesis; applications; surface functionality; biocompatibility; low toxicity; bioimaging; applications

## 1. Introduction

Nanoparticles are microscopic particles with a size range of 1–100 nm. During the past decade, considerable research was conducted on the fabrication and application of nanoparticles in many fields. Based on their unique properties, nanoparticles have a substantial impact in various industries, including health, cosmetics, energy, pharmaceuticals, and food.

Enormous work was completed in recent years to design nanostructured materials with specific characteristics that will ultimately influence their function and application. In this era of carbon nanotechnology, special emphasis is laid on the organic functionality of nanomaterials or organic nanomaterials, including graphene, carbon nanotubes, and fullerenes. Because of their biocompatibility, ease of fabrication, and fascinating features, especially their water solubility fluorescence emission, carbon nanodots with a size in the range of 1–10 nm have taken the central stage of materials research. Carbon nanodots (CDs) are known to have zero dimension with almost spherical geometry. This material has become a rising star in the field of luminescent nanomaterials [1]. Due to their desirable qualities, such as hydrophilicity, ease of functionalization, outstanding biocompatibility, bright luminescence, good solubility, high chemical inertness, and low toxicity, they are potent candidates for various applications in solar cells, biosensors [2–10], bioimaging, and optoelectronic devices, etc. CNDs exhibit many remarkable properties including outstanding photoinduced electron transfer, stable chemical

inertness, low cytotoxicity [11–15], good biocompatibility, and efficient light harvesting. Dots made of carbon, such as carbon nanodots (CDs) and graphene quantum dots (GQDs), are a brand-new carbonaceous nanomaterial with zero dimensions [16–22]. Until now, a lot of work has been carried out, and substantial advances have been made in the synthesis and uses of carbon-based dots [23–27].

This class of carbon-based nanomaterials was initially found by the top-down approach of minimizing huge carbon nanomaterials, and has recently advanced at a startling rate. Their purity standards, classification, and fluorescence mechanisms have made an impact on the research community, as evident from the rapid pace of research and publications in this area. The emphasis of the review is mainly on the synthesis and applications of carbon nanodots. A schematic of the methods of synthesis of CNDs and their structure was depicted in Figures 1 and 2, respectively. They usually have the inner hybridization of $sp^2$ and outer hybridization of $sp^3$, and these hybridized structures tend to have functional groups containing oxygen atoms (such as -OH, -COOH, -CO, and many more).

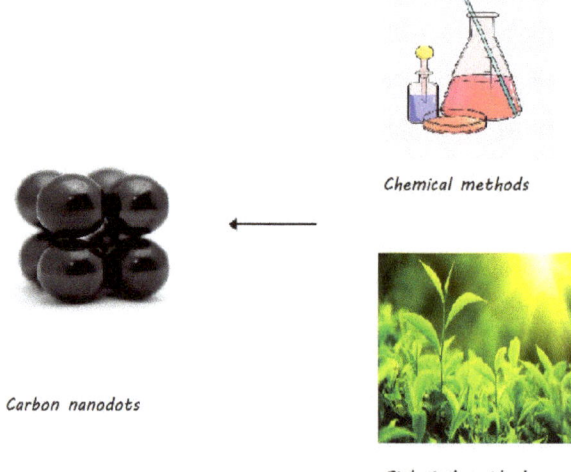

Figure 1. Methods used for the synthesis of Carbon nanodots (1–10 nm).

Due to their excellent characteristics, CNDs are prospective replacement probes for bioimaging and bioassay [28–33]. So far, a number of methods have been investigated to synthesize CNDs, including thermal oxidation, chemical oxidation, and arc discharge, laser ablation of graphite, electrochemical synthesis, microwave synthesis, and ultrasonic methods [34–40]. The use of most of these techniques are, however, constrained since they sometimes require an expensive carbon source, intricate reactions, lengthy process time, and post-treatment steps. Therefore, there is a significant demand for synthetic methods that are easy to use, sustainable, and ecologically friendly for the mass production of high-quality CNDs [41].

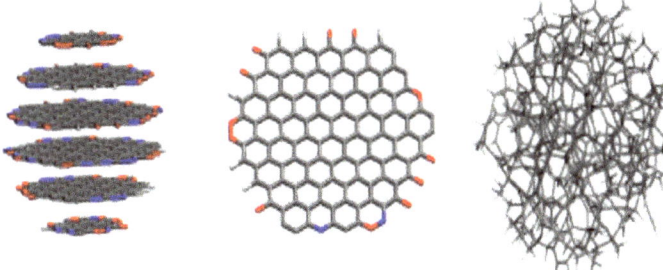

**Figure 2.** Commonly available carbon nanodot structures: spherical particles, nanosheets of graphene, and amorphous structures. Reproduced with the permission from Ref. [41], 2019, American Chemical Society.

There are not many reviews focused explicitly on the synthetic pathways, characteristics, and uses of the CDs, despite the fact that they have all been thoroughly summarized elsewhere [42–45]. For improved clarity and understanding, a general evaluation of the most recent developments in the synthetic methods of CNDs is presented. Moreover, the advances in the understanding of the properties of carbon dots and the resulting applications were highlighted, and a new insight is provided by correlating the synthetic strategy, property, and application [46–50]. The organization of the review comprises a discussion of the approaches for the fabrication of carbon nanodots for various sustainable resources (including biomass), with an emphasis on cost-effectiveness and eco-friendliness. Afterward, the major applications of CNDs were highlighted, with an emphasis on bioimaging and photocatalysis.

## 2. Novel Methods for the Synthesis of Carbon Nanodots

During the last ten years, namely 2013–2023, many methods were developed to synthesize carbon nanodots with attractive features and applications in a specific field. These well-known CD synthesis techniques are typically categorized into "top-down" and "bottom-up" categories. The top-down techniques involve the exfoliation of nanomaterials chemically, laser ablation, electrical and chemical oxidation, arc discharge, and ultrasonic synthesis. Graphene quantum dots, or two-dimensional nanomaterials, are often produced via a "top-down" technique by exfoliating and cutting the macroscale framework of carbon species, such as carbon rods, tubes, graphite powder, activated carbon, carbon black, carbon soot, and carbon fibers possessing the graphene lattices. Top-down strategies typically demand a lengthy processing time, challenging reaction environments, and expensive materials and machinery [51–53], and these methods work well for the mass production of CNDs. On the contrary, the bottom-up method is used for producing carbonized polymer dots and carbon quantum dots (3D nanoparticles with spherical centers) by polymerizing molecular precursors, including glucose, sucrose, and citric acid, using processes such as chemical vapor deposition, plasma treatment, microwave pyrolysis, and solvothermal reactions, with a high degree of controllability. Of course, these methods are not flawless altogether [54].

Hawrylak et al. [55] synthesized carbon nanodots (CNDs) for the first time by surprise in 2004 in their effort to purify single-walled carbon nanotubes. Arc-discharge soot was used in the experiment as a source of carbon nanotubes. The components of the soot suspension were separated during the procedure using gel-electrophoresis, which exposed a brand-new band of fluorescent material. Currently, CNDs can be synthesized by chemical or physical processes. Thermal therapy, electrochemistry, acidic or hydrothermal oxidation, or ultrasonic treatment are examples of chemical processes. Arc discharge, plasma therapy, and laser ablation are examples of physical approaches [56]. The processes for synthesizing CNDs can be roughly categorized as top-down or bottom-up syntheses, as mentioned earlier [57–66]. Top-down techniques often include etching, intercalation, hydrothermal or solvothermal cutting, chemical oxidation, and laser ablation to break down a larger carbon structure into progressively smaller pieces. In bottom-up methods, CNDs are synthesized by carbonizing organic precursors. These include dehydration with sulfuric acid, microwave pyrolysis with solvent mediation, and refluxing pyrolysis. Using solvothermal or direct thermal breakdown, organic precursors such as allotropic carbon forms, natural gas, and carbohydrates are transformed into CNDs using direct thermal or solvothermal breakdown [67]. To achieve uniformity, the final product can be processed through electrophoresis, chromatography, centrifugation, dialysis, or through some other processes [68]. The carbon precursor, preparation technique, and experimental circumstances have a significant impact on the shape and structure of CNDs. For instance, depending on the parent material, CNDs made using top-down techniques have a different size and shape (coal, graphite powder, or graphene nanosheets). UV radiation or hydrothermal treatments can be used in a single preparation technique such as etching, and both would yield different outcomes. CNDs produced from the top-down approach would typically have dimensions below 10 nm, a spherical or sheet-like shape, and a size below 3 nm. Data indicate that bottom-up synthesized CNDs may be layered with a size of less than 10 nm.

Arc discharge and laser ablation are likely the most well-known top-down techniques for producing carbon-based nanomaterials. The term "arc discharge" refers to producing a current between two electrodes, often graphite rods, which causes them to vaporize. As a result, soot is formed, which may contain various nanoparticles of carbon. In contrast, the technique of laser ablation comprises applying a pulse of laser energy to a solid surface, resulting in carbon nanomaterials. Laser ablation in solution (LAS) has attracted interest as a top-down, single-step method for producing nanomaterials quickly and affordably, as shown in Figure 3 [69].

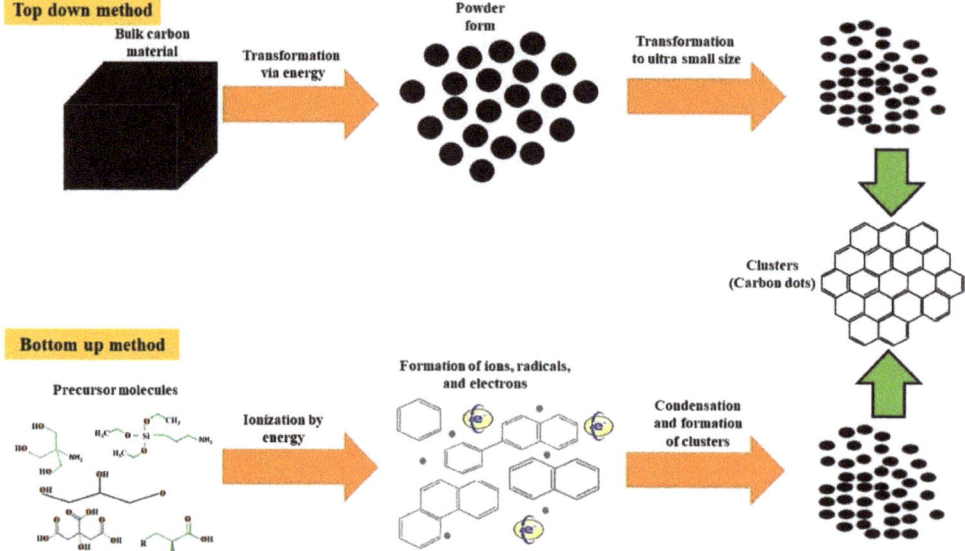

**Figure 3.** General methods of synthesis carbon nanodots (CNDs: Bottom-up approach: CNDs were synthesized from smaller carbon units (small organic molecules) by applying energy (electrochemical/chemical, thermal. laser, microwave, etc.). The source molecules will get ionized, dissociated, evaporated, or sublimated and then condensed to form CNDs; Top-down approach: CNDs are synthesized by transformation of larger carbon structures into ultra-small fragments by applying energy (thermal, mechanical, chemical, ultrasonic, etc.). Reproduced with the permission from Ref. [69], 2019, Springer Nature.

Amorphous (a) nanoparticles (a-CDs) are often produced at relatively low temperatures (<300 °C), whereas graphitized (g) structures are produced at higher temperatures (g-CDs). The surface functional groups of the resulting nanoparticles are significantly influenced by the precursors used. The most prevalent surface functional groups are amine and carboxylate, using precursors such as citric acid or polyamines. Such functionality can be generated via post-synthetic functionalization reactions. At the beginning of the research in this area, researchers found that proteins were the suitable precursors for the synthesis of carbon nanodots because they were readily available, affordable, and capable of undergoing dehydration and decarboxylation reactions to produce CNDs with heteroatom doping [70,71]. N-doping is accountable for better photoluminescence quantum yields, red-shifted absorption, and more favorable optoelectronic features. Citric acid in combination with amino acids (arginine, Arg) was also explored, leading to the formation of carbogenic nanoparticles' molecular precursor, which benefits from its distinct reactive behavior and capacity to serve as a "passivating agent" or "capping agent" for the outer surface [72,73].

The hydrothermal approach, aided by microwave heating, is one among numerous potential synthesis processes. It has been widely employed for manufacturing a variety of carbon materials. Hydrothermal synthesis offers a minimal toxicological impact on materials and processes [74]. The use of hydrothermal conditions causes the reagent's solubility to rise or change, enhances their chemical and physical interactions, and makes it easier for the carbonaceous structures to form. The production of nanomaterials with higher amounts of carbon, such as graphitic carbon compounds, and nanotubes at higher temperatures is a reliable process. Microwave-assisted methods have also grown in popularity as a method for synthesizing nanomaterials. Issues with the conventional heating process used for preparing nanomaterials, such as the tendency for insoluble compounds to cause heterogeneous heating, leading to an increase in the size of nanomaterials, is solved by microwave heating [75,76]. Due to its high energy consumption efficiency, MW irradiation offers a safe, inexpensive, and practical mode of heating, producing higher yields of the desired products [77]. As a result, the MW-assisted hydrothermal approach, which combines the benefits of both MW and hydrothermal processes, has become essential for the production of carbon dots.

The carbonization of small molecule precursors is achieved in the bottom-up fabrication of carbon dots. The most commonly available methods for the fabrication of carbon dots via the bottom-up method include a mixture of molecules having nitrogen atoms (such as urea) and citric acid [78–84]. The pyrolysis of these molecular precursors in an autoclave or microwave forms a black nanopowder of CDs. These CDs are easily dissolved in water and have exceptional fluorescent characteristics. These CDs are capable of emitting blue [85], green [86], and red emissions depending on their surface properties and circumstances (excitation source of radiation) [87,88], albeit a thorough purification is frequently required to separate the carbon dots [89]. When it comes to top-down methods, the starting materials include carbon structures such as amorphous activated carbon, carbon fibers, graphite, nanotubes, and fullerenes that are physically or chemically fragmented to produce very small carbon nanoparticles [90–93]. One such instance is graphitic oxidation in an extremely acidic environment [94,95], enabling the surface to be functionalized and the bulk-precursor to be broken up, resulting in the optical characteristics that are typical of CDs. Top-down synthetic methods frequently produce CDs with lower quantum yields of emission. These top-down synthetic methods of CDs are far more complex and time-consuming. However, they enable better structural control and end-product purity [96,97].

*2.1. Sonochemical/Ultra-Sonic Fabrication of CDs*

Sonochemistry is exploited for the synthesis of nanostructured materials. In this acoustic activation technique, severe physical and chemical conditions are generated as a result of the use of high-intensity ultrasound [98–106]. As the molecular dimensions are smaller than acoustic wavelengths, the chemical features of the resulting materials are not a result of the interaction between the ultrasonic and chemical species in the liquid state. Sonochemical fabrication is the result of the high compression heating of gas and vapor, resulting in incredibly high temperature and pressure conditions [107].

For the first time, Zhuo et al. [108] described the synthesis of graphene quantum dots using the ultra-sonic exfoliation of graphene. A simple sonochemical approach for the production of extremely photoluminescent CDs was devised by Wei et al. [109] in 2014. High-intensity ultrasound forms collapsing bubbles that serve as microreactors and offer intense, momentary conditions ideal for the pyrolysis of carbon precursors. Sono-chemically produced CDs in the presence of surface passivation agents have a high quantum yield and outstanding photostability.

The ultrasonic technique has merits of being inexpensive and easy to operate for the synthesis of carbon dots. The ultrasound method involves alternate high-pressure and low-pressure waves, which cause small bubbles in liquid to form and break. Thus, by means of powerful hydrodynamic shear forces resulting from the cavitation of tiny bubbles, macroscopic carbon materials were reduced to nanoscale CDs. Generally, the ultrasonic power, reaction time, and solvent and carbon source ratio were altered to produce CDs with various properties.

Huang et al. [110] used a direct ultrasonic exfoliation method to produce the chlorine-infused graphene quantum dots. Park et al. [111] conducted a typical experiment in which they first produced water-soluble CQDs from food waste. From ethanol and food waste mixture, approximately 120 g of carbon dots with an average diameter of 2–4 nm can be produced. The benefits of the as-prepared CDs for in vitro bioimaging include photostability, low cytotoxicity, and good PL characteristics.

Jiang et al. [112] reported a one-pot eco-friendly fabrication of silver nanoparticles supported on carbon from carbon and silver nitrate liquid solution without additional capping or reducing agents. Simply altering the molar ratio of carbon nanodots and $AgNO_3$ would change the size of the silver nanoparticles supported on carbon. The simple method used to create amorphous carbon-supported Ag NPs was significant because it is synthesized in the absence of reducing or capping agents. Moreover, the stability and electrical and catalytic activities of the silver nanoparticles for the electrocatalytic reduction of $H_2O_2$ were also enhanced. The great sensitivity and low detection limit of the Ag/C nanocomposites make them excellent non-enzymatic $H_2O_2$ sensors. The Ag/C nanocomposites acted as a non-enzymatic $H_2O_2$ sensor due to their high sensitivity and selectivity, as shown in Scheme 1.

Manoharan et al. [113] used an easy and affordable technique to turn coconut water into vivid eco-friendly fluorescent carbon nanodots. Coconut water, as an environmentally friendly and less expensive carbon precursor, is used to fabricate finely dispersible carbon nanodots with both the amorphous and nano-crystalline carbon-phase. High-resolution transmission electron microscopy was used to demonstrate the monodispersed CNDs' spherical shape, ($4 \pm 1$ nm) nm particle size. FT-IR measurements revealed extensive surface functionalization. Using UV-visible absorption and photoluminescence spectroscopic techniques, the eco-friendly luminescent properties of the carbon nanodots were assessed. The fluorescence quantum yield of the carbon nanodots having a core of carbon with extensive surface functionalization was found to be 60.18%. These CNDs, fabricated from tender coconut water, are more favorable than those from other resources due to their stability, high quality, fast reaction rate, and fine dispersion.

Currently, microalgae productivity has been increasing worldwide. To take advantage of the situation, Choi et al. [114] set out to show that the aqueous-type biofriendly luminescent carbon nanodots (C-paints) could be successfully applied for enhancing the growth rate of microalgae, *Haematococcus pluviali*. A straightforward procedure of ultrasonic irradiation with the passivating agent, polyethylene glycol, was used to prepare C-paints. The end product, called a C-paint, has a carbonyl-rich surface, outstanding particle size homogeneity, high water-solubility, photo-stability, fluorescence efficacy, and biocompatibility.

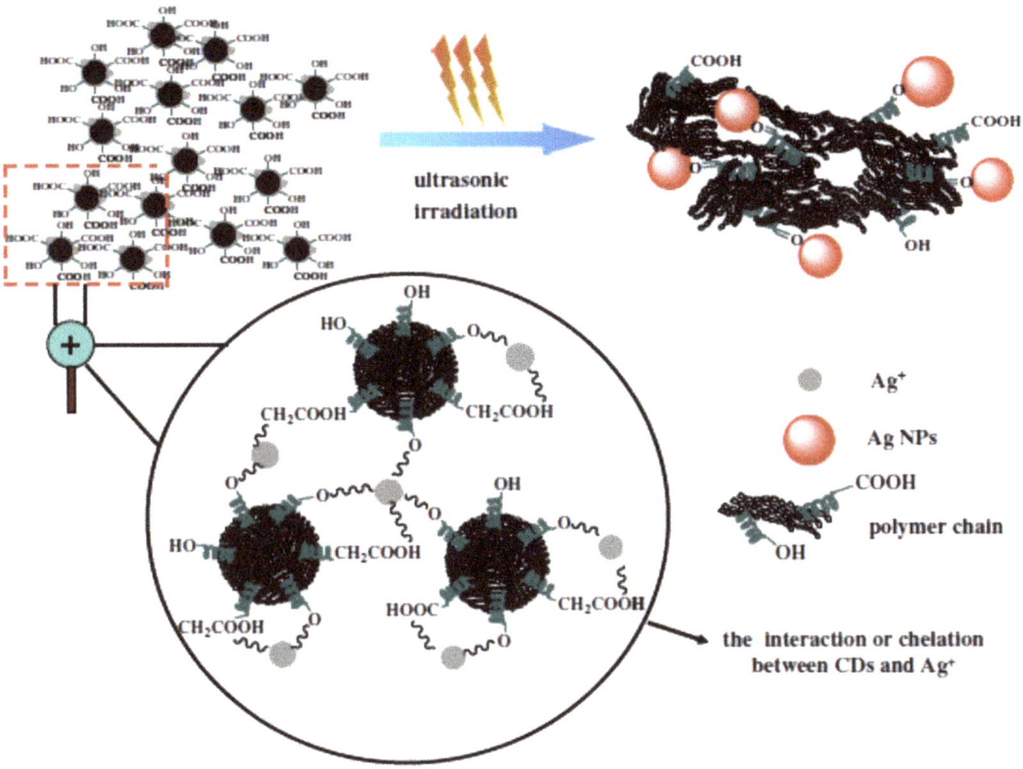

**Scheme 1.** Schematic view of the formation of carbon-supported silver nanoparticles. Reproduced with permission from Ref. [112]. 2014, Elsevier.

Using polymer dots enclosed in NIR emissive hydrophobic carbon nanodots, Huang et al. [115] proposed the first endoplasmic reticulum (ER) focused nearinfrared (NIR) nanosensor for detecting $Cu^{2+}$ in biosystems. With a detection limit of 13 nM, this nanosensor with stable fluorescence can be utilized to quantify $Cu^{2+}$ in a linear range from 0.25 to 9.0 M. It responded quickly to $Cu^{2+}$ (120 s). In addition, compared to other metal ions and amino acids, the nanosensor's fluorescence fluctuations are extremely selective to $Cu^{2+}$. Moreover, the developed nanosensor showed low cytotoxicity, superior biocompatibility, and ER targeting capability [116].

*2.2. Hydrothermal Synthesis*

The hydrothermal method for producing CDs is inexpensive and non-toxic. This is an easy method for creating carbon quantum dots compared to other synthetic approaches. Teflon lined stainless steel autoclaves are used as reaction vessels for the aqueous solution of the CD precursor and chemical agents. The autoclave is then placed in an air oven where the contents are hydrothermally reacted at a high pressure and high temperature to form the CDs [117–123]. Mehta et al. [124] developed a plant-based (sugarcane juice) source for producing luminescent carbon quantum dots, which are soluble in an aqueous medium with a size of less than 5 nm. These CQDs were used for the sensitive and specific detection of $Cu^{2+}$. In their proposal, Lu et al. [125] suggested the production of carbon quantum dots via this method from pomelo peel having the size of less than 5 nm. The fabricated carbon quantum dots, which had remarkable yield, were used for the sensitive detection of $Hg^{2+}$ at lower concentrations for the examination of water samples collected from the lake.

Li et al. [126] successfully manufactured CDs in 2018 using a one-step hydrothermal process that was ecologically friendly, easy, and affordable. Oxidation resistance, stability, excellent solubility, and high quantum yield (18.67%) were all features of the CDs. It was discovered that a charge transfer procedure could cause picric acid (PA) to quench the CDs' early fluorescence. These CDs worked well as fluorescent probes to identify PA in our study. The technique was successfully used on actual and laboratory-derived tap water samples, and it was found to have beneficial properties, such as an outstanding selective nature, excellent sensitive nature, and lower detection limit of 10 nM.

Gao et al. [127] proposed a simple and inexpensive technique in which they showed the coupling of graphene quantum dots with carbon nitride (hexagonal-structure) via freeze-drying. The result showed enhanced photocatalytic activity, improved absorption in the visible region, and the effective separation of photon-generated electron-hole pairs. We have successfully completed the simple synthesis of B/N co-doped, fluorescent surface passivated carbon nanodots with a high quantum yield at a low cost, as reported by Jahan et al. [128]. Further employing these carbon dots results in the production of supramolecular moieties, which turns off fluorescence before being turned on by SR III.

For the first time, Soni et al. [129] have shown the precise source of light absorption and its emission of carbon nanodots. They demonstrated that molecular fluorophore, which is generally found in the fabrication mixture as a by-product, is the true source of the emission in red emissive carbon nanodots.

Using p-phenylenediamine and urea, Ding et al. [130] developed multiple multicolor-emitting carbon nanodots via this process. After being purified using column chromatography, the carbon nanodots were obtained without excitation, causing fluorophores to show different colors. Along with a progressive increase in the red-shifted fluorescence emission, the oxidation on the surface of the carbon nanodots also became enhanced. The band width narrows as the oxygen content of the carbon nanodots' surface increases; as a result, the higher level of surface oxidation causes the red-shifted emission.

Bakier et al. [131] proposed a new turn-off fluorescent chemical sensor for the ultra-sensitive detection of aniline in the liquid phase, via the formation of colloidal carbon nanodots supported on nitrogen. They demonstrated a susceptible fluorescent aniline liquid sensor based on incredibly tiny carbon dots supported on nitrogen (Figure 4).

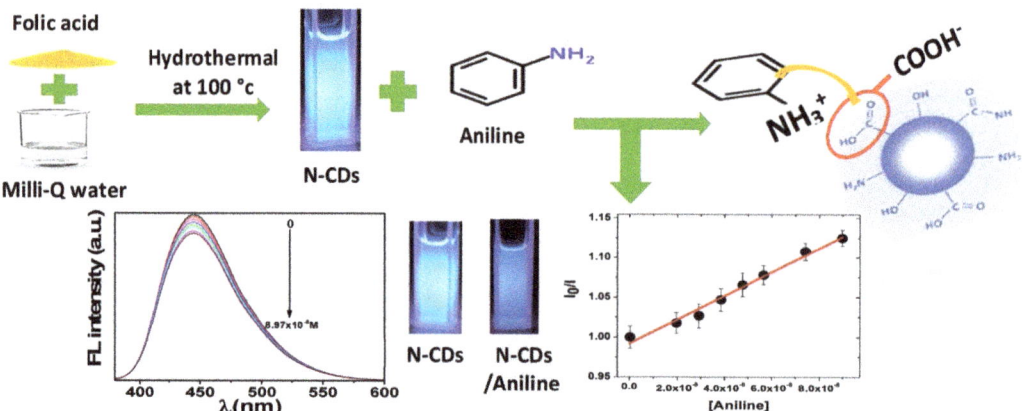

**Figure 4.** Fluorescent chemical sensor for ultra-sensitive detection of aniline in the liquid phase, via formation of colloidal carbon nanodots supported on nitrogen. Reproduced with permission from Ref. [131]. 2021, Elsevier.

The carbon dots supported on nitrogen were produced using folic acid that had undergone ultrasonic processing at lower temperatures. They further discovered that the sensor's operation followed the static N-CD fluorescence quenching caused by electrostatic contact with aniline. The detection limit for aniline via any method was 3.75 nM (0.332 ppb), which is the sensor's detection limit. Furthermore, real sample analysis was investigated using the N-CDs' nano-probe with real tap water, and excellent results were obtained with 99.7–101% recovery (Figure 5). Hence, this proposal could prove to be helpful in developing a simple and environmentally benign nano-sensing process with excellent sensitivity, outstanding selectivity, and good quantitative value to monitor harmful aniline against the degradation of the environment [131].

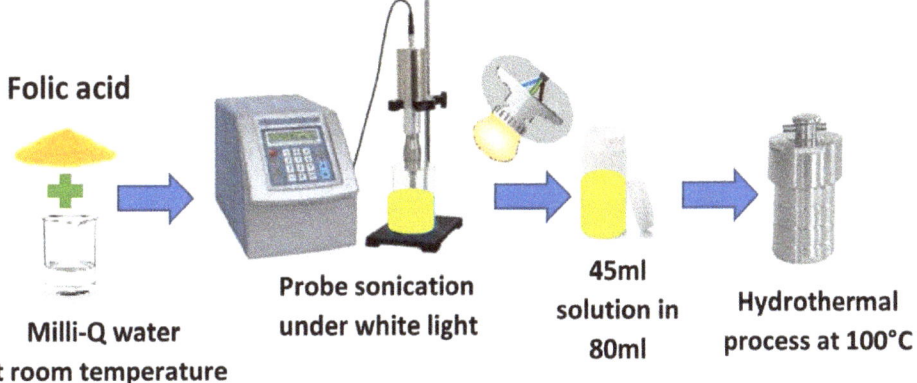

**Figure 5.** Schematic view of the experimental procedure. Reproduced with permission from Ref. [131]. 2021, Elsevier.

## 2.3. Carbonization/Pyrolysis

In recent years, pyrolysis has emerged as a potent technique for producing fluorescent CDs by using precursors that are microscopic carbon structures. Short reaction times, minimal costs, simple procedures, the absence of any solvent, and high quantum yields are all benefits of this technology. Under high temperatures, the following basic processes of heating, dehydrating, degrading, and carbonization are essential for converting the molecules with organic carbon into carbon quantum dots. During the pyrolysis process, strong concentrations of alkali perform the cleavage of carbon initiators into carbon nanoparticles.

Ma et al. [132] produced nitrogen-doped graphene quantum dots by directly carbonizing ethylene diamine tetra acetic acid at 260–280 °C, and this study also offered a growth mechanism for GQDs. It is important to note that ion doping has been found to produce a variety of CQD kinds. Li and colleagues created chlorine-doped graphene quantum dots by using HCl and fructose as precursors in a typical experiment. The average size of the quantum dots was found to be 5.4 nm. They altered the color of the emission by alternating the excitation wavelength from 300 to 600 nm, and the color changed from blue to red, respectively [133]. The fluorescent carbon quantum dots were also made by Praneerad et al. [134] by carbonizing the durian peel biomass waste. The produced CQDs were used to develop a composite-based electrode that displayed a significantly greater specific capacitance value as compared to the electrode made of pure carbon. According to 135.Zhang et al. [135], the quantum dots of carbon with increased sulphur and nitrogen contents were created by carbonizing hair fiber combined with $H_2SO_4$ through sonication.

Gunjal et al. [136] used a straightforward carbonization process to create waste tea residue carbon dots from surplus and inexpensive kitchen waste biomass, so that it is cheaper, greener, and more environmentally friendly than previous techniques. As soon as they are created, waste tea residue carbon dots exhibit excitation dependent emission and are very stable in ionic media. Furthermore, due to the oxidative nature of the ion, it has demonstrated excellent fluorescence quenching for ClO⁻. The fabricated sensor has the advantage in that it is highly sensitive and selective in comparison to the other 21 common interfering ions which were tested against it. Its detection limit is comparatively lower than other biomass-made carbon dots. Its quick rate of reaction enables easy and feasible ClO⁻ detection in real samples with excellent precision and reliability. A simple method to covalently immobilize nanoscale carbon dots upon conducting carbon surface for sensing purposes is reported by Gutiérrez-Sánchez et al. [137]. The carbon nanodots (N-CD) containing amine functionalization on the surface can be electro-grafted upon the electrodes of carbon, where they are then readily covalently immobilized. They were made using a carbonization approach with microwave aid and cost-effective, biocompatible initiators, such as D-fructose as the primary carbon source and urea as the N-donor reagent, to produce peripheral enhanced nitrogen CD. It has been determined through various methods of analysis that the synthesized nanomaterial comprises regular-sized amorphous structures that glow blue when exposed to UV light. Through the relatively stable immobilization of nitrogen carbon dots onto the electrode surfaces through electrografting, hybrid electrodes with higher relative surface areas and enhanced electron transfer capacities are generated, holding great potential for electrochemical sensing. Because of their conductive nature, electrical properties, abundant edges sites, and high catalytic activity, N-CDs that are immobilized on carbon electrodes efficiently amplify the electro-chemiluminiscence (ECL) signal from the luminophore $[Ru(bpy)_3]^{2+}$ in a taurine sensor.

*2.4. Electrochemical Synthesis*

The one-pot electrochemical method is used to controllably synthesize fluorescent or luminescent carbon nanodots (C-dots) from small molecular alcohols as a single carbon source for the first time. By adjusting the applied potential, it is possible to control the size of the resulting C-dots, which can then be used to image cells using luminescence microscopy.

A titanium tube cathode and a pure graphite loop electrode were assembled in the center, according to Pender et al. [138]. Distilled water was used for synthesizing both the electrolyte and the cathode while the anode was isolated from them by an insulating O-ring. Luminescent blue-colored carbon dots were broadly employed in pure water thanks to the use of electronic voltage and ultrasonic control, which eliminated the need for laborious cleaning. The amount yield was 8.9%, while the size of the synthesized C-dots was 2–3 nm. The C-dots offered good fluorescent properties and thermodynamic constancy in the aqueous phase. Fluorescent CDs were made from ethanol by electrochemical carbonization, according to Miao et al. [139]. The synthesizing procedure is easy, economical, and environmentally benign. The synthesized carbon dots were amorphous, spherical, and easily dispensable in water, making them ideal for analytical uses. A strong fluorescence intensity with a QY of 10.04% was attained in the absence of a surface passivation reagent. By identifying $Fe^{3+}$ induced fluorescence quenching, carbon nanodots were successfully used for the $Fe^{3+}$ test.

Keerthana and Ashraf [140] highlighted the hydrothermal carbonization approach for synthesizing carbon dots from chitosan. Chitosan was totally transformed into carbon dots, according to an analysis using UV-Visible spectroscopy. With one step microwave synthesis, Arvapalli et al. [141] were able to synthesize carbon nanodots that had remarkable selectivity and sensitivity for the detection of Fe (III) ions. The synthesized carbon nanodots exhibit excellent stability, high photoluminescence, and strong water solubility. Bright blue fluorescence from carbon nanodots was successfully internalized inside endothelial cells, and when the cells were nurtured with iron, the fluorescence quenching phenomenon was seen, demonstrating the possibility of sensing iron in living cells. The transfer of charge specifically between the carbon nanodots and iron was responsible for the fluorescence quenching of the carbon nanodots, and cyclic voltammetry experiments have further confirmed this.

Tyrosinase was immobilized on carbon-based nanoparticles and cysteamine (electrically active layer) covering the gold electrode in the small gold-epinephrine biosensor reported by Baluta et al. [142]. This sensor system made use of the differential pulse and cyclic voltammetry voltammetric methods to monitor the oxidation of norepinephrinetonorepinephrine-quinone via catalysis.

*2.5. Microwave-Assisted Synthesis*

The electromagnetic wave known as the microwave has a vast wavelength range of 1 mm to 1 m and is frequently employed in daily life and scientific study. Microwaves can also deliver high energy to breakdown the chemical bonds in a substrate, just like lasers can. It is believed that using a microwave to create CDs is an energy-efficient method. Additionally, the reaction time may be significantly reduced. The substrate is typically pyrolyzed and the surface functionalized during microwave-aided synthesis [143,144].

The CDs are synthesized more quickly using a green, economical microwave-aided method. For the creation of CDs, microwave irradiation can deliver consistent heat. For the first time, Li et al. [145] produced green-colored luminescent graphene quantum dots via the cleavage of graphene oxide sheets chemically in the presence of acids under microwave conditions. They have an emission peak of 500 nm when excited at 260 nm and 340 nm. For the first time, electrochemiluminescence has been observed from the graphene quantum dots and is highly applicable in imaging and bio-sensing.

Liu et al. [146] produced carbon dots under microwave conditions. They used glutaraldehyde as a cross-linking agent for the fluorescent system. The luminescent emissions of the carbon dots come in a range on the basis of the amount of glutaraldehyde used. The as-prepared carbon dots showed remarkable luminescent characteristics and were less toxic, highly stable, and water soluble.

A simple microwave-assisted hydrothermal was used to produce CDs from *Mangifera indica* leaves [147]. The resulting carbon dots were employed as temperature sensors inside the cells, had good biocompatibility, and strong photostability. To create carbon dots from raw cashew gum, Pires et al. [148] devised a heating method that is microwave-assisted and has dual steps. The carbon quantum dots have an average size of nearly 9 nm. The synthesis involves two steps: the first step is the partial depolymerization, i.e., autohydrolysis of the gum and production of 5-hydroxymethyl furfural, while the second step involves poly-condensation for the production of the polyfuranic structure, accompanied by carbonization and nucleation. The generated carbon quantum dots have been used in the cell imaging of live cells because they exhibit good biocompatibility and low cytotoxicity.

Simsek et al. [149] showed that under different physical conditions, a quick and one-step green synthesis of carbon nanodots from *Nerium oleander* leaves may be achieved using a household oven and a microwave-assisted hydro-thermal synthesizer (Figure 6). The effects of the synthesizer system, the kind of extract based on the plant extraction of the plant solvents, and the synthetic conditions, including the time of reaction, temperature of reaction, surface-passivation reagent inclusion into the reaction medium, physical and chemical properties, and optical characteristics of carbon dots, were examined.

Ren et al. [150] reported the synthesis of 5.6 nm-diameter N-doped graphene quantum dots using microwave-assisted heat. The resulting N-doped graphene quantum dots were used in metal ion detection and exhibit strong and constant blue fluorescence emission with an 8% quantum yield.

Sendao et al. [151] suggested a microwave method to synthesize blue-emitting carbon quantum dots and looked into photoluminescent emission features. They discovered that the synthesis technique created green-emitting molecular fluorophore that can hide the photoluminescent emission of the carbon dots. It is important to note that in the same solution, these fluorophores and the carbon dots do not function as different species with independent emissions. Instead, their interaction results in a hybrid luminescence which is seen. This method demonstrates that the reactive nature and the characteristics in the excited-state are indistinct in comparison to their individual characteristics. The impurities of the fluorescence generated from its formation have formed a critical drawback in the investigation of the photoluminescent property of the carbon quantum dots (Tables 1 and 2).

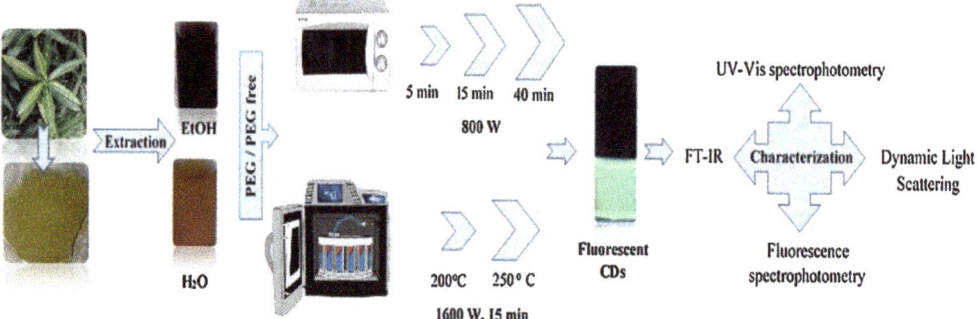

**Figure 6.** Schematic illustration of the fabrication process. Reproduced with permission from Ref. [149]. 2019, Elsevier.

**Table 1.** Advantages and disadvantages of different synthetic methods of carbon nanodots.

| Synthetic Method | Advantages | Disadvantages | References |
| --- | --- | --- | --- |
| Sonochemical/Ultra-sonic fabrication | Easy-operation | Expensive cost of energy | [108–116] |
| Hydrothermal synthesis | Cost-effective, environmentally benign, non-toxic | No uniformity in size | [124–131] |
| Carbonization/Pyrolysis | No solvent required, cost-effective, bulk-production | No uniformity in distribution of size | [132–136] |
| Electrochemical synthesis | Easy, cost-effective, environmentally benign | Uniformity in size distribution | [137–142] |
| Microwave-assisted synthesis | Fast, cost-effective, environmentally benign | No uniformity in distribution of size | [143–151] |

**Table 2.** Methods for the conversion of biobased and chemical feedstock into functionalized carbon nanodots.

| Sources | Synthetic Methods | References |
| --- | --- | --- |
| Carbon, Silver nitrate liquid solution | Sonochemical synthesis | [112] |
| *Saccharum officinarum* (Sugarcane) | Hydrothermal synthesis | [124] |
| Coconut water | Ultrasonication synthesis | [113] |
| *o*-phenylenediamine | Hydrothermal synthesis | [129] |
| Waste biomass | Carbonization synthesis | [136] |
| Ethanol | Electrochemical carbonization synthesis | [139] |
| Graphene | Sonochemical synthesis | [108] |
| *p*-phenylenediamine and urea | Hydrothermal synthesis | [130] |
| *Mangifera indica* (Mango) | Microwave-assisted hydrothermal synthesis | [147] |
| Chitosan | Hydrothermal carbonization synthesis | [140] |

## 3. Applications of Carbon Dots (CDs)

There are various applications which are associated with carbon dots. CDs also show a number of biomedical applications. The application of CDs is shown in Figure 7.

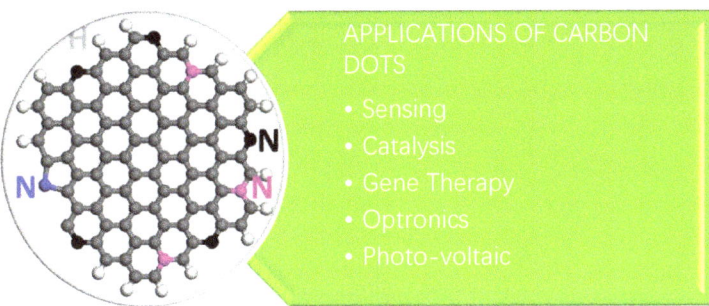

**Figure 7.** Application of Carbon dots.

*3.1. Sensing*

One of the most common and potentially significant uses of CDs is sensing [152–154]. Due to their superior optical qualities, high fluorescence sensitivity to the surrounding environment [155,156], and ability to function as effective electron donors [157–159], CDs are frequently suggested as detectors for a variety of harmful substances, including heavy metals such as mercury [160–162], copper, and iron [163–166]. To make CDs more sensitive to one or more of these analytes, persistent work is being conducted in this direction. Only a handful of studies, however, have attempted to examine the interactions of CDs with metal ions at a more fundamental level; for example, Goncalves and colleagues demonstrated that the fluorescence emissions of both CQD solution and CQDs immobilized in sol–gel are sensitive to the presence of $Hg^{2+}$ [167]. In their study, laser-ablated and $NH_2$-$PEG_{200}$ and N-acetyl-L-cysteine-passivated CQDs were used as fluorescent probes. It was observed that the fluorescence intensity of the CQDs is efficiently quenched by micro molar amounts of $Hg^{2+}$ with a Stern–Volmer constant of $1.3 \times 10^5$ $M^{-1}$. Therefore, judging from the relatively large magnitude of the Stern–Volmer constant [168], the quenching provoked by $Hg^{2+}$ is probably due to the static quenching arising from the formation of a stable non-fluorescent complex between CQD and $Hg^{2+}$. A substantial improvement in the sensitivity down to nanomolars was later realized by replacing the laser-ablated CQDs with N-CQDs. Again, static quenching is thought to be responsible for the quenching of fluorescence, but with a much larger Stern–Volmer constant of $1.4 \times 10^7$ $M^{-1}$, two orders of magnitude higher than that of the previous system [169]. It was suggested that the presence of the nitrogen element in the N-CQDs, most probably -CN groups on the N-CQD surface, is responsible for the much-improved performance of $Hg^{2+}$ sensing.

*3.2. Bio Imaging Probes*

An intriguing application of C dots is their use as a potential agent for in vivo and in vitro bioimaging of cells and species due to their photoluminescence, which is an important property of C dots [170–172]. The bioimaging of cells and tissues is an important part of the diagnosis of many diseases, particularly cancer. Various fluorescent systems for diagnostic purposes have been reported, ranging from organic and inorganic dyes to the most recent nanoparticle-based systems.

To be considered suitable for use as an imaging probe, a bioimaging agent must have excellent biocompatibility, a tunable emission spectrum, and be free of cytotoxicity. Rapid progress in implementing a new class of nanoparticles has resulted in a material that meets these criteria and can be used for both diagnostic and therapeutic purposes. Chemical functionalization is used to successfully conjugate the required drug molecule to the fluorescent nanoprobes for these theranostic applications. Sahu et al. [173] reported the synthesis of C dots from orange juice hydrothermal treatment. This was one of the first examples of making fluorescent C dots from readily available natural resources [174,175]. The C dots were non-cytotoxic and efficiently taken up by MG-63 human osteosarcoma cells for cellular imaging.

The "central dogma" states that genetic information flows from DNA to RNA to proteins. Researchers investigated the physiological activity of RNA during cancer research by using RNA dynamics in cellular functions and the real-time monitoring of their temporo-spatial distribution. The experiments were carried out using fluorescent carbon dots created by the one-pot hydrothermal treatment of o-, m-, or p-phenylenediamines with triethylenetetramine by Chen et al. [176]. Because carbon has excellent biocompatibility and negligible cytotoxicity, there has been a lot of interest in using carbon nanodots as bioimaging probes instead of other types of nanoparticles. C dots are ideal candidates for theranostic applications due to their ease of synthesis, acceptable emission spectra, high photostability, and lack of cytotoxicity.

Tao et al. [177] used a mixed acid treatment to create C dots from carbon nanotubes (CNTs) and graphite. Under UV light, the C dots emit a strong yellow fluorescence with no cellular toxicity. They also demonstrated in vivo bioimaging in the near-infrared region using a rat model, and this experiment exemplified the possibilities for the development of fluorescent imaging probes in both the ultraviolet (UV) and infrared (IR) range spectra.

*3.3. Photodynamic Therapy*

Photodynamic therapy is a relatively new advancement in biomedical nanotechnology that uses energy transfer to destroy damaged cells and tissues. This method is useful in dealing with cancer cells because it effectively targets and destroys malignant tissue while leaving normal, healthy tissue alone. This targeted destruction in photodynamic therapy can be accomplished with fluorescent C dots that have adequate photostability [178].

Shi et al. [179] used the hydrothermal method to create N-doped C dots from rapeseed flowers and bee pollen. The authors demonstrated that C dots had no cytotoxic effect up to a limiting concentration of 0.5 mg/mL after this successful large-scale synthesis. Human colon carcinoma cells were imaged successfully in this study, and the C dots were found to have good photostability and biocompatibility.

Wang et al. [180] reported C dot synthesis from the condensation carbonization of linear polyethylenic amine (PEA) analogues and citric acid (CA) of different ratios. The authors successfully demonstrated that the extent of conjugated π-domains with CN in the carbon backbone was correlated with their photoluminescence quantum yield. The main conclusion from this study is that the emission arises not only from the $sp^2/sp^3$ carbon core and surface passivation of C nanodots, but also from the molecular fluorophores integrated into the C dot framework. This work provided an insight into the excellent biocompatibility, low cytotoxicity, and enhanced bioimaging properties of N-doped C dots, which opens the possibilities for new bioimaging applications.

Bankoti et al. [181] fabricated C dots from onion peel powder waste using the microwave method and studied cell imaging and wound healing aspects. The C dots exhibited stable fluorescence at an excitation wavelength of 450 nm and an emission wavelength of 520 nm at variable pH, along with the ability to scavenge free radicals, which can be further explored for antioxidant activity. The radical scavenging ability leads to an enhanced wound healing ability in a full-thickness wound in a rat model.

*3.4. Photocatalysis*

There has been significant research interest in photocatalysts over the past decade due to the scenario of environmental safety and sustainable energy. The applications of nanomaterials for the efficient fabrication of photocatalysts made the journey fast and effective.

Ming et al. [182] successfully developed C dots using a one-pot electrochemical method that only used water as the main reagent. This is an extremely promising synthetic methodology because it is a green protocol that is also cost-effective, with good photocatalytic activity of C dots for methyl orange degradation.

Song et al. [183] devised a two-step hydrothermal method for the creation of a C dot–$WO_2$ photocatalyst. The authors used this system to photocatalytically degrade rhodamine

B. It is worth noting that the reaction rate constant reported in this study is 0.01942 min$^{-1}$, which is approximately 7.7 times higher than the catalytic rate using $WO_2$ alone.

For photocatalytic hydrogen generation, a C-dot/g-$C_3N_4$ system was used. The authors created C dots from rapeseed flower pollen and hydrothermally incorporated them into g-$C_3N_4$. Under visible light irradiation, this system was able to photocatalytically generate hydrogen via sound with an output greater than that of bulk g-$C_3N_4$.

*3.5. Biological Sensors and Chemical Sensors*

There is great interest in using nanoparticles as biochemical sensors because C dots have been found to be useful in sensing chemical compounds or elements. Based on the properties of C dots, particularly their fluorescence properties and surface-functionalized chemical groups, various sensors for biological and chemical applications have been developed.

Qu et al. [184] developed ratiometric fluorescent nano-sensors using C dots in a single step of microwave-assisted synthesis. This research is significant in C-dot sensor research because the developed nanosensors are multi-sensory and can detect temperature, pH, and metal ions such as Fe (III). Because it can detect and estimate multiple metabolic parameters at the same time, this exciting feature is proving to be widely applicable in the biological environment. The sensory mechanism is non-cytotoxic and based on ratiometric fluorescence, which is a promising feature for future research.

Vedamalai et al. [185] developed C dots that are highly sensitive to copper (II) ions in cancer cells. They used a relatively simple hydrothermal synthesis method based on ortho-phenylenediamine (OPD). The orange color was caused by the formation of the $Cu(OPD)_2$ complex on the surface of the C dots. Further investigation revealed that the C dots were highly water dispersible, photostable, chemically stable, and biocompatible.

Shi et al. [186] used C dots to detect Cu(II) ions in living cells as well. The hydrothermal pyrolysis of leeks resulted in blue and green fluorescent C dots. In a single step of hydrothermal carbonization, the C dots were modified with boronic acid using phenylboronic acid as the precursor. This C-dot-based sensor successfully detected blood sugar levels and demonstrated good selectivity with minimal chemical interference from other species [187].

Nie et al. [188] used a novel bottom-up method to develop a pH sensor out of C dots. This method yielded C dots with high crystallinity and stability. The procedure involved a one-pot synthesis with high reproducibility using chloroform and diethylamine. The authors were able to use the technique for cancer diagnosis after successfully implementing the pH detection of two C dots with different emission wavelengths.

Wang et al. [189] described an intriguing C-dot sensor for hemoglobin detection (Hb). The C dots were developed from glycine using an electrochemical method that included multiple steps, such as electro-oxidation, electro-polymerization, carbonization, and passivation. The authors successfully validated the sensitivity of Hb detection and discovered that the luminescence intensity varied inversely with Hb concentration in the 0.05–250 nM range.

*3.6. Drug Delivery*

Carbon dots' excellent biocompatibility and clearance from the body meet the requirements for in vivo applications. Carbon dots with rich and tunable function groups, such as amino, carboxyl, or hydroxyl, can carry therapeutic agents, resulting in theranosticnanomedicines [190–195]. The bright emission of carbon dots allows for the dynamic and real-time monitoring of drug distribution and response. Zheng et al. [196] used carbon dots synthesized through the thermal pyrolysis of citric acid and polyene polyamine to transport oxaliplatin, a platinum-based drug, because platinum-based drugs are the most effective anticancer drugs and are used in more than 50% of clinical cancer patients' chemotherapeutic treatments.

## 3.7. Micro-Fluidic Marker

The study of fluidic physics at the micro-scale is now best conducted using microfluidic systems. Because of their considerably higher surface-to-volume ratio, surface tension and viscosity dominate those of inertia, making the fluid easier to control. Static laminar flows and dynamic droplet formation are typical microfluidic situations. Both exhibit many advantages, including minimal reagent use, high sensitivity, and high output, which leads to a wide range of applications in bioassays, chemical reactions, drug delivery, etc. The majority of applications rely on the microfluidic circuit's ability to visualize fluid flow. However, the biocompatibility and cheap cost of the fluorescent materials currently in use cannot be balanced, which is a critical issue for microfluidic applications, particularly for bio-applications. Sun's colleagues used carbon dots, synthesized by heating glucose and urea in a microwave, to visualize microfluid flows for the first time to address this problem [197–199]. The scientists used carbon dots dissolved in the deionized water as a fluorescent marker to investigate the dynamics of the mixture of glycerol and deionized water. When the interface is ruptured by an electric field above a threshold, fast mixing occurs at the microscale. In addition to laminar flow, the authors also synthesized mono-dispersed droplets in a flow focusing system, where the continuous phase was mineral oil while the aqueous solution of carbon dots appeared as the dispersed phase. The diameter of the droplets will shrink because a higher capillary number results in a greater interfacial shear force. Additionally, the authors successfully demonstrated the multiple component droplet, merged droplet, and double emulsion, each of which has a distinct core-shell structure. To more accurately determine the speed of the flow field, luminescent seeding carbon dots were made via a mixture of carbon dots (liquid state) and polystyrene microparticles [200–202].

## 3.8. Bioimaging

Carbon dots have significant advantages over fluorescent organic dyes and genetically engineered fluorescent proteins, such as high PL quantum yield, photostability, and resistance to metabolic degradation, which endows them with enormous potential for use in bioapplications. While the toxicity testing of carbon dots is required before exploring their bioapplications, Yang et al. [203] used human breast cancer MCF-7 cells and human colorectal adenocarcinoma HT-29 cells (previously reported by other scientists, Yang modified and used it) to assess the in vitro toxicity of carbon dots synthesized by the laser ablation of graphite powder and cement with PEG1500N [204–206] as a surface passivation agent. All the observations of cell proliferation, mortality, and viability from both cell lines indicated that the carbon dots exhibited superior biocompatibility, even at concentrations as high as 50 mg/mL, which is much higher than the practical application demand, for example, in living cell imaging.

## 3.9. Carbon Dots Chiral Photonics

Chirality is essential in a number of practical application fields, such as chiral drug recognition, chiral molecular biology, and chiral chemistry [207–209]. As a result, as previously proposed by M. Va'zquez-Nakagawa et al. [210], chirality and carbon dots can be combined to form intriguing chiral optics based on carbon dots. The carbon dots used in their groundbreaking research were created by chemically exfoliating graphite with strong sulfuric and nitric acids. The carbon dots' surface carboxylic acid groups were subsequently converted to acid chlorides using thionyl chloride. When the acid chlorides and the (R) or (S)-2-phenyl-1-propanol reacted simultaneously, enantiomerically pure esters and chiral carbon dots were created (chiral molecular). Enantiomerically esters and chiral carbon dots were formed, and their formation was verified using $^{13}$C-NMR and FTIR spectroscopy. The presence of phenyl substituents was suggested by the appearance of peaks in the $^{13}$C–NMR. The recent work in this field is the most notable development in the chiral regulation of bioreactions for chiral carbon dots. Xin et al. [211] described the destruction of the cell walls of gram-positive and gram-negative bacteria via carbon dots in the presence of D-glutamic acid, which resulted in the fatality of bacteria. In contrast, the carbon dots formed in the presence of L-glutamic acid demonstrated an insignificant effect on bacterial cells. This implied that antimicrobial nanoagents with chirality can be synthesized from carbon dots. The D-form and L-form of cysteine-based carbon dots were used to regulate the chirality of the enzyme. For instance, L-form cysteine carbon dots reduce the enzymatic activity while D-form cysteine carbon dots enhance the enzymatic activity of the enzyme. According to Li et al. [212], these cysteine-based carbon nanodots have the capacity to affect cellular energy metabolism. We anticipate that other chiral carbon dots-based applications will be investigated in the future [213], and that carbon dots with chirality will emerge as a novel but exciting topic because of their wide applications.

## 4. Conclusions

Carbon dots have drawn rigorous attention since they possess outstanding photoluminescence, fluorescence, biocompatibility, sensing and imaging, photostability, excellent colloidal stability, eco-friendly synthesis, low toxicity, and are cost-effective. In this review, widespread synthesis procedures have been discussed in detail, including bottom-up and top-down methods, along with biological and eco-friendly synthetic ways. This concludes numerous synthesizing routes that could be helpful to many scientific and research areas, since carbon nanodots can be easily synthesized for various applications. Earlier, the synthetic methods were limited because of unreliable quantum yields. However, in recent years, the synthesizing methods have seen a remarkable lift in yield, hence enhancing their use in different fields for varied applications. Despite many advancements in the field of carbon nanodots, there is still room for improvement in its synthetic methods. Several bio-related fields are left undiscovered and need special attention.

**Author Contributions:** Writing—original draft, A.B.; review and editing, S.G.; Funding acquisition, S.J.; investigation, formal analysis, data curation, N.K.J.: conceptualization, methodology, A.S.; Chemistry and English language editing, I.N.P.; methodology, data curation, J.D.; Supervision, M.S. All authors have read and agreed to the published version of the manuscript.

**Funding:** This work is financially supported by the Department of Science & Technology—Fund for Improvement of S&T Infrastructure in Universities and Higher Educational Institutions (DST—FIST), India (Order No. SR/FST/CS-II/2022/252).

**Data Availability Statement:** Data is available upon request.

**Acknowledgments:** The authors are thankful to the Department of Chemistry, Banasthali Vidyapith for providing the necessary infrastructure. We are thankful to publishing houses, namely ACS and Elsevier, Springer Nature for providing copyright permissions for the figures used in this review article.

**Conflicts of Interest:** This research received no external funding.

# References

1. Xu, X.Y.; Ray, R.; Gu, Y.L.; Ploehn, H.J.; Gearheart, L.; Raker, K.; Scrivens, W.A. Electrophoretic Analysis and Purification of Fluorescent Single-Walled Carbon Nanotube Fragments. *J. Am. Chem. Soc.* **2004**, *126*, 12736–12737. [CrossRef] [PubMed]
2. Sun, Y.P.; Zhou, B.; Lin, Y.; Wang, W.; Fernando, K.A.S.; Pathak, P.; Meziani, M.J.; Harruff, B.A.; Wang, X.; Wang, H.; et al. QuantumSized Carbon Dots for Bright and Colorful Photoluminescence. *J. Am. Chem. Soc.* **2006**, *128*, 7756–7757. [CrossRef] [PubMed]
3. Zheng, L.; Chi, Y.; Dong, Y.; Lin, J.; Wang, B. Electrochemiluminescence of Water-Soluble Carbon Nanocrystals Released Electrochemically from Graphite. *J. Am. Chem. Soc.* **2009**, *131*, 4564–4565. [CrossRef] [PubMed]
4. Baker, S.N.; Baker, G.A. Luminescent Carbon Nanodots: Emergent Nanolights. *Angew. Chem. Int. Ed.* **2010**, *49*, 6726–6744. [CrossRef] [PubMed]
5. Li, H.T.; He, X.D.; Kang, Z.H.; Huang, H.; Liu, Y.; Liu, J.L.; Lian, S.Y.; Tsang, C.C.A.; Yang, X.B.; Lee, S.T. Water-Soluble Fluorescent Carbon Quantum Dots and Catalyst Design. *Angew. Chem. Int. Ed.* **2010**, *49*, 4430–4434. [CrossRef] [PubMed]
6. Ma, Z.; Zhang, Y.L.; Wang, L.; Ming, H.; Li, H.T.; Zhang, X.; Wang, F.; Liu, Y.; Kang, Z.H.; Lee, S.T. Bioinspired Photoelectric Conversion System Based on Carbon Quantum Dot Doped Dye—Semiconductor Complex. *ACS Appl. Mater. Interfaces* **2013**, *5*, 5080–5084. [CrossRef] [PubMed]

7. Mu, X.W.; Wu, M.X.; Zhang, B.; Liu, X.; Xu, S.M.; Huang, Y.B.; Wang, X.H.; Sun, Q. A sensitive "off-on" carbon dots-Ag nanoparticles fluorescent probe for cysteamine detection via the inner filter effect. *Talanta* **2021**, *221*, 121463. [CrossRef]
8. Huang, H.L.; Ge, H.; Ren, Z.P.; Huang, Z.J.; Xu, M.; Wang, X.H. Controllable Synthesis of Biocompatible Fluorescent Carbon Dots from Cellulose Hydrogel for the Specific Detection of $Hg^{2+}$. *Front. Bioeng. Biotechnol.* **2021**, *9*, 617097. [CrossRef]
9. Wang, Y.; Zhuang, Q.; Ni, Y. Facile Microwave Assisted Solid Phase Synthesis of Highly Fluorescent Nitrogen Sulfur Co-doped Carbon Quantum Dots for Cellular Imaging Applications. *Chem. Eur. J.* **2015**, *21*, 13004–13011. [CrossRef]
10. Li, D.; Li, W.; Zhang, H.; Zhang, X.; Zhuang, J.; Liu, Y.; Hu, C.; Lei, B. Far-Red Carbon Dotsas Efficient Light-Harvesting Agents for Enhanced Photosynthesis. *ACS Appl. Mater. Interfaces* **2020**, *12*, 21009–21019. [CrossRef]
11. Wang, J.; Li, R.S.; Zhang, H.Z.; Wang, N.; Zhang, Z.; Huang, C.Z. Highly fluorescent carbon dots as selective and visual probes for sensing copper ions in living cells via an electron transfer process. *Biosens. Bioelectron.* **2017**, *97*, 157–163. [CrossRef] [PubMed]
12. Jana, J.; Lee, H.J.; Chung, J.S.; Kim, M.H.; Hur, S.H. Blue emitting nitrogen-doped carbon dots as a fluorescent probe for nitrite ion sensing and cell-imaging. *Anal. Chim. Acta* **2019**, *1079*, 212–219. [CrossRef]
13. Hu, J.; Tang, F.; Jiang, Y.; Liu, C. Rapid screening and quantitative detection of *Salmonella* using a quantum dot nanobead-based biosensor. *Analyst* **2020**, *145*, 2184–2190. [CrossRef] [PubMed]
14. Liu, M.L.; Chen, B.B.; Li, C.M.; Huang, C.Z. Carbon Dots: Synthesis, Formation Mechanism, Fluorescence Origin and Sensing Applications. *Green Chem.* **2019**, *21*, 449–471. [CrossRef]
15. Qin, K.H.; Zhang, D.F.; Ding, Y.F.; Zheng, X.D.; Xiang, Y.Y.; Hua, J.H.; Zhang, Q.; Ji, X.L.; Li, B.; Wei, Y.L. Applications of hydrothermal synthesis of *Escherichia coli* derived carbon dots in in vitro and in vivo imaging and p-nitrophenol detection. *Analyst* **2020**, *145*, 177–183. [CrossRef]
16. Cui, L.; Wang, J.; Sun, M.T. Graphene plasmon for optoelectronics. *Rev. Phys.* **2021**, *6*, 100054. [CrossRef]
17. Yuan, F.; Wang, Y.K.; Sharma, G.; Dong, Y.; Zheng, X.; Li, P.; Johnston, A.; Bappi, G.; Fan, J.Z.; Kung, H. Bright High-Colour Purity Deep-Blue Carbon Dot Light-Emitting Diodes via Efficient Edge Amination. *Nat. Photonics* **2020**, *14*, 171–176. [CrossRef]
18. Hu, C.; Li, M.Y.; Qiu, J.S.; Sun, Y.P. Design and fabrication of carbon dots for energy conversion and storage. *Chem. Soc. Rev.* **2019**, *48*, 2315–2337. [CrossRef]
19. Cao, Y.; Cheng, Y.; Sun, M.T. Graphene-based SERS for sensor and catalysis. *Appl. Spectrosco. Rev.* **2021**, *58*, 1–38. [CrossRef]
20. Hsu, P.C.; Shih, Z.Y.; Lee, C.H.; Chang, H.T. Synthesis and analytical applications of photoluminescent carbon nanodots. *Green Chem.* **2012**, *14*, 917–920. [CrossRef]
21. Chandra, S.; Pathan, S.H.; Mitra, S.; Modha, B.H.; Goswami, A.; Pramanik, P. Tuning of photoluminescence on different surface functionalized carbon quantum dots. *RSC Adv.* **2012**, *2*, 3602–3606. [CrossRef]
22. Liu, M.; Chen, W. Green synthesis of silver nanoclusters supported on carbon nanodots: Enhanced photoluminescence and high catalytic activity for oxygen reduction reaction. *Nanoscale* **2013**, *5*, 12558–12564. [CrossRef] [PubMed]
23. Kottam, N.; Smrithi, S.P. Luminescent carbon nanodots: Current prospects on synthesis, properties and sensing applications. *Methods Appl. Fluoresc.* **2021**, *9*, 012001. [CrossRef] [PubMed]
24. Xu, A.; Wang, G.; Li, Y.; Dong, H.; Yang, S.; He, P.; Ding, G. Carbon-based quantum dots with solid-state photoluminescent: Mechanism, implementation, and application. *Small* **2020**, *16*, 2004621. [CrossRef]
25. Roy, P.; Chen, P.C.; Periasamy, A.P.; Chen, Y.N.; Chang, H.T. Photoluminescent carbon nanodots: Synthesis, physicochemical properties and analytical applications. *Mater. Today* **2015**, *18*, 447–458. [CrossRef]
26. Li, H.; Kang, Z.; Liu, Y.; Lee, S.T. Carbon nanodots: Synthesis, properties and applications. *J. Mat. Chem.* **2012**, *22*, 24230–24253. [CrossRef]
27. Philippidis, A.; Stefanakis, D.; Anglos, D.; Ghanotakis, D. Microwave heating of arginine yields highly fluorescent nanoparticles. *J. Nanopart. Res.* **2013**, *15*, 1–9. [CrossRef]
28. Zhu, L.; Yin, Y.; Wang, C.F.; Chen, S. Plant leaf-derived fluorescent carbon dots for sensing, patterning and coding. *J. Mater. Chem. C* **2013**, *1*, 4925–4932. [CrossRef]
29. Tang, J.; Zhang, Y.; Kong, B.; Wang, Y.; Da, P.; Li, J.; Elzatahry, A.A.; Zhao, D.; Gong, X.; Zheng, G. Solar-driven photoelectrochemical probing of nanodot/nanowire/cell interface. *Nano Lett.* **2014**, *14*, 2702–2708. [CrossRef]
30. Li, Q.; Ohulchanskyy, T.Y.; Liu, R.; Koynov, K.; Wu, D.; Best, A.; Kumar, R.; Bonoiu, A.; Prasad, P.N. Photoluminescent carbon dots as biocompatible nanoprobes for targeting cancer cells in vitro. *J. Phys. Chem. C* **2010**, *114*, 12062–12068. [CrossRef]
31. Da Silva, J.C.E.; Gonçalves, H.M. Analytical and bioanalytical applications of carbon dots. *TrAC Trends Anal. Chem.* **2011**, *30*, 1327–1336. [CrossRef]
32. Sanderson, K. Quantum dots go large: A small industry could be on the verge of a boom. *Nature* **2009**, *459*, 760–762. [CrossRef]
33. Cao, L.; Wang, X.; Meziani, J.M.; Lu, F.; Wang, H.; Luo, P.G.; Lin, Y.; Harruff, B.A.; Veca, L.M.; Murray, D.; et al. Carbon Dots for Multiphoton Bioimaging. *J. Am. Chem. Soc.* **2007**, *129*, 11318–11319. [CrossRef] [PubMed]
34. Yang, S.T.; Cao, L.; Luo, P.G.; Lu, F.S.; Wang, X.; Wang, H.F.; Meziani, M.J.; Liu, Y.F.; Qi, G.; Sun, Y.P. Carbon Dots for Optical Imaging in Vivo. *J. Am. Chem. Soc.* **2009**, *131*, 11308–11309. [CrossRef] [PubMed]
35. Sun, D.; Ban, R.; Zhang, P.H.; Wu, G.H.; Zhang, J.R.; Zhu, J.J. Hair Fiber as a Precursor for Synthesizing of Sulfur- and Nitrogen-Codoped Carbon Dots with Tunable Luminescence Properties. *Carbon* **2013**, *64*, 424–434. [CrossRef]
36. Miao, P.; Han, K.; Tang, Y.; Wang, B.; Lin, T.; Cheng, W. Recent advances in carbon nanodots: Synthesis, properties and biomedical applications. *Nanoscale* **2015**, *7*, 1586–1595. [CrossRef]

37. Wang, H.; Boghossian, A.A. Covalent conjugation of proteins onto fluorescent single-walled carbon nanotubes for biological and medical applications. *Mater. Adv.* **2023**, *4*, 823–834. [CrossRef]
38. Hu, S.L.; Niu, K.Y.; Sun, J.; Yang, J.; Zhao, N.Q.; Du, X.W. One-Step Synthesis of Fluorescent Carbon Nanoparticles by Laser Irradiation. *J. Mater. Chem.* **2009**, *19*, 484–488. [CrossRef]
39. Ma, Z.; Ming, H.; Huang, H.; Liu, Y.; Kang, Z. One-step ultrasonic synthesis of fluorescent N-doped carbon dots from glucose and their visible-light sensitive photocatalytic ability. *New J. Chem.* **2012**, *36*, 861–864. [CrossRef]
40. Dong, Y.; Pang, H.; Ren, S.; Chen, C.; Chi, Y.; Yu, T. Etching Single-Wall Carbon Nanotubes into Green and Yellow Single-Layer Graphene Quantum Dots. *Carbon* **2013**, *64*, 245–251. [CrossRef]
41. Cadranel, A.; Margraf, J.T.; Strauss, V.; Clark, T.; Guldi, D.M. Carbon nanodots for charge-transfer processes. *Acc. Chem. Res.* **2019**, *52*, 955–963. [CrossRef] [PubMed]
42. Li, H.; He, X.; Liu, Y.; Huang, H.; Lian, S.; Lee, S.T.; Kang, Z. One-Step Ultrasonic Synthesis of Water-Soluble Carbon Nanoparticles with Excellent Photoluminescent Properties. *Carbon* **2011**, *49*, 605–609. [CrossRef]
43. Li, H.T.; He, X.D.; Liu, Y.; Yu, H.; Kang, Z.H.; Lee, S.T. Synthesis of Fluorescent Carbon Nanoparticles Directly from Active Carbon via a One-Step Ultrasonic Treatment. *Mater. Res. Bull.* **2011**, *16*, 147–151. [CrossRef]
44. Dong, Y.; Pang, H.; Yang, H.B.; Guo, C.; Shao, J.; Chi, Y.; Li, C.M.; Yu, T. Carbon-based Dots Co-doped with Nitrogen and Sulfur for High Quantum Yield and Excitation-Independent Emission. *Angew. Chem. Int. Ed.* **2013**, *52*, 7800–7804. [CrossRef]
45. Tabaraki, R.; Abdi, O. Microwave assisted synthesis of N-doped carbon dots: An easy, fast and cheap sensor for determination of aspartic acid in sport supplements. *J. Fluoresc.* **2019**, *29*, 751–756. [CrossRef] [PubMed]
46. Tian, M.; Wang, Y.T.; Zhang, Y. Synthesis of fluorescent nitrogen-doped carbon quantum dots for selective detection of picric acid in water samples. *J. Nanosci. Nanotechnol.* **2018**, *18*, 8111–8117. [CrossRef]
47. Luo, J.B.; Sun, Z.S.; Zhou, W.Y.; Mo, F.W.; Wu, Z.C.; Zhang, X.G. Hydrothermal synthesis of bright blue-emitting carbon dots for bioimaging and fluorescent determination of baicalein. *Opt. Mater.* **2021**, *113*, 110796. [CrossRef]
48. Li, C.; Sun, X.Y.; Li, Y.; Liu, H.L.; Long, B.B.; Xie, D.; Chen, J.J.; Wang, K. Rapid and Green Fabrication of Carbon Dots for Cellular Imaging and Anti-Counterfeiting Applications. *ACS Omega* **2021**, *6*, 3232–3237. [CrossRef]
49. Prasannan, A.; Imae, T. One-Pot Synthesis of Fluorescent Carbon Dots from Orange Waste. *Ind. Eng. Chem. Res.* **2013**, *52*, 15673–15678. [CrossRef]
50. Aji, M.P.; Wiguna, P.A.; Susanto-Rosita, N.; Sucingingtyas, S.A. Sulhadi Performance of Photocatalyst Based Carbon Nanodots from Waste Frying Oil in Water Purification. *AIP Conf. Proc.* **2016**, *1725*, 79–94.
51. Achilleos, D.S.; Kasap, H.; Reisner, E. Photocatalytic hydrogen generation coupled to pollutant utilization using carbon dots produced from biomass. *Green Chem.* **2020**, *22*, 2831–2839. [CrossRef]
52. Aggarwal, R.; Saini, D.; Singh, B.; Kaushik, J.; Garg, A.K.; Sonkar, S.K. Bitter apple peel derived photoactive carbon dots for the sunlight induced photocatalytic degradation of crystal violet dye. *Sol. Energy* **2020**, *197*, 326–331. [CrossRef]
53. Tyagi, A.; Tripathi, K.M.; Singh, N.; Choudhary, S.; Gupta, R.K. Green synthesis of carbon quantum dots from lemon peel waste: Applications in sensing and photocatalysis. *RSC Adv.* **2016**, *6*, 72423–72432. [CrossRef]
54. Guclu, A.D.; Potasz, P.; Korkusinski, M.; Hawrylak, P. *Graphene Quantum Dots*; Springer: Berlin/Heidelberg, Germany, 2014.
55. Hawrylak, P.; Peeters, F.; Ensslin, K. Carbononics-integrating electronics, photonics and spintronics with graphene quantum dots. *Phys. Status Solidi RRL* **2016**, *10*, 11–12. [CrossRef]
56. Hai, K.; Feng, J.; Chen, X.W.; Wang, J.H. Tuning the optical properties of graphene quantum dots for biosensing and bioimaging. *J. Mater. Chem. B* **2018**, *6*, 3219–3234. [CrossRef]
57. Bisker, G.; Bakh, N.A.; Lee, M.A.; Ahn, J.; Park, M.; O'Connell, E.B.; Strano, M.S. Insulin detection using a corona phase molecular recognition site on single-walled carbon nanotubes. *ACS Sens.* **2018**, *3*, 367–377. [CrossRef]
58. Xu, Z.Q.; Lan, J.Y.; Jin, J.C.; Dong, P.; Jiang, F.L.; Liu, Y. Highly Photoluminescent Nitrogen-Doped Carbon Nanodots and Their Protective Effects against Oxidative Stress on Cells. *ACS Appl. Mater. Interfaces* **2015**, *7*, 28346–28352. [CrossRef]
59. Zhao, L.; Zhao, L.; Li, H.; Ma, J.; Bian, L.; Wang, X.; Pu, Q. Controlled synthesis of fluorescent carbon materials with the assistance of capillary electrophoresis. *Talanta* **2021**, *228*, 122224. [CrossRef]
60. Yuvali, D.; Narin, I.; Soylak, M.; Yilmaz, E. Green synthesis of magnetic carbon nanodot/graphene oxide hybrid material ($Fe_3O_4$@C-nanodot@GO) for magnetic solid phase extraction of ibuprofen in human blood samples prior to HPLC-DAD determination. *J. Pharm. Biomed. Anal.* **2020**, *179*, 113001. [CrossRef]
61. Yang, D.; Li, L.; Cao, L.; Chang, Z.; Mei, Q.; Yan, R.; Ge, M.; Jiang, C.; Dong, W.F. Green Synthesis of Lutein-Based Carbon Dots Applied for Free-Radical Scavenging within Cells. *Materials* **2020**, *13*, 4146. [CrossRef]
62. Xu, Y.; Li, P.; Cheng, D.; Wu, C.; Lu, Q.; Yang, W.; Zhu, X.; Yin, P.; Liu, M.; Li, H. Group IV nanodots: Synthesis, surface engineering and application in bioimaging and biotherapy. *J. Mater. Chem. B* **2020**, *8*, 10290–10308. [CrossRef]
63. Wang, Y.; Zhang, Y.; Yan, J.; Yu, J.; Ding, B. One-step synthesis of a macroporous Cu-g/$C_3N_4$ nanofiber electrocatalyst for efficient oxygen reduction reaction. *Chem. Commun.* **2020**, *56*, 14087–14090. [CrossRef]
64. Qin, X.; Liu, J.; Zhang, Q.; Chen, W.; Zhong, X.; He, J. Synthesis of Yellow-Fluorescent Carbon Nano-dots by Microplasma for Imaging and Photocatalytic Inactivation of Cancer Cells. *Nanoscale Res. Lett.* **2021**, *16*, 14. [CrossRef]
65. Ha, J.; Seo, Y.; Kim, Y.; Lee, J.; Lee, H.; Kim, S.; Choi, Y.; Oh, H.; Lee, Y.; Park, E. Synthesis of nitrogen-doped carbon nanodots to destroy bacteria competing with *Campylobacter jejuni* in enrichment medium, and development of a monoclonal antibody to detect *C. jejuni* after enrichment. *Int. J. Food Microbiol.* **2021**, *339*, 109014. [CrossRef]

66. Guo, F.; Bao, L.; Wang, H.; Larson, S.L.; Ballard, J.H.; Knotek-Smith, H.M.; Zhang, Q.; Su, Y.; Wang, X.; Han, F. A simple method for the synthesis of biochar nanodots using hydrothermal reactor. *MethodsX* **2020**, *7*, 101022. [CrossRef] [PubMed]
67. Sharma, V.; Tiwari, P.; Mobin, S.M. Sustainable carbon-dots: Recent advances in green carbon dots for sensing and bioimaging. *J. Mater. Chem. B* **2017**, *5*, 8904–8924. [CrossRef] [PubMed]
68. Yang, N.; Jiang, X.; Pang, D.W. *Carbon Nanoparticles and Nanostructures*; Springer International Publishing: Cham, Switzerland, 2016; p. 360.
69. Sharma, A.; Das, J. Small molecules derived carbon dots: Synthesis and applications in sensing, catalysis, imaging, and biomedicine. *J. Nanobiotechnol.* **2019**, *17*, 92. [CrossRef]
70. Jiang, J.; He, Y.; Li, S.; Cui, H. Amino Acids as the Source for Producing Carbon Nanodots: Microwave Assisted One-Step Synthesis, Intrinsic Photoluminescence Property and Intense Chemiluminescence Enhancement. *Chem. Commun.* **2012**, *48*, 9634–9636. [CrossRef]
71. Mazzier, D.; Favaro, M.; Agnoli, S.; Silvestrini, S.; Granozzi, G.; Maggini, M.; Moretto, A. Synthesis of Luminescent 3D Microstructures Formed by Carbon Quantum Dots and Their Self Assembly Properties. *Chem. Commun.* **2014**, *50*, 6592–6595. [CrossRef]
72. Krysmann, M.J.; Kelarakis, A.; Dallas, P.; Giannelis, E.P. Formation Mechanism of Carbogenic Nanoparticles with Dual Photoluminescence Emission. *J. Am. Chem. Soc.* **2012**, *134*, 747–750. [CrossRef] [PubMed]
73. Zhu, S.; Meng, Q.; Wang, L.; Zhang, J.; Song, Y.; Jin, H.; Zhang, K.; Sun, H.; Wang, H.; Yang, B. Highly Photoluminescent Carbon Dots for Multicolor Patterning, Sensors, and Bioimaging. *Angew. Chem. Int. Ed.* **2013**, *52*, 3953–3957. [CrossRef]
74. Hu, B.; Wang, K.; Wu, L.; Yu, S.-H.; Antonietti, M.; Titirici, M.-M. Engineering Carbon Materials from the Hydrothermal Carbonization Process of Biomass. *Adv. Mater.* **2010**, *22*, 813–828. [CrossRef] [PubMed]
75. Schwenke, A.M.; Hoeppener, S.; Schubert, U.S. Synthesis and Modification of Carbon Nanomaterials Utilizing Microwave Heating. *Adv. Mater.* **2015**, *27*, 4113–4141. [CrossRef]
76. Vazquez, E.; Giacalone, F.; Prato, M. Non-Conventional Methods and Media for the Activation and Manipulation of Carbon Nanoforms. *Chem. Soc. Rev.* **2014**, *43*, 58–69. [CrossRef]
77. Vazquez, E.; Prato, M. Carbon Nanotubes and Microwaves: Interactions, Responses, and Applications. *ACS Nano* **2009**, *3*, 3819–3824. [CrossRef]
78. Liu, Q.; Zhang, N.; Shi, H.; Ji, W.; Guo, X.; Yuan, W.; Hu, Q. One-step microwave synthesis of carbon dots for highly sensitive and selective detection of copper ions in aqueous solution. *New. J. Chem.* **2018**, *42*, 3097. [CrossRef]
79. Schneider, J.; Reckmeier, C.J.; Xiong, Y.; von Seckendorff, M.; Susha, A.S.; Kasák, P.; Rogach, A.L. Molecular fluorescence in citric acid-based carbon dots. *J. Phys. Chem. C* **2017**, *121*, 2014. [CrossRef]
80. Cayuela, A.; Soriano, M.L.; Valcàrcel, M. Strong luminescence of carbon dots induced by acetone passivation: Efficient sensor for a rapid analysis of two different pollutants. *Anal. Chim. Acta* **2013**, *804*, 246. [CrossRef]
81. Ðordevi'c, L.; Arcudi, F.; Prato, M. Preparation, functionalization and characterization of engineered carbon nanodots. *Nat. Protoc.* **2019**, *14*, 2931–2953. [CrossRef]
82. Sciortino, A.; Mauro, N.; Buscarino, G.; Sciortino, L.; and Popescu, R.; Schneider, R.; Giammona, G.; Gerthsen, D.; Cannas, M.; Messina, F. β-$C_3N_4$ nanocrystals: Carbon dots with extraordinary morphological, structural, and optical homogeneity. *Chem. Mater.* **2018**, *30*, 1695. [CrossRef]
83. Ji, Z.; Yin, Z.; Jia, Z.; Wei, J. Carbon nanodots derived from urea and citric acid in living cells: Cellular uptake and antioxidation effect. *Langmuir* **2020**, *36*, 8632. [CrossRef]
84. Shan, X.; Chai, L.; Ma, J.; Qian, Z.; Chen, J.; Feng, H. B-doped carbon quantum dots as a sensitive fluorescence probe for hydrogen peroxide and glucose detection. *Analyst* **2014**, *139*, 2322. [CrossRef]
85. Malfatti, L.; Innocenzi, P. Sol-gel chemistry for carbon dots. *Chem. Rev.* **2018**, *18*, 1192. [CrossRef]
86. Ðordevi'c, L.; Arcudi, F.; Prato, M. Synthesis, separation, and characterization of small and highly fluorescent nitrogen-doped carbon nanodots. *Angew. Chem. Int. Ed.* **2016**, *55*, 2107. [CrossRef] [PubMed]
87. Gan, Z.; Wu, X.; Hao, Y. The mechanism of blue photoluminescence from carbon nanodots. *Cryst. Eng. Comm.* **2014**, *16*, 4981. [CrossRef]
88. Sai, L.; Yinzi, C.; Chun, L.; Jiali, F.; Weidong, X.; Xiaojuan, L. Carbon nanodots with intense emission from green to red and their multifunctional applications. *J. Alloys Compd.* **2018**, *742*, 212.
89. Holá, K.; Sudolská, M.; Kalytchuk, S.; Nachtigallová, D.; Rogach, A.L.; Otyepka, M.; Zboril, R. Graphitic nitrogen triggers red fluorescence in carbon dots. *ACS Nano* **2017**, *11*, 12402. [CrossRef]
90. Reckmeier, C.J.; Schneider, J.; Xiong, Y.; Häusler, J.; Kasák, P.; Schnick, W.; Rogach, A.L. Aggregated molecular fluorophores in the ammonothermal synthesis of carbon dots. *Chem. Mater.* **2017**, *29*, 10352. [CrossRef]
91. Bao, L.; Liu, C.; Zhang, Z.L.; Pang, D.W. Photoluminescence-tunable carbon nanodots: Surface-state energy-gap tuning. *Adv. Mater.* **2015**, *27*, 1663. [CrossRef]
92. Le Croy, G.E.; Sonkar, S.K.; Yang, F.; Monica Veca, L.; Wang, P.; Tackett II, K.N.; Yu, J.-J.; Vasile, E.; Qian, H.; Liu, Y. Towards structurally defined carbon dots as ultracompact fluorescent probes. *ACS Nano* **2014**, *8*, 4522–4529. [CrossRef] [PubMed]
93. Campuzano, S.; Yáñez-Sedeño, P.; Pingarrón, J.M. Carbon dots and graphene quantum dots in electrochemical biosensing. *Nanomaterials* **2019**, *9*, 634. [CrossRef] [PubMed]

94. Fresco-Cala, B.; Soriano, M.L.; Sciortino, A.; Cannas, M.; Messina, F.; Cardenas, S. One-pot synthesis of graphene quantum dots and simultaneous nanostructured self-assembly via a novel microwave-assisted method: Impact on triazine removal and efficiency monitoring. *RSC Adv.* **2018**, *8*, 29939. [CrossRef]
95. Sun, Y.; Wang, S.; Li, C.; Luo, P.; Tao, L.; Wei, Y.; Shi, G. Large scale preparation of graphene quantum dots from graphite with tunable fluorescence properties. *Phys. Chem. Chem. Phys.* **2013**, *15*, 9907. [CrossRef] [PubMed]
96. Wang, L.; Wang, Y.; Xu, T.; Liao, H.; Yao, C.; Liu, Y.; Li, Z.; Chen, Z.; Pan, D.; Sun, L. Gram-scale synthesis of single-crystalline graphene quantum dots with superior optical properties. *Nat. Commun.* **2014**, *5*, 5357. [CrossRef] [PubMed]
97. Kwon, W.; Lee, G.; Do, S.; Joo, T.; Rhee, S.W. Size-controlled soft-template synthesis of carbon nanodots toward versatile photoactive materials. *Small* **2014**, *10*, 506. [CrossRef]
98. Bang, J.H.; Suslick, K.S. Applications of ultrasound to the synthesis of nanostructured materials. *Adv. Mater.* **2010**, *22*, 1039–1059. [CrossRef]
99. Shchukin, D.G.; Radziuk, D.; Möhwald, H. Ultrasonic fabrication of metallic nanomaterials and nanoalloys. *Ann. Rev. Mater. Res.* **2010**, *40*, 345–362. [CrossRef]
100. Vinodgopal, K.; Neppolian, B.; Lightcap, I.V.; Grieser, F.; Ashokkumar, M.; Kamat, P.V. Sonolytic design of graphene–Au nanocomposites. Simultaneous and sequential reduction of graphene oxide and Au (III). *Phys. Chem. Lett.* **2010**, *1*, 1987. [CrossRef]
101. Geng, J.; Jiang, L.; Zhu, J. Crystal formation and growth mechanism of inorganic nanomaterials in sonochemical syntheses. *Sci. China Chem.* **2012**, *55*, 2292–2310. [CrossRef]
102. Skrabalak, S.E. Ultrasound-assisted synthesis of carbon materials. *Phys. Chem. Chem. Phys.* **2009**, *11*, 4930. [CrossRef]
103. Gedanken, A. Using sonochemistry for the fabrication of nanomaterials. *Ultrason. Sonochem.* **2004**, *11*, 47–55. [CrossRef]
104. Gao, T.; Wang, T. Sonochemical synthesis of $SnO_2$ nanobelt/CdS nanoparticle core/shell heterostructures. *Chem. Commun.* **2004**, *22*, 2558–2559. [CrossRef]
105. Hasin, P.; Wu, Y. Sonochemical synthesis of copper hydride (CuH). *Chem. Commun.* **2012**, *48*, 1302–1304. [CrossRef] [PubMed]
106. Xu, H.; Zeiger, B.W.; Suslick, K.S. Sonochemical synthesis of nanomaterials. *Chem. Soc. Rev.* **2013**, *42*, 2555–2567. [CrossRef] [PubMed]
107. Suslick, K.S.; Flannigan, D.J. Inside a collapsing bubble: Sonoluminescence and the conditions during cavitation. *Annu. Rev. Phys. Chem.* **2008**, *59*, 659–683. [CrossRef]
108. Zhuo, S.; Shao, M.; Lee, S.T. Upconversion and downconversion fluorescent graphene quantum dots: Ultrasonic preparation and photocatalysis. *ACS Nano* **2012**, *6*, 1059–1064. [CrossRef]
109. Wei, K.; Li, J.; Ge, Z.; You, Y.; Xu, H. Sonochemical synthesis of highly photoluminescent carbon nanodots. *RSC Adv.* **2014**, *4*, 52230–52234. [CrossRef]
110. Huang, H.Y.; Cui, Y.; Liu, M.Y.; Chen, J.Y.; Wan, Q.; Wen, Y.Q.; Wei, Y. A one-step ultrasonic irradiation assisted strategy for the preparation of polymer-functionalized carbon quantum dots and their biological imaging. *J. Colloid Interface Sci.* **2018**, *532*, 767–773. [CrossRef]
111. Park, S.Y.; Lee, H.U.; Park, E.S.; Lee, S.C.; Lee, J.W.; Jeong, S.W.; Chi, H.K.; Lee, Y.C.; Yun, S.H.; Lee, J. Photoluminescent Green Carbon Nanodots from Food-Waste-Derived Sources: Large-Scale Synthesis, Properties, and Biomedical Applications. *ACS Appl. Mater. Interfaces* **2014**, *6*, 3365–3370. [CrossRef]
112. Jiang, D.; Zhang, Y.; Huang, M.; Liu, J.; Wan, J.; Chu, H.; Chen, M. Carbon nanodots as reductant and stabilizer for one-pot sonochemical synthesis of amorphous carbon-supported silver nanoparticles for electrochemical nonenzymatic $H_2O_2$ sensing. *J. Electroanal. Chem.* **2014**, *728*, 26–33. [CrossRef]
113. Manoharan, P.; Dhanabalan, S.C.; Alagan, M.; Muthuvijayan, S.; Ponraj, J.S.; Somasundaram, C.K. Facile synthesis and characterisation of green luminescent carbon nanodots prepared from tender coconut water using the acid-assisted ultrasonic route. *Micro Nano Lett.* **2020**, *15*, 920–924. [CrossRef]
114. Choi, S.A.; Jeong, Y.; Lee, J.; Huh, Y.H.; Choi, S.H.; Kim, H.S.; Cho, D.H.; Lee, J.S.; Kim, H.; An, H.R.; et al. Biocompatible liquid-type carbon nanodots (C-paints) as light delivery materials for cell growth and astaxanthin induction of *Haematococcus pluvialis*. *Mater. Sci. Eng. C* **2020**, *109*, 110500. [CrossRef] [PubMed]
115. Huang, H.; Li, S.; Chen, B.; Wang, Y.; Shen, Z.; Qiu, M.; Pan, H.; Wang, W.; Wang, Y.; Li, X. Endoplasmic reticulum-targeted polymer dots encapsulated with ultrasonic synthesized near-infrared carbon nanodots and their application for in vivo monitoring of $Cu^{2+}$. *J. Colloid Interface Sci.* **2022**, *627*, 705–715. [CrossRef]
116. Bandi, R.; Gangapuram, B.R.; Dadigala, R.; Eslavath, R.; Singh, S.S.; Guttena, V. Facile and green synthesis of fluorescent carbon dots from onion waste and their potential applications as sensor and multicolour imaging agents. *RSC Adv.* **2016**, *6*, 28633–28639. [CrossRef]
117. Sachdev, A.; Gopinath, P. Green synthesis of multifunctional carbon dots from coriander leaves and their potential application as antioxidants, sensors and bioimaging agents. *Analyst* **2015**, *140*, 4260–4269. [CrossRef]
118. Zhao, S.; Lan, M.; Zhu, X.; Xue, H.; Ng, T.W.; Meng, X.; Lee, C.S.; Wang, P.; Zhang, W. Green synthesis of bifunctional fluorescent carbon dots from garlic for cellular imaging and free radical scavenging. *ACS Appl. Mater. Interfaces* **2015**, *7*, 17054–17060. [CrossRef]
119. Wang, N.; Wang, Y.; Guo, T.; Yang, T.; Chen, M.; Wang, J. Green preparation of carbon dots with papaya as carbon source for effective fluorescent sensing of Iron (III) and *Escherichia coli*. *Biosens. Bioelectron.* **2016**, *85*, 68–75. [CrossRef]

120. Kasibabu, B.S.B.; D'Souza, S.L.; Jha, S.; Singhal, R.K.; Basu, H.; Kailasa, S.K. One-step synthesis of fluorescent carbon dots for imaging bacterial and fungal cells. *Anal. Methods* **2015**, *7*, 2373–2378. [CrossRef]
121. Zhang, J.; Yuan, Y.; Liang, G.; Yu, S.H. Scale-up synthesis of fragrant nitrogen-doped carbon dots from bee pollens for bioimaging and catalysis. *Adv. Sci.* **2015**, *2*, 1500002. [CrossRef]
122. Latief, U.; ul Islam, S.; Khan, Z.M.; Khan, M.S. A facile green synthesis of functionalized carbon quantum dots as fluorescent probes for a highly selective and sensitive detection of Fe3+ ions. *Spectrochim. Acta Part A Mol. Biomol. Spectrosc.* **2021**, *262*, 120132. [CrossRef] [PubMed]
123. Xu, J.Y.; Zhou, Y.; Liu, S.X.; Dong, M.T.; Huang, C.B. Low-cost synthesis of carbon nanodots from natural products used as a fluorescent probe for the detection of ferrum(III) ions in lake water. *Anal. Methods* **2014**, *6*, 2086. [CrossRef]
124. Mehta, V.N.; Jha, S.; Kailasa, S.K. One-pot green synthesis of carbon dots by using Saccharum officinarum juice for fluorescent imaging of bacteria (*Escherichia coli*) and yeast (*Saccharomyces cerevisiae*) cells. *Mater. Sci. Eng. C* **2014**, *38*, 20–27. [CrossRef] [PubMed]
125. Lu, W.B.; Qin, X.Y.; Liu, S.; Chang, G.H.; Zhang, Y.W.; Luo, Y.L.; Asiri, A.M.; Al-Youbi, A.O.; Sun, X.P. Economical, Green Synthesis of Fluorescent Carbon Nanoparticles and Their Use as Probes for Sensitive and Selective Detection of Mercury(II) Ions. *Anal. Chem.* **2012**, *84*, 5351–5357. [CrossRef]
126. Li, J.; Zhang, L.; Li, P.; Zhang, Y.; Dong, C. One step hydrothermal synthesis of carbon nanodots to realize the fluorescence detection of picric acid in real samples. *Sens. Actuator B Chem.* **2018**, *258*, 580–588. [CrossRef]
127. Gao, Y.; Hou, F.; Hu, S.; Wu, B.G.; Wang, Y.; Zhang, H.Q.; Jiang, B.J.; Fu, H.G. Graphene Quantum-Dot-Modified Hexagonal Tubular Carbon Nitride for Visible-Light Photocatalytic Hydrogen Evolution. *RSC Adv.* **2018**, *10*, 1330–1335. [CrossRef]
128. Jahan, S.; Mansoor, F.; Naz, S.; Lei, J.; Kanwal, S. Oxidative synthesis of highly fluorescent boron/nitrogen co-doped carbon nanodots enabling detection of photosensitizer and carcinogenic dye. *Anal. Chem.* **2013**, *85*, 10232–10239. [CrossRef]
129. Soni, N.; Singh, S.; Sharma, S.; Batra, G.; Kaushik, K.; Rao, C.; Verma, N.C.; Mondal, B.; Yadav, A.; Nandi, C.K. Absorption and emission of light in red emissive carbon nanodots. *Chem. Sci.* **2021**, *12*, 3615–3626. [CrossRef]
130. Ding, H.; Yu, S.B.; Wei, J.S.; Xiong, H.M. Full-color light-emitting carbon dots with a surface-state-controlled luminescence mechanism. *ACS Nano* **2016**, *10*, 484–491. [CrossRef]
131. Bakier, Y.M.; Ghali, M.; Elkun, A.; Beltagi, A.M.; Zahra, W.K. Static interaction between colloidal carbon nano-dots and aniline: A novel platform for ultrasensitive detection of aniline in aqueous media. *Mater. Res. Bull.* **2021**, *134*, 111119. [CrossRef]
132. Ma, C.; Zhu, Z.; Wang, H.; Huang, X.; Zhang, X.; Qi, X.; Zhang, H.L.; Zhu, Y.; Deng, X.; Peng, Y. A General Solid-State Synthesis of Chemically-Doped Fluorescent Graphene Quantum Dots for Bioimaging and Optoelectronic Applications. *Nanoscale* **2015**, *7*, 10162–10169. [CrossRef]
133. Li, X.; Lau, S.; Tang, L.; Ji, R.; Yang, P. Multicolour light emission from chlorine-doped graphene quantum dots. *J. Mater. Chem. C* **2013**, *1*, 7308–7313. [CrossRef]
134. Praneerad, J.; Neungnoraj, K.; In, I.; Paoprasert, P. Environmentally friendly supercapacitor based on carbon dots from durian peel as an electrode. *Key Eng. Mater.* **2019**, *803*, 115–119. [CrossRef]
135. Zhang, Y.Q.; Ma, D.K.; Zhuang, Y.; Zhang, X.; Chen, W.; Hong, L.L.; Huang, S.M. One-pot synthesis of N-doped carbon dots with tunable luminescence properties. *J. Mater. Chem.* **2012**, *22*, 16714–16718. [CrossRef]
136. Gunjal, D.B.; Naik, V.M.; Waghmare, R.D.; Patil, C.S.; Shejwal, R.V.; Gore, A.H.; Kolekar, G.B. Sustainable carbon nanodots synthesised from kitchen derived waste tea residue for highly selective fluorimetric recognition of free chlorine in acidic water: A waste utilization approach. *J. Taiwan Inst. Chem. Eng.* **2019**, *95*, 147–154. [CrossRef]
137. Gutiérrez-Sánchez, C.; Mediavilla, M.; Guerrero-Esteban, T.; Revenga-Parra, M.; Pariente, F.; Lorenzo, E. Direct covalent immobilization of new nitrogen-doped carbon nanodots by electrografting for sensing applications. *Carbon* **2020**, *159*, 303–310. [CrossRef]
138. Pender, J.P.; Jha, G.; Youn, D.H.; Ziegler, J.M.; Andoni, I.; Choi, E.; Heller, A.; Dunn, B.S.; Weiss, P.S.; Penner, R.M. Electrode Degradation in Lithium-Ion Batteries. *ACS Nano* **2020**, *14*, 1243–1295. [CrossRef]
139. Miao, P.; Tang, Y.; Han, K.; Wang, B. Facile synthesis of carbon nanodots from ethanol and their application in ferric (III) ion assay. *J. Mater. Chem. A* **2015**, *3*, 15068–15073. [CrossRef]
140. Keerthana, A.K.; Ashraf, P.M. Carbon nanodots synthesized from chitosan and its application as a corrosion inhibitor in boat-building carbon steel BIS2062. *Appl. Nano.* **2020**, *10*, 1061–1071. [CrossRef]
141. Arvapalli, D.M.; Sheardy, A.T.; Alapati, K.C.; Wei, J. High quantum yields fluorescent carbon nanodots for detection of Fe (III) Ions and electrochemical study of quenching mechanism. *Talanta* **2020**, *209*, 120538. [CrossRef]
142. Baluta, S.; Lesiak, A.; Cabaj, J. Simple and cost-effective electrochemical method for norepinephrine determination based on carbon dots and tyrosinase. *Sensors* **2020**, *20*, 4567. [CrossRef]
143. Zhai, X.; Zhang, P.; Liu, C.; Bai, T.; Li, W.; Dai, L.; Liu, W. Highly luminescent carbon nanodots by microwave-assisted pyrolysis. *Chem. Commun.* **2012**, *48*, 7955–7957. [CrossRef] [PubMed]
144. Zhang, P.; Li, W.; Zhai, X.; Liu, C.; Dai, L.; Liu, W. A facile and versatile approach to biocompatible "fluorescent polymers" from polymerizable carbon nanodots. *Chem. Commun.* **2012**, *48*, 10431–10433. [CrossRef] [PubMed]
145. Li, L.; Ji, J.; Fei, R.; Wang, C.; Lu, Q.; Zhang, J.; Jiang, L.P.; Zhu, J.J. A facile microwave avenue to electrochemiluminescenttwo-color graphene quantum dots. *Adv. Funct. Mater.* **2012**, *22*, 2971–2979. [CrossRef]

146. Liu, H.; He, Z.; Jiang, L.P.; Zhu, J.J. Microwave-assisted synthesis of wavelength-tunablephotoluminescent carbon nanodots and their potential applications. *ACS Appl. Mater. Interfaces* **2015**, *7*, 4913–4920. [CrossRef]
147. Kumawat, M.K.; Thakur, M.; Gurung, R.B.; Srivastava, R. Graphene quantum dots from mangiferaindica: Application in near-infrared bioimaging and intracellular nanothermometry. *ACS Sustain. Chem. Eng.* **2017**, *5*, 1382–1391. [CrossRef]
148. Pires, N.R.; Santos, C.M.W.; Sousa, R.R.; De Paula, R.C.M.; Cunha, P.L.R.; Feitosa, J.P.A. Novel and Fast Microwave-Assisted Synthesis of Carbon Quantum Dots from Raw Cashew Gum. *J. Brazil. Chem. Soc.* **2015**, *26*, 1274–1282. [CrossRef]
149. Simsek, S.; Alas, M.O.; Ozbek, B.; Genc, R. Evaluation of the physical properties of fluorescent carbon nanodots synthesized using Nerium oleander extracts by microwave-assisted synthesis methods. *J. Mater. Res. Technol.* **2019**, *8*, 2721–2731. [CrossRef]
150. Ren, Q.; Ga, L.; Ai, J. Rapid synthesis of highly fluorescent nitrogendoped graphene quantum dots for effective detection of ferric ions and as fluorescent ink. *ACS Omega* **2019**, *4*, 15842–15848. [CrossRef]
151. Sendao, R.M.S.; Crista, J.D.M.A.; Afonso, A.C.P.; Yuso, M.V.M.; Algarra, M.; Silva, J.C.E.; Silva, L.P. Insight into the hybrid luminescence showed by carbon dots and molecular fluorophores in solution. *Phys. Chem. Chem. Phys.* **2019**, *21*, 20919. [CrossRef]
152. Luo, X.; Huang, G.; Bai, C.; Wang, C.; Yu, Y.; Tan, Y.; Tang, C.; Kong, J.; Huang, J.; Li, Z. A versatile platform for colorimetric, fluorescence and photothermal multi-mode glyphosate sensing by carbon dots anchoring ferrocene metal-organic framework nanosheet. *J. Hazard. Mat.* **2023**, *443*, 130277. [CrossRef]
153. Wang, C.; Xu, Z.; Zhang, C. Polyethyleneimine-functionalized fluorescent carbon dots: Water stability, pH sensing, and cellular imaging. *ChemNanoMat* **2015**, *1*, 122–127. [CrossRef]
154. Mandal, P.; Sahoo, D.; Sarkar, P.; Chakraborty, K.; Das, S. Fluorescence turn-on andturn-off sensing of pesticides by carbon dot-based sensor. *New J. Chem.* **2019**, *43*, 12137–12151. [CrossRef]
155. Papaioannou, N.; Marinovic, A.; Yoshizawa, N.; Goode, A.E.; Fay, M.; Khlobystov, A.; Titirici, M.M.; Sapelkin, A. Structure and solvents effects on the optical properties of sugar-derivedcarbon nanodots. *Sci. Rep.* **2018**, *8*, 6559. [CrossRef]
156. Wang, J.C.; Violette, K.; Ogunsolu, O.O.; Hanson, K. Metal ion mediated electrontransfer at dye–semiconductor interfaces. *Phys. Chem. Chem. Phys.* **2017**, *19*, 2679–2682. [CrossRef]
157. Zheng, H.; Wang, Q.; Long, Y.; Zhang, H.; Huang, X.; Zhu, R. Enhancing theluminescence of carbon dots with a reduction pathway. *Chem. Commun.* **2011**, *47*, 10650–10652. [CrossRef]
158. Deshmukh, S.; Deore, A.; Mondal, S. Ultrafast dynamics in carbon dots as photosensitizers: A review. *ACS Appl. Nano Mater.* **2021**, *4*, 7587–7606. [CrossRef]
159. Sciortino, A.; Madonia, A.; Gazzetto, M.; Sciortino, L.; Rohwer, E.J.; Feurer, T.; Gelardi, F.M.; Cannas, M.; Cannizzo, A.; Messina, F. The interaction of photoexcited carbon nanodotswith metal ions disclosed down to the femtosecond scale. *Nanoscale* **2017**, *9*, 11902–11911. [CrossRef]
160. Liu, R.; Li, H.; Kong, W.; Liu, J.; Liu, Y.; Tong, C.; Zhang, X.; Kang, Z. Ultra-sensitiveand selective $Hg^{2+}$ detection based on fluorescent carbon dots. *Mater. Res. Bull.* **2013**, *48*, 2529–2534. [CrossRef]
161. Huang, H.; Lv, J.J.; Zhou, D.L.; Bao, N.; Xu, Y.; Wang, A.J.; Feng, J.J. One-pot greensynthesis of nitrogen-doped carbon nanoparticles as fluorescent probes for mercury ions. *RSC Adv.* **2013**, *3*, 21691–21696. [CrossRef]
162. Wang, X.; Zhang, J.; Zou, W.; Wang, R. Facile synthesis of polyaniline/carbon dotnanocomposites and their application as a fluorescent probe to detect mercury. *RSC Adv.* **2015**, *5*, 41914–41919. [CrossRef]
163. Liu, C.; Tang, B.; Zhang, S.; Zhou, M.; Yang, M.; Liu, Y.; Zhang, Z.L.; Zhang, B.; Pang, D.W. Photoinduced electron transfer mediated by coordination between carboxyl on carbon nanodotsand $Cu^{2+}$ quenching photoluminescence. *J. Phy. Chem. C* **2018**, *122*, 3662–3668. [CrossRef]
164. Zhu, A.; Qu, Q.; Shao, X.; Kong, B.; Tian, Y. Carbon-dot-based dual-emissionnanohybrid produces a ratiometric fluorescent sensor for in vivo imaging of cellular copperions. *Angew. Chem. Int. Ed.* **2012**, *51*, 7185–7189. [CrossRef] [PubMed]
165. Mohammed, L.J.; Omer, K.M. Dual functional highly luminescence B, N Co-dopedcarbon nanodots as nanothermometer and $Fe^{3+}$/$Fe^{2+}$ sensor. *Sci. Rep.* **2020**, *10*, 3028. [CrossRef] [PubMed]
166. Sekar, A.; Yadav, R.; Basavaraj, N. Fluorescence quenching mechanism and the application of green carbon nanodots in the detection of heavy metal ions: A review. *New J. Chem.* **2021**, *45*, 2326–2360. [CrossRef]
167. Goncalves, H.M.; Duarte, A.J.; da Silva, J.C.E. Optical fiber sensor for Hg (II) based on carbon dots. *Biosen. Bioelect.* **2010**, *26*, 1302–1306. [CrossRef] [PubMed]
168. Gharat, P.M.; Pal, H.; Choudhury, S.D. Photophysics and luminescence quenching ofcarbon dots derived from lemon juice and glycerol. *Spectrochim. Acta Part A Mol. Biomol. Spectro.* **2019**, *209*, 14–21. [CrossRef]
169. Karali, K.K.; Sygellou, L.; Stalikas, C.D. Highly fluorescent N-doped carbon nanodotsas an effective multi-probe quenching system for the determination of nitrite, nitrate and ferric ions infood matrices. *Talanta* **2018**, *189*, 480–488. [CrossRef] [PubMed]
170. Guo, J.; Liu, D.; Filpponen, I.; Johansson, L.S.; Malho, J.M.; Quraishi, S.; Liebner, F.; Santos, H.A.; Rojas, O.J. Photoluminescent hybrids of cellulose nanocrystals and carbon quantum dots as cytocompatible probes for in vitro bioimaging. *Biomacromolecules* **2017**, *18*, 2045–2055. [CrossRef]
171. Cronican, J.J.; Thompson, D.B.; Beier, K.T.; McNaughton, B.R.; Cepko, C.L.; Liu, D.R. Potent delivery of functional proteins into Mammalian cells in vitro and in vivo using a superchargedprotein. *ACS Chem. Biol.* **2010**, *5*, 747–752. [CrossRef]
172. Pan, L.; Sun, S.; Zhang, L.; Jiang, K.; Lin, H. Near-infrared emissive carbon dots for two-photon fluorescence bioimaging. *Nanoscale* **2016**, *8*, 17350–17356. [CrossRef]

173. Sahu, S.; Behera, B.; Maiti, T.K.; Mohapatra, S. Simple one-step synthesis of highly luminescent carbon dots from orange juice: Application as excellent bio-imaging agents. *Chem. Commun.* **2012**, *48*, 8835–8837. [CrossRef] [PubMed]
174. Sivasankarapillai, V.S.; Jose, J.; Shanavas, M.S.; Marathakam, A.; Uddin, M.; Mathew, B. Silicon quantum dots: Promising theranostic probes for the future. *Curr. Drug Targets* **2019**, *20*, 1255–1263. [CrossRef] [PubMed]
175. Liu, C.; Zhang, P.; Tian, F.; Li, W.; Li, F.; Liu, W. One-step synthesis of surfacepassivated carbon nanodots by microwave assisted pyrolysis for enhanced multicolour photoluminescence and bioimaging. *J. Mater. Chem.* **2011**, *21*, 13163–13167. [CrossRef]
176. Chen, B.B.; Wang, X.Y.; Qian, R.C. Rolling "wool-balls": Rapid live-cell mapping of membrane sialic acids via poly-p-benzoquinone/ethylenediamine nanoclusters. *Chem. Commun.* **2019**, *55*, 9681–9684. [CrossRef] [PubMed]
177. Tao, H.; Yang, K.; Ma, Z.; Wan, J.; Zhang, Y.; Kang, Z.; Liu, Z. In vivo NIR fluorescenceimaging, biodistribution, and toxicology of photoluminescent carbon dots produced from carbonnanotubes and graphite. *Small* **2012**, *8*, 281–290. [CrossRef]
178. Choi, Y.; Kim, S.; Choi, M.H.; Ryoo, S.R.; Park, J.; Min, D.H.; Kim, B.S. Highly biocompatible carbon nanodots for simultaneous bioimaging and targeted photodynamic therapy invitro and in vivo. *Adv. Functional Mater.* **2014**, *24*, 5781–5789. [CrossRef]
179. Shi, Q.Q.; Li, Y.H.; Xu, Y.; Wang, Y.; Yin, X.B.; He, X.W.; Zhang, Y.K. High-yield and high-solubility nitrogen-doped carbon dots: Formation, fluorescence mechanism and imaging application. *RSC Adv.* **2014**, *4*, 1563–1566. [CrossRef]
180. Wang, J.; Zhang, P.; Huang, C.; Liu, G.; Leung, K.C.F.; Wáng, Y.X.J. High performancephotoluminescent carbon dots for in vitro and in vivo bioimaging: Effect of nitrogen dopingratios. *Langmuir* **2015**, *31*, 8063–8073. [CrossRef]
181. Bankoti, K.; Rameshbabu, A.P.; Datta, S.; Das, B.; Mitra, A.; Dhara, S. Onion derived carbon nanodots for live cell imaging and accelerated skin wound healing. *J. Mater. Chem. B* **2017**, *5*, 6579–6592. [CrossRef]
182. Ming, H.; Ma, Z.; Liu, Y.; Pan, K.; Yu, H.; Wang, F.; Kang, Z. Large scaleelectrochemical synthesis of high-quality carbon nanodots and their photocatalytic property. *Dalton Trans.* **2012**, *41*, 9526–9531. [CrossRef]
183. Song, B.; Wang, T.; Sun, H.; Shao, Q.; Zhao, J.; Song, K.; Hao, L.; Wang, L.; Guo, Z. Two-step hydrothermally synthesized carbon nanodots/WO$_3$ photocatalysts with enhanced photocatalytic performance. *Dalton Trans.* **2017**, *46*, 15769–15777. [CrossRef] [PubMed]
184. Qu, S.; Chen, H.; Zheng, X.; Cao, J.; Liu, X. Ratiometric fluorescent nanosensor based on water soluble carbon nanodots with multiple sensing capacities. *Nanoscale* **2013**, *5*, 5514–5518. [CrossRef] [PubMed]
185. Vedamalai, M.; Periasamy, A.P.; Wang, C.W.; Tseng, Y.T.; Ho, L.C.; Shih, C.C.; Chang, H.T. Carbon nanodots prepared from o-phenylenediamine for sensing of Cu$^{2+}$ ions in cells. *Nanoscale* **2014**, *6*, 13119–13125. [CrossRef] [PubMed]
186. Shi, L.; Li, Y.; Li, X.; Zhao, B.; Wen, X.; Zhang, G.; Dong, C.; Shuang, S. Controllable synthesis of green and blue fluorescent carbon nanodots for pH and Cu$^{2+}$ sensing in living cells. *Biosen. Bioelect.* **2016**, *77*, 598–602. [CrossRef]
187. Xu, B.; Zhao, C.; Wei, W.; Ren, J.; Miyoshi, D.; Sugimoto, N.; Qu, X. Aptamer carbon-nanodot sandwich used for fluorescent detection of protein. *Analyst* **2012**, *137*, 5483–5486. [CrossRef]
188. Nie, H.; Li, M.; Li, Q.; Liang, S.; Tan, Y.; Sheng, L.; Shi, W.; Zhang, S.X.A. Carbon dots with continuously tunable full-color emission and their application in ratiometric pH sensing. *Chem. Mater.* **2014**, *26*, 3104–3112. [CrossRef]
189. Wang, C.I.; Wu, W.C.; Periasamy, A.P.; Chang, H.T. Electrochemical synthesis of photoluminescent carbon nanodots from glycine for highly sensitive detection of hemoglobin. *Green Chem.* **2014**, *16*, 2509–2514. [CrossRef]
190. Gogoi, N.; Chowdhury, D. Novel carbon dot coated alginate beads with superior stability, swelling and pH responsive drug delivery. *J. Mater. Chem. B* **2014**, *2*, 4089–4099. [CrossRef]
191. Karthik, S.; Saha, B.; Ghosh, S.K.; Singh, N.P. Photoresponsivequinoline tethered fluorescent carbon dots for regulated anticancer drug delivery. *Chem. Commun.* **2013**, *49*, 10471–10473. [CrossRef]
192. Chen, B.; Yang, Z.; Zhu, Y.; Xia, Y. Zeolitic imidazolate framework materials: Recent progress in synthesis and applications. *J. Mater. Chem. A* **2014**, *2*, 16811–16831. [CrossRef]
193. He, L.; Wang, T.; An, J.; Li, X.; Zhang, L.; Li, L.; Li, G.; Wu, X.; Su, Z.; Wang, C. Carbon nanodots@ zeolitic imidazolate framework-8 nanoparticles for simultaneous pH-responsive drug delivery and fluorescence imaging. *Cryst. Eng. Commun.* **2014**, *16*, 3259–3263. [CrossRef]
194. Kim, Y.; Jang, G.; Lee, T.S. New fluorescent metal-ion detection using a paper-based sensor strip containing tethered rhodamine carbon nanodots. *ACS Appl. Mater. Interfaces* **2015**, *7*, 15649–15657. [CrossRef]
195. Wang, J.; Zhang, Z.; Zha, S.; Zhu, Y.; Wu, P.; Ehrenberg, B.; Chen, J.Y. Carbon nanodots featuring efficient FRET for two-photon photodynamic cancer therapy with a low fs laser power density. *Biomaterials* **2014**, *35*, 9372–9381. [CrossRef] [PubMed]
196. Zheng, M.; Liu, S.; Li, J.; Qu, D.; Zhao, H.; Guan, X.; Hu, X.; Xie, Z.; Jing, X.; Sun, Z. Integrating oxaliplatin with highly luminescent carbon dots: An unprecedented theranostic agent for personalized medicine. *Adv. Mat.* **2014**, *26*, 3554–3560. [CrossRef] [PubMed]
197. Huang, Y.; Chen, J.; Wong, T.; Liow, J.L. Experimental and theoretical investigations of non-Newtonian electro-osmotic driven flow in rectangular microchannels. *Soft Matter* **2016**, *12*, 6206–6213. [CrossRef]
198. Atencia, J.; Beebe, D.J. Controlled microfluidic interfaces. *Nature* **2005**, *437*, 648–655. [CrossRef]
199. Stewart, M.P.; Sharei, A.; Ding, X.; Sahay, G.; Langer, R.; Jensen, K.F. In vitro and ex vivo strategies for intracellular delivery. *Nature* **2016**, *538*, 183–192. [CrossRef]
200. Song, H.; Chen, D.L.; Ismagilov, R.F. Reactions in droplets in microfluidic channels. *Angew. Chem. Int. Ed.* **2006**, *45*, 7336–7356. [CrossRef]
201. Huang, Y.; Xiao, L.; An, T.; Lim, W.; Wong, T.; Sun, H. Fast dynamic visualizations in microfluidics enabled by fluorescent carbon nanodots. *Small* **2017**, *13*, 1700869. [CrossRef]

202. Lin, H.; Storey, B.D.; Oddy, M.H.; Chen, C.H.; Santiago, J.G. Instability of electrokinetic microchannel flows with conductivity gradients. *Phys. Fluids* **2004**, *16*, 1922–1935. [CrossRef]
203. Yang, S.T.; Wang, X.; Wang, H.; Lu, F.; Luo, P.G.; Cao, L.; Meziani, M.J.; Liu, J.H.; Liu, Y.; Chen, M.; et al. Carbon Dots as Nontoxic and High-Performance Fluorescence Imaging Agents. *J. Phys. Chem. C* **2009**, *113*, 18110–18114. [CrossRef] [PubMed]
204. Zhu, A.; Luo, Z.; Ding, C.; Li, B.; Zhou, S.; Wang, R.; Tian, Y. A two-photon "turn-on" fluorescent probe based on carbon nanodots for imaging and selective biosensing of hydrogen sulfide in live cells and tissues. *Analyst* **2014**, *139*, 1945–1952. [CrossRef] [PubMed]
205. Chandra, A.; Deshpande, S.; Shinde, D.B.; Pillai, V.K.; Singh, N. Mitigating the cytotoxicity of graphene quantum dots and enhancing their applications in bioimaging and drug delivery. *ACS Macro Lett.* **2014**, *3*, 1064–1068. [CrossRef] [PubMed]
206. Song, Y.; Shi, W.; Chen, W.; Li, X.; Ma, H. Fluorescent carbon nanodots conjugated with folic acid for distinguishing folate-receptor-positive cancer cells from normal cells. *J. Mater. Chem.* **2012**, *22*, 12568–12573. [CrossRef]
207. Milton, F.P.; Govan, J.; Mukhina, M.V.; Gun'ko, Y.K. The chiral nano-world: Chiroptically active quantum nanostructures. *Nano. Hori.* **2016**, *1*, 14–26. [CrossRef]
208. Liu, X.L.; Tsunega, S.; Jin, R.H. Self-directing chiral information in solid–solid transformation: Unusual chiral-transfer without racemization from amorphous silica to crystalline silicon. *Nano. Hori.* **2017**, *2*, 147–155. [CrossRef]
209. Walker, R.; Pociecha, D.; Abberley, J.P.; Martinez-Felipe, A.; Paterson, D.A.; Forsyth, E.; Lawrence, G.B.; Henderson, P.A.; Storey, J.M.; Gorecka, E.; et al. Spontaneous chirality through mixing achiral components: A twist-bend nematic phase driven by hydrogen-bonding between unlike components. *Chem. Commun.* **2018**, *54*, 3383–3386. [CrossRef]
210. Vázquez-Nakagawa, M.; Rodríguez-Pérez, L.; Herranz, M.A.; Martín, N. Chirality transfer from graphene quantum dots. *Chem. Commun.* **2016**, *52*, 665–668. [CrossRef]
211. Xin, Q.; Liu, Q.; Geng, L.; Fang, Q.; Gong, J.R. Chiral nanoparticles as a new efficient antimicrobial nanoagent. *Adv. Healthcare Mater.* **2017**, *6*, 1601011. [CrossRef]
212. Li, F.; Li, Y.; Yang, X.; Han, X.; Jiao, Y.; Wei, T.; Yang, D.; Xu, H.; Nie, G. Highly fluorescent chiral N-S-doped carbon dots from cysteine: Affecting cellular energy metabolism. *Angew. Chem.* **2018**, *130*, 2401–2406. [CrossRef]
213. Malishev, R.; Arad, E.; Bhunia, S.K.; Shaham-Niv, S.; Kolusheva, S.; Gazit, E.; Jelinek, R. Chiral modulation of amyloid beta fibrillation and cytotoxicity by enantiomeric carbon dots. *Chem. Commun.* **2018**, *54*, 7762–7765. [CrossRef] [PubMed]

**Disclaimer/Publisher's Note:** The statements, opinions and data contained in all publications are solely those of the individual author(s) and contributor(s) and not of MDPI and/or the editor(s). MDPI and/or the editor(s) disclaim responsibility for any injury to people or property resulting from any ideas, methods, instructions or products referred to in the content.

*Review*

# Carbon Quantum Dots: Synthesis, Structure, Properties, and Catalytic Applications for Organic Synthesis

Pradeep Kumar Yadav [1], Subhash Chandra [2], Vivek Kumar [3], Deepak Kumar [4] and Syed Hadi Hasan [3],*

[1] Department of Chemistry, Jagatpur P.G. College, Varanasi 221301, India
[2] Department of Chemistry, Bappa Sri Narain Vocational P.G. College, Lucknow 226001, India
[3] Nano Material Research Laboratory, Department of Chemistry, Indian Institute of Technology (BHU), Varanasi 221005, India
[4] Department of Chemistry, Chhadami Lal Jain P.G. College, Firozabad 283203, India
* Correspondence: shhasan.apc@itbhu.ac.in; Tel.: +91-9839089919

**Abstract:** Carbon quantum dots (CQDs), also known as carbon dots (CDs), are novel zero-dimensional fluorescent carbon-based nanomaterials. CQDs have attracted enormous attention around the world because of their excellent optical properties as well as water solubility, biocompatibility, low toxicity, eco-friendliness, and simple synthesis routes. CQDs have numerous applications in bioimaging, biosensing, chemical sensing, nanomedicine, solar cells, drug delivery, and light-emitting diodes. In this review paper, the structure of CQDs, their physical and chemical properties, their synthesis approach, and their application as a catalyst in the synthesis of multisubstituted 4H pyran, in azide-alkyne cycloadditions, in the degradation of levofloxacin, in the selective oxidation of alcohols to aldehydes, in the removal of Rhodamine B, as H-bond catalysis in Aldol condensations, in cyclohexane oxidation, in intrinsic peroxidase-mimetic enzyme activity, in the selective oxidation of amines and alcohols, and in the ring opening of epoxides are discussed. Finally, we also discuss the future challenges in this research field. We hope this review paper will open a new channel for the application of CQDs as a catalyst in organic synthesis.

**Keywords:** carbon quantum dots; synthetic methods; fluorescence; optical properties; catalyst

## 1. Introduction

Recently, carbon-based nanomaterials such as graphene [1], fullerenes [2], nanodiamonds [3], carbon nanotubes (CNTs) [4], and carbon quantum dots (CQDs) have attracted great attention because of their distinctive structural dimensions, as well as their outstanding chemical and physical properties [5]. It was found that the preparation and separation of nanodiamonds are complicated, while other nanomaterials such as graphene, fullerenes, and CNTs do not display good water solubility and also do not exhibit strong fluorescence in the visible region. These limitations prevent their applications in different areas [6]. Although semiconductor quantum dots (SQDs) exhibit good fluorescence properties, because of the presence of heavy metals, they are toxic in nature. This prevents their biological applicationin biosensors, bio-imaging, and drug delivery. In contrast, fluorescent CQDs are nontoxic and, thus, have attracted enormous interest over other carbon-based nanomaterials [7]. Xu et al. in 2004 accidentally discovered CQDs using gel electrophoresis during the purification of single-walled carbon nanotubes [8]. However, the name CQDs was given by Sun et al. in 2006 during the synthesis of carbon nanomaterials of different sizes [9]. Subsequently, CQDs became rising stars among various carbon-based nanoparticles and are considered an extremely precious asset of nanotechnology. CQDs are also known as carbon nano-lights because of their strong luminescence properties [10]. CQDs have attractive features such as ease of synthesis, good water solubility, high photostability, high photoresponse, low cytotoxicity, facile surface functionalization,

**Citation:** Yadav, P.K.; Chandra, S.; Kumar, V.; Kumar, D.; Hasan, S.H. Carbon Quantum Dots: Synthesis, Structure, Properties, and Catalytic Applications for Organic Synthesis. *Catalysts* **2023**, *13*, 422. https://doi.org/10.3390/catal13020422

Academic Editors: Indra Neel Pulidindi, Archana Deokar and Aharon Gedanken

Received: 31 December 2022
Revised: 10 February 2023
Accepted: 14 February 2023
Published: 16 February 2023

**Copyright:** © 2023 by the authors. Licensee MDPI, Basel, Switzerland. This article is an open access article distributed under the terms and conditions of the Creative Commons Attribution (CC BY) license (https://creativecommons.org/licenses/by/4.0/).

good catalysis properties, and tunable excitation–emission [11–17]. Due to these characteristic properties, CQDs are widely utilized in photovoltaic devices, medical diagnosis, sensing, drug delivery, catalysts, photocatalysis, optronic devices, bio-imaging, laser, single electron transistors, solar cells, and LEDs [18–29]. However, very few reports have been investigated regarding the application of CQDs as a catalyst in photochemical water splitting [30] the preparation of substituted 4H pyran with indole moieties [31], azide-alkyne cycloadditions [32], the degradation of levofloxacin [33], the selective oxidation of alcohols to aldehydes [34], the removal of Rhodamine B [35], the selective oxidation of amines and imine [36], high-efficiency cyclohexane oxidation [37], H-bond catalysis in Aldol condensations [38], intrinsic peroxidase-mimetic enzyme activity [39], and the ring opening of epoxides [40]. In this review paper, we explain the synthetic approach, structure, optical properties, and applications of CQDs as a catalyst. Finally, we also discuss their future prospects.

## 2. Synthesis Approach

Since the discovery of carbon quantum dots (CQDs), several convenient, cost-effective, size-controlled, and large-scale production approaches have been developed. For the synthesis of CQDs, two general categories, top-down and bottom-up, approaches are utilized (Figure 1). Even though CQDs synthesis is facile, there are definite challenges related to their synthesis, such as an aggregation of nanomaterials, the tuning of surface properties, and controlling the size and uniformity [41]. To adjust the functional groups present on the surface and achieve better CQDs performance, post-treatment can be conducted in both approaches. Quantum yields (QYs) of CQDs can be enhanced after surface passivation, which eliminates the emissive traps from the surface. CQDs doped with heteroatoms (N and P) or metals such as Au or Mg improve solubility and electrical conductivity [42]. Even though for the synthesis of CQDs, both the top-down and bottom-up approaches have been used, the environmentally and cost-effective bottom-up approach is most commonly used [43].

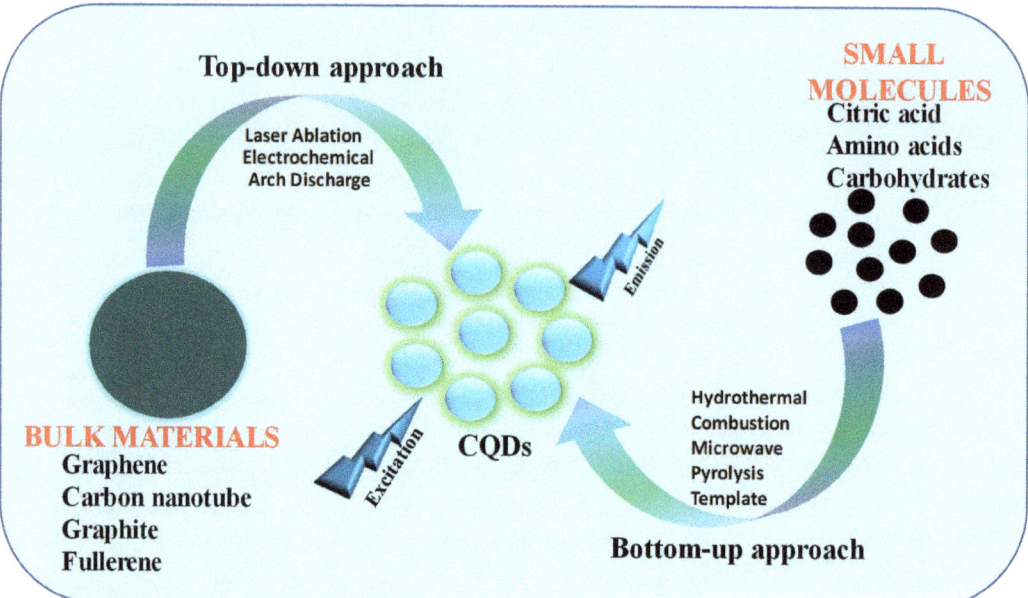

**Figure 1.** The typical approaches for the synthesis of CQDs.

## 2.1. Top-Down Approach

In a top-down approach, the larger carbon resources such as carbon nanotubes, fullerene, graphite, graphene, carbon soot, activated carbon, etc., are broken down into smaller constituents with the help of different techniques such as laser ablation and electrochemical and arch discharge [28,44–47]. Carbon structures with $sp^2$ hybridization that lack efficient energy gaps or band gaps are commonly used as starting materials for top-down processes. Although the top-down approach is extremely helpful and suitable for microsystem industries, it has some limitations, such as the fact that pure nanomaterials cannot be obtained from the large carbon precursor; their purification is costly and also unable to accurately control the morphology and size distribution of CQDs [48].

### 2.1.1. Laser Ablation Method

Sun and co-workers in 2006 first reported a laser ablation technique. In this technique, the CQDs are synthesized by irradiating a target surface with a high-energy laser pulse [9]. Recently, Li et al. synthesized ultra-small CQDs with uniform sizes by using the laser ablation method. They utilized fluorescent CQDs for cell imaging applications [49]. Cui and co-workers have also synthesized homogeneous CQDs by an ultrafast, highly efficient dual-beam pulsed laser ablation method for bio-imaging applications along with high QYs [50]. Buendia and co-workers also used laser ablation techniques to synthesize fluorescent CQDs for cell labeling [51]. The CQDs synthesized by this technique are usually non-fluorescent in nature, have heterogeneity in size, and have low quantum yield, which influences different potential applications of CQDs. Therefore, to increase the fluorescence properties and quantum yield, pre-treatments such as surface passivation (doping) and oxidation are required.

### 2.1.2. Electrochemical Method

The electrochemical method was first described by Zhou and coworkers in 2007. They used tetra-butyl ammonium perchlorate solution as the electrolyte to fabricate the first blue luminescent CQDs from multiwall carbon nanotubes (CNTs) [52]. In this method, larger carbon precursors are cut down into smaller parts by electrochemical oxidation in the presence of a reference electrode. Zhao et al. prepared fluorescent carbon nanomaterial by electrochemical oxidation with the help of a graphite rod as a working electrode [53]. Subsequently, Zheng and colleagues developed water soluble CQDs with tunable luminescence using graphite as an electrode material and buffering the pH with phosphate [54]. Using the oxidation method, Deng and coworkers synthesized the CQDs from low-molecular-weight alcohol. According to them, the most straightforward and convenient way to create CQDs is to conduct it under ambient pressure and temperature [55]. Hou and colleagues manufactured bright blue emitting CQDs in 2015 by treating urea and sodium citrate electrochemically in de-ionized water [56]. The electrochemical method has a few benefits; for example, it requires no surface passivation, is low cost, and has a simple purification process [42]. However, the limitation of this method is that for the synthesis of CQDs, it allows only a few little molecular precursors and has a tedious purification process. Therefore, it is the least frequently used technique [41].

### 2.1.3. Arch Discharge Method

Fluorescent carbon quantum dots were first discovered by Xu and coworkers accidentally during the separation and purification of a single-wall carbon nanotube by the arch discharge method. In this process, nitric acid was used as an oxidizing agent to oxidize arch ash, which formed the different functional groups on the surface, due to which aqueous solubility increased. The QYs obtained were 1.66% at a 366 nm excitation wavelength [8]. An additional experiment demonstrated that the surface of CQDs was attached to hydrophilic carboxyl groups. In the discharge process, carbon particles of different sizes are produced. CQDs obtained using this method are highly water soluble, having a wide distribution of particle sizes. Furthermore, an electronic flash method was used to separate fluorescent nanomaterials from neat carbon nanostructures and carbon nanostructures oxidized with nitric acid [57,58]. Zhang et al. synthesized CQDs with up-conversion fluorescence using arc-synthesized carbon by-products, and Hamid Delavariet al. synthesized CQDs by arc discharge in water [59,60]. However, CQDs synthesized by this technique have some impurities that are difficult to eliminate because of their complex composition [28].

*2.2. Bottom-Up Approach*

In a bottom-up approach, the smaller carbon resources such as amino acids, polymers, carbohydrates, and waste materials combine to form CQDs by a variety of techniques such as hydrothermal/solvothermal, combustion, pyrolysis, and microwave irradiation. In this method, the size and structure of CQDs depend on a variety of factors such as solvent, precursor molecular structures, and conditions of the reaction (temperature, pressure, reaction time, etc.). The conditions of the reaction are necessary, since they influence the reactants and the extremely casual nucleation and escalation procedure of CQDs. This approach strengthens the material chemistry because of its ease of operation, lower cost, and easier implementation for production in a large scale [61].

The precursor used for the synthesis of CQDs may be both chemical and biological, i.e., natural. The chemical precursors include glucose, sucrose, citric acid, lactic acid, ascorbic acid, glycerol, ethylene glycol, etc. [62–68]. The natural sources include Artocarpous lakoocha seeds, rice husks, Azadirachta indica leaves, pomelo peel, the latex of Ficus benghalensis, aloe vera, etc. (Figure 2) [69–72].

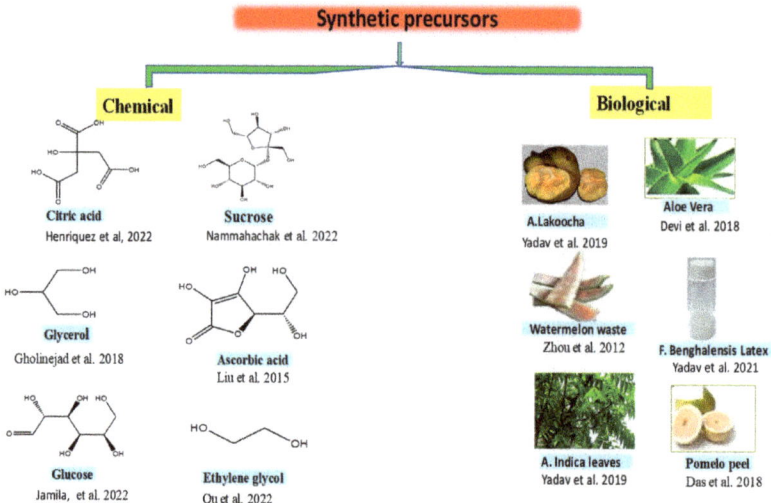

**Figure 2.** Chemical and biological precursors utilized for the synthesis of CQDs [61–72].

2.2.1. Hydrothermal Method

The hydrothermal method was first reported by Zhang et al., for the synthesis of CQDs from the precursor L-ascorbic acid (carbon source) without any chemical action or other surface passivation. The average size of the synthesized CQDs was ~2 nm, and the QY obtained was 6.79%. They utilized four different solvents (water, ethyl acetate, acetone, and ethanol) for the synthesis of bright blue emission CQDs and observed that the water soluble CQDs were very stable at room temperature over 6 months. Additionally, the fluorescence intensity of CQDs was stable in a wide pH range and highly ionic salt conditions (2 M NaCl) [73]. In the hydrothermal process, the precursor molecules are dissolved in water, set aside in a Teflon-lined stainless steel autoclave, and placed in the hydrothermal chamber at high temperature and pressure for a few hours [66].The precursor molecules utilized for the synthesis include proteins, polymers, amino acids, polyols, glucose, some wastes, and natural products [13,74]. In recent years, the hydrothermal method has attracted great attention around the world because of its single step, ease of operation, nontoxicity, low cost, and ecofriendliness. CQDs prepared from the hydrothermal treatment have a range of beneficial properties, such as being highly homogeneous, watersoluble, monodispersed, and photostable, having salt tolerance and a controlled particle size, and exhibiting an elevated QY with no surface passivation. Similar to the hydrothermal method, for the synthesis of CQDs, a solvothermal method is also utilized using ammonia, alcohol, and other organic and inorganic solvents as a substitute for water [63,75–77].

2.2.2. Combustion Method

In 2007, Liu et al. first reported the combustion method to synthesize CQDs. This method involves oxidative acid treatments which aggregate smaller carbon resources into CQDs, enhance the aqueous solubility, and control the fluorescence properties. Liu and coworkers explained that candle ashes were obtained by partial combustion of a candle with aluminum foil and refluxing it in nitric acid solution. When the candle ashes were dissolved in a neutral medium followed by centrifugation and a dialysis method, the pure CQDs were obtained [78]. The CQDs synthesized by the combustion method had low QY but displayed good fluorescence without doping [70].

2.2.3. Pyrolysis Method

The pyrolysis method is the thermal decomposition of the precursor at an elevated temperature (typically over 430 °C) and under pressure in the absence of oxygen. Additionally, the carbon precursor cleavages into nanoscale colloidal particles in the presence of an alkali and strong acid concentration as a catalyst. The advantageous properties of this method include practicability, repeatability, and simplicity, as well as having a high QY. However, it is challenging to separate small precursors from raw materials.

In 2009, Liu et al. first described a novel method for the preparation of CQDs through pyrolysis using resol (as a carbon source) and surfactant-modified silica spheres. The synthesized CQDs exhibited blue fluorescence and were amorphous, with sizes ranging from 1.5 to 2.5 nm, and the QY obtained was 14.7%. Moreover, the CQDs were stable in a broad pH range (pH 5–9) [79]. After that, several investigations were carried out for the preparation of CQDs using the pyrolysis method. Pan et al., in 2010, synthesized extremely blue fluorescent CQDs from ethylenediamine-tetraacetic acid (EDTA) salts using the pyrolysis method. The average size of the synthesized CQDs was 6 nm. The quantum yield (QY) obtained was 40.6% [80]. With the help of the pyrolysis of citric acid at 180 °C, Martindale and coworkers (in 2015) synthesized fluorescent CQDs with an average size of 6 nm, and at the excitation of 360 nm, the calculated QY was 2.3% [81]. Rong and coworkers in 2017 also prepared fluorescent N-CQDs by the pyrolysis of citric acid and guanidinium chloride without organic solvent, acid, alkali, or further modification and passivation, resulting in N-CQDs with a size of 2.2 nm and a QY of 19.2%. They utilized N-CQDs intensively in the detection of metal-ion ($Fe^{3+}$) and in bio-imaging [82]. Lately, several CQDs were synthesized using the pyrolysis method and utilized in different fields [41,83,84].

2.2.4. Microwave Irradiation Method

Microwave synthesis is a faster and cost-effective method for the synthesis of CQDs via microwave heating. Compared to other techniques, this is a simple and convenient method because it requires less time for the synthesis of CQDs, with an improved quantum yield. Zhu et al. first synthesized fluorescent CQDs under the microwave (500W) by heating poly (ethylene glycol) (PEG-200) and saccharide for 2–10 min [48]. This method is rapid, novel, green, and energy efficient in synthesizing CQDs. However, there are some limitations, such as difficulty in the separation procedure and purification, and that non-uniform particle sizes of CQDs restrict their prospective applications [85,86]. Recently, various investigations were carried out for the preparation of CQDs using microwave irradiation, utilizing them for different applications [87–91].

2.2.5. Template Method

Bourlinos and coworkers first synthesized fluorescent CQDs using the template method [92]. The template method involves two steps: (i) The preparation of CQDs in the appropriate template or silicon sphere by calcinations. (ii) The etching process occurs to eliminate the supporting materials. Some advantageous properties of the template method are that it is straightforward, the equipment is easily obtainable, it is suitable for the surface passivation of CQDs, it prevents the particles from agglomerating, and it controls the size of CQDs. The disadvantageous property of the template method is the difficulty in the separation of the CQDs from the template, which may affect the purity, particle size, fluorescence property, and QY.

## 3. Structure of CQDs

Tang et al. reported that CQDs have core–shell structures which are either amorphous (mixed $sp^2/sp^3$) or graphitic crystalline ($sp^2$), depending upon the extent of the occurrence of $sp^2$ carbon in the core [93]. Graphitic crystalline ($sp^2$) cores were reported by several researchers [94–96]. The size of cores is very small (2–3 nm), with a characteristic lattice spacing of ~0.2 nm [97]. The cores are categorized depending on the technique utilized for the synthesis and the precursors used, as well as other synthetic parameters (such as duration, temperature, pH, etc.) [98]. Generally, the graphitization ($sp^2$) structure is obtained at over 300 °C reaction temperatures, while amorphous cores are obtained at lower temperatures, unless $sp^2/sp^3$-hybridized C is present in the precursor [99]. To determine the core structure of CQDs, various instrumental techniques such as Transmission Electron Microscopy (TEM) or High Resolution (HR) TEM, Scanning Electron Microscopy (SEM), Raman spectroscopy, and X-ray diffraction (XRD) are utilized. To measure the size and morphology of the CQDs, TEM or SEM are carried out [100]. The selected area electron diffraction (SAED) patterns reveal the amorphous or crystalline nature of CQDs [101]. The XRD pattern also determines the crystal structure of CQDs. The broad peak at 2θ 23° indicates the amorphous nature of CQD, while the occurrence of two broad peaks at 2θ 25° and 44° specifies a low-graphitic carbon structure analogous to (002) and (100) diffraction [102]. The general structure and presence of different functional groups on the surface of CQDs are determined using Fourier transform infrared (FT-IR) spectroscopy, X-ray photoelectron spectroscopy (XPS), elemental analysis (EA), and nuclear magnetic resonance (NMR) [103,104]. Using nitrogen sorption analysis, the surface area of the carbon nanoparticles is calculated [103]. To decide the optical properties and qualitative information regarding the presence of C=C and C=O in CQDs, UV-Vis absorption spectroscopy is carried out [105]. To determine the positive or negative charge on the surface of CQDs and the extent of the electrostatic interaction between them, zeta potential is conceded [106,107].

Figure 3 is the typical structure of carbon quantum dots (CQDs), which reveals the presence of different functional groups (such as carbonyl, carboxyl, hydroxyl, amino, etc.) on the surface of CQDs. The presence of these functional groups was confirmed by instrumental techniques such as FTIR and XPS [108].

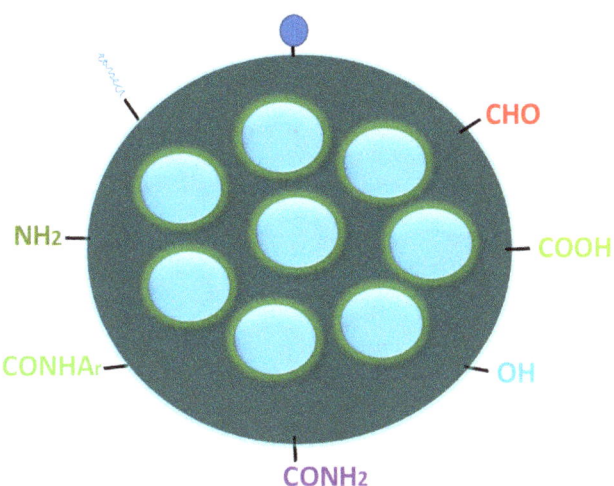

**Figure 3.** Typical structure of CQDs with different functional groups on the surface.

## 4. Optical Properties of Carbon Quantum Dots (CQDs)

### 4.1. Absorbance

CQDs generally exhibit two absorption bands in the visible region around 280 nm and 350 nm, alongside a tail broadly in the UV region. Hu et al. reported that an absorption band at 280 nm is due to a pi-pi* ($\pi$-$\pi$*) transition of a C=C bond, and the one at 350 nm is due to an n-$\pi$* transition of the C=O bond [109]. Figure 4 is the typical UV-visible absorption spectrum of fluorescent CQDs. The absorption properties of CQDs can be influenced by surface modification or surface passivation [110–113]. Depending on the raw precursor and synthesis methodology, the positions of these absorption bands are different to some extent. Doping in CQDs can also alter the absorption wavelength.

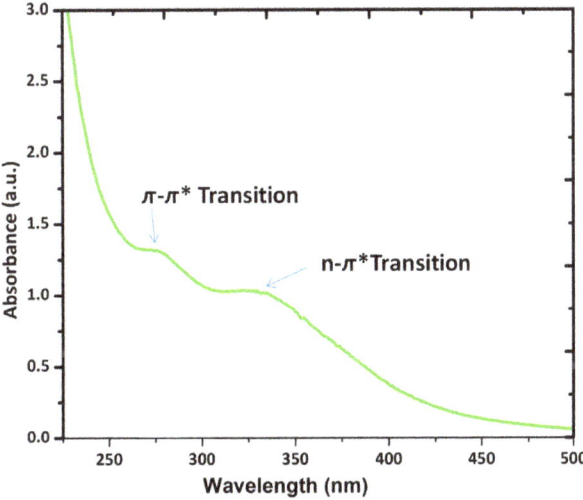

**Figure 4.** The UV-visible absorption spectrum of fluorescent CQDs.

The optical properties of CQDs can be customized by doping/co-doping with heteroatoms, functional groups, and surface passivation [114]. In the process of surface passivation, a slim insulating (protecting) layer of covering materials such as thiols, thionyl chloride, spiropyrans, and oligomers (polyethylene glycol (PEG), etc.) is formed on the CQDs surface. The important functions of such types of protective layers are to shield CQDs from the adhesion of impurities and to provide stability [115]. CQDs with surface-passivating agents become extremely optically active, demonstrating considerable fluorescence from the visible to the near-IR region [116]. The quantum yields (QYs) of CQDs can also be enhanced up to 55–60% by surface passivation [114]. The absorbance of CQDs improved to longer wavelengths (350–550 nm) after surface passivation with 4,7,10-trioxa-1,13-tridecanediamine (TTDDA) [117]. Particle size is associated with the absorption wavelength. As the size of the CQDs increases, absorption wavelength also increases [118,119]. The CQDs are viable for covalent bonding with functionalizing agents [114]. Different functional groups such as amines, carboxyl, hydroxyl, carbonyl, etc., were introduced on the surface of CQDs by surface functionalization. The functionalized CQDs revealed good biocompatibility, high stability, outstanding photoreversibility, and low toxicity compared to undoped CQDs. The efficient technique to modify the CQDs absorption spectrum is doping/co-doping with heteroatoms (such as boron (B), nitrogen (N), fluorine (F), phosphorous (P), and sulfur (S)). The dopant adjusts the bandgap, electronic structure, and, consequently, the optical properties of CQDs by altering the $\pi$-$\pi$* energy level (related through the core-$sp^2$ carbon system) [120]. On increasing N-dopant concentration, a gradual increase in the band gap of the CQDs from 2.2 to 2.7 eV was observed [121]. In contrast, it was also found that the doping of N in CQDs results in a reduction in size [122]. The CQDs established innovative electronic states, resulting in a reduction in the bandgap of CQDs (about ~48–57%) [123]. Zuo et al. synthesized F-doped CQDs using a hydrothermal method which exhibited higher QYs and enhanced the electron transfer and acted as a superior photocatalyst [124].

*4.2. Photoluminescence*

The emission of light from a substance upon the absorption of light (photon) is called photoluminescence (PL). Photoluminescence includes two types, namely fluorescence and phosphorescence. Fluorescent materials emit absorbed light from the lowest singlet excited state ($S_1$) to the singlet ground state ($S_0$). This process is very fast and has a nanosecond lifetime. The transitions that occur among two electronic states in the fluorescence process are allowed because it has the same spin multiplicity. In contrast, in phosphorescence, the transition occurs from the lowest triplet excited state ($T_1$) to singlet ground state ($S_0$), i.e., a forbidden transition occurs according to the spin selection rule.

4.2.1. Fluorescence

The fluorescence properties of CQDs have attracted great attention among researchers because of their several sensing and analytical applications. Numerous mechanisms have been reported to gain deep insight into the cause of fluorescence in CQDs [125–130]. Among them, the following two have been found more prominent. The first is that the fluorescence mechanism is due to band gaps' transitions arising from the $\pi$-conjugated domains ($sp^2$-hybridized), which is similar to aromatic molecules employing definite energy band gaps in favor of absorptions and emissions [131]. The second cause of fluorescence is related to the surface defects, quantum size effect, carbon core state, surface passivation/functionalization effect, and different emissive traps on the surface of CQDs [132–134].

The main reason for the surface defects in CQDs is an unsymmetrical allocation of $sp^2$- and $sp^3$-hybridized carbon atoms, and the existence of heteroatoms such as B, N, P, and S [126,135]. When this surface defect is independently incorporated into the solid host, it creates surroundings similar to aromatic molecules. These molecules can attract UV light and display various color emissions [131,136]. CQDs show two types of emission, i.e., excitation-dependent emission (tunable emission) and excitation-independent emission. The tunable emission is due to the presence of various emission sites on the surface of CQDs along with particle size distribution; because of this, most CQDs exhibit tunable emissions [137]. The excitation-independent emission is due to the extremely ordered graphitic structure of CQDs [118]. CQDs exhibit extensive and unremitting excitation spectra which are highly photostable and have steady fluorescence, in contrast to traditional organic dye [95,138,139].

4.2.2. Phosphorescence

In CQDs, the phosphorescence property is also observed, which was first described by De et al. via dispersing CQDs to polyvinyl alcohol matrix at RT and exciting them with ultraviolet light. The maximum emission obtained was 500 nm, with an average lifetime of 380 ns at a 325 nm excitation [140]. Phosphorescence in CQDs arises when the singlet and triplet states of an aromatic carbonyl group in CQDs and polyvinyl alcohol matrix are close in energy to assist spin–orbit coupling, which increases the intersystem crossing (ISC). By using microwave synthesis, Lu et al. synthesized ultra-long phosphorescent carbon quantum dots (P-CQDs). When P-CQDs were excited at 354 nm, they displayed yellow-green phosphorescence (525 nm) for up to 9 s. They concluded that as the pH increases, the phosphorescence intensity of P-CQDs gradually decreases. The reason is that protonation dissociates the hydrogen bonds and distresses the phosphorescent sources. By introducing the tetracyclines (TCs), the phosphorescence of P-CQDs was quenched. They applied P-CQDs as biological and chemical sensing and time-resolved imaging [141]. Figure 5 is the typical excitation (black line) and emission (red line) spectrum of fluorescent CQDs.

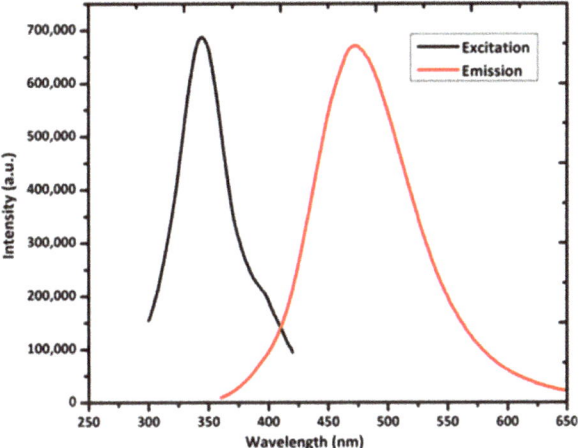

Figure 5. Excitation and emission spectrum of CQDs.

## 5. Application of Carbon Quantum Dots as a Catalyst

The presence of different functional groups such as -OH, -COOH, -NH$_2$, etc., on carbon quantum dots' (CQDs) surface provides vigorous coordination sites to bind with transition metal ions. The CQDs doped with multiple heteroatoms might further improve the catalytic activity by encouraging electron transfer via interior interactions. The presence of more active catalytic reaction sites offered by CQDs and favorable charge transfers during the catalytic process is also responsible for the application of CQDs as a catalyst (Figure 6) [6,142–144].

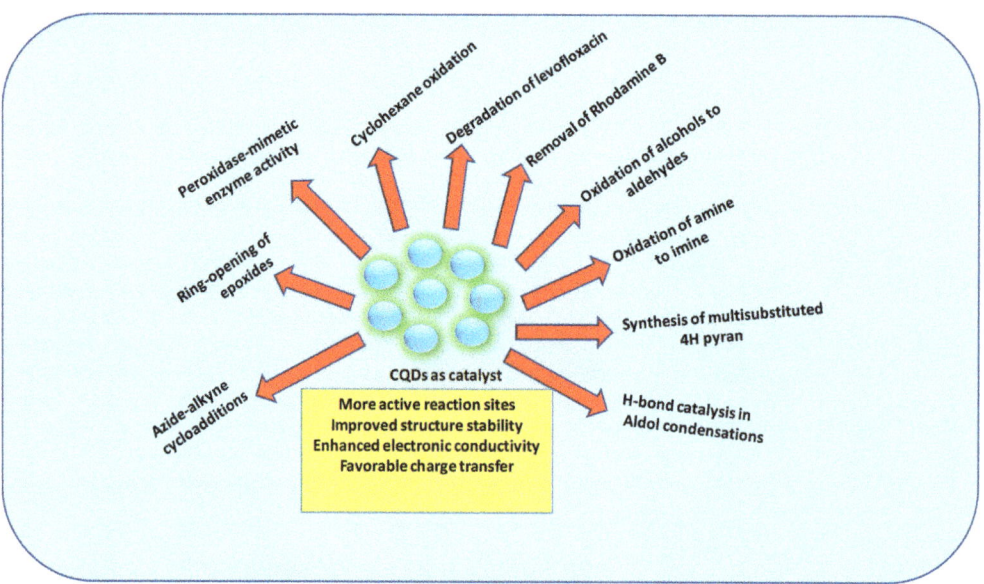

**Figure 6.** Catalytic applications of CQDs.

*5.1. CQDs as a Catalyst for the Peroxidase-Mimetic Enzyme Activity*

Natural enzymes such as peroxidase can catalyze a variety of reactions with high catalytic activity and excessive surface specificity [145]. Because of this, they are broadly utilized in different fields such as the pharmaceuticals industry, medicine, agriculture, etc. [146]. However, they possess some limitations such as high cost, short storage life, rigorous storage conditions, and poor thermal stability [147]. Therefore, to point out these limitations, carbon-based nanomaterials were found very suitable for intrinsic peroxidase-mimetic catalytic activity. Yadav et al. have synthesized fluorescent CQDs from leaf extracts of neem (*Azadirachtaindica*) by using a one-pot hydrothermal method. The as-prepared Neem-Carbon Quantum Dots (N-CQDs) exhibited peroxidase-mimetics catalytic activity in an extensive pH range for the oxidation of peroxidase substrate 3,3′,5,5′- tetramethylbenzidine (TMB) in the presence of hydrogen peroxide ($H_2O_2$). The peroxidase-mimetic catalytic activity of N-CQDs was confirmed by taking UV–visible absorption spectra of N-CQDs in the presence and absence of $H_2O_2$ with TMB in an acetate buffer. When the mixtures of TMB and N-CQDs were taken, no absorbance at 652 nm was observed, revealing no oxidation of TMB. Additionally, when the mixture of TMB and $H_2O_2$ reacted, a less intense peak at 652 nm was obtained, enlightening the partial oxidation of TMB with the existence of a partial blue color. Interestingly, in the presence of N-CQDs, TMB, and $H_2O_2$, the absorbance at 652 was found at a maximum, with the color changing from colorless to blue, revealing the complete oxidation of TMB. These results powerfully confirmed that N-CQDs act as a catalyst for peroxidase-mimetic activity. To determine the intermediate reaction, the active species trapping experiment with isopropyl alcohol (IPA) and methyl alcohol (MA) was carried out. The IPA and MA are hydroxyls radical (•OH) scavengers. When these scavengers were added to the oxidized blue-colored solution of TMB, a decrease in the absorption at 652 nm was observed, enlightening the incomplete oxidation of TMB because the IPA and MA consumed the •OH radical. This examination specifies that in the presence of N-CQDs, the •OH radicals were generated during a peroxidase-like catalytic reaction, which oxidized TMB via a one-electron transfer to produce a blue-colored solution. Additionally, the high surface area, small size, and presence of a negative-charge density on the N-CQDs surface were also responsible for this catalytic activity (Figure 7) [39].

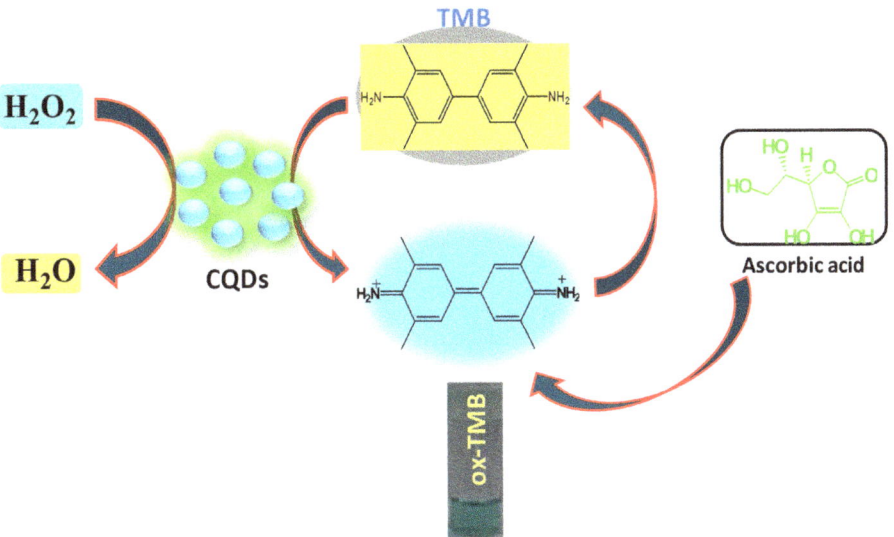

**Figure 7.** Showing the oxidation of TMB along with $H_2O_2$ in the presence of CQDs as a catalyst.

*5.2. CQDs as a Catalyst for Selective Oxidation of Alcohols to Aldehydes*

Aldehydes are highly demanded as a crucial intermediate for the production of an extensive range of materials, such as pesticides, toiletries, dyes, and perfumes, in the pharmaceuticals and agribusiness industries. The popular method for the synthesis of aldehydes is catalytic alcohol oxidation, but establishing an ecofriendly method with high-yield production and selectivity is still a major challenge for researchers [148–150]. Rezaie et al. developed a multifunctional tungstate-decorated CQDs base catalyst, A-CQDs/W, by using a one-pot hydrothermal technique, and utilized it for the oxidation of a variety of alcoholic substrates into analogous aldehydes with the help of $H_2O_2$ as an oxidant and an ultrasound effect as a green activation method. Before investigating the catalytic activity, the oxidizing potential of an amphiphilic multifunctional catalyst was examined, and they observed that A-CQDs/W were capable of oxidizing a wide range of alcoholic substances into corresponding aldehydes with 100% selectivity and above 95% yield. This achievement was because of the synergic effect among ultrasound irradiation and the suitable design of the catalyst. The proposed mechanism for this oxidation reaction firstly involves the reaction between $H_2O_2$ and A-CQDs/W, resulting in the production of bisperoxo tungstate, which is immobilized on A-CQDs via hydrophilic groups. This is able to diffuse into the organic alcoholic phase and trigger the oxidation reaction with the assistance of an ultrasound wave. Finally, aldehyde was fabricated after inserting the alcoholic ligand on A-CQDs/W, followed by a ligand exchange reaction [34,151,152].

*5.3. CQDs as a Catalyst for Selective Oxidation of Amine to Imine*

Imines are valuable for the preparation of biologically active molecules, such as oxazolidines, chiral amines, amides, nitrones, aminonitriles, and hydroxylamines. Additionally, β-lactams complexes are also synthesized using imine intermediates [153–155]. Several materials were used as a catalyst for the selective oxidation of amine to imine, but the carbon-based materials such as CQDs, mesoporous carbon, graphene oxide (GO), amorphous carbon, graphitic carbon nitride, and carbon nanotubes (CNTs) have been recognized as potential catalysts compared to conventional metal-based catalysts because of their relatively low cost and natural abundance [156–158].

Ye et al. prepared oxygen-rich carbon quantum dots (O-CQDs) from fullerenes ($C_{60}$) and utilized them as nanocatalysts (metal-free) for the oxidation of amines to imine with an excellent 98% yield. The mechanism behind this catalytic oxidation reveals that the molecular oxygen and amine molecules are trapped and activated by carboxylic functional groups present on the surface of CQDs, along with the unpaired electrons, resulting in the conversion of amine. For the oxidative coupling of amine to imine, the catalytic performance of O-CQDs was further improved by heat treatment. The aerobic oxidation of amines was probably because of the occurrence of several carboxyl functional groups, which coupled with spins of π-electrons from the atoms situated at the surface of O-CQDs [36].

### 5.4. CQDs as a Catalyst in the Synthesis of Multisubstituted 4H Pyran with Indole Moieties

Indole scaffolds have attracted much attention among researchers because of their applications in the field of pharmacology, such as antihypertensive, antiproliferative, anticholinergic, antifungal, cardiovascular, optimal inhibitory, antibacterial, antiviral, and anticonvulsant activities [159–161]. Additionally, there are some pharmaceutically significant compounds and natural products which have anticancer, hypoglycemic, anti-inflammatory, antipyretic, and antitumor properties, and contain indole scaffolds in their structures [162,163]. 4H-pyrans are an important family of oxygen-containing heterocyclic compounds with a wide spectrum of biological properties such as antioxidant, anticoagulant, diuretic, spasmolytic, anti-anaphylactic, and anticancer activities [164,165]. Rasooll et al. synthesized a novel heterogeneous nano-catalyst from CQDs and phosphorus acid moieties by using ultrasonic irritation followed by a hydrothermal method and named it CQDs–N(CH$_2$PO$_3$H$_2$)$_2$. The instrumental techniques such as transmission electron microscopy (TEM), energy-dispersive X-ray (EDX) spectroscopy, X-ray diffraction (XRD), FT-IR spectroscopy, scanning electron microscopy (SEM), fluorescence, and thermogravimetric (TG) analysis were utilized to characterize this catalyst. An efficient catalyst, CQDs–N (CH$_2$PO$_3$H$_2$)$_2$, was effectively applied for the preparation of 2-amino-6-(2-methyl-1H-indol-3-yl)-4-phenyl-4H-pyran-3,5-dicarbonitriles, with the help of a variety of aromatic aldehydes, 3-(1H-indol3-yl)-3-oxopropanenitrile derivatives, and malononitrile. The principal advantages of this catalytic activity include fresh and mild reaction conditions, little reaction time, and the recycling of the catalyst.

The anticipated mechanism for this catalytic reaction is that, firstly, the acidic proton of CQDs–N(CH$_2$PO$_3$H$_2$)$_2$ activates the aldehyde group, followed by the reaction with malononitrile, and intermediate (I) is formed by the loss of one molecule of H$_2$O. In the next step, 3-(1Hindol-3-yl)-3-oxopropanenitrile reacts with intermediate (I) to provide intermediate (II) following tautomerization. Finally, after intramolecular cyclization, the desired product is obtained from intermediate (II) with the loss of another molecule of H$_2$O [31].

### 5.5. As a Photocatalyst for High-Efficiency Cyclohexane Oxidation

In the 21st century, the highly efficient and highly selective catalytic oxidation of cyclohexane under mild conditions is the principle objective of catalysis chemistry. Liu et al. synthesized fluorescent CQDs and gold (Au) nanoparticle composites (Au/CQDs composites). The CQDs were prepared through the electrochemical ablation method using graphite. A chemical reduction method was used to synthesize AuNPs by an aqueous solution of HAuCl$_4$ and trisodium citrate, which resulted in a pink color immediately after the addition of the NaBH$_4$ solution. When in the solution of CQDs, a HAuCl4 solution was added, and the solution turned red, revealing the formation of a composite (Au/CQDs composites). Interestingly, they utilized this composite as a tunable photocatalyst for the selective oxidation of cyclohexane to cyclohexanone with the help of an oxidant H$_2$O$_2$ (30%). The conversion efficiency was 63.8% and selectivity was over 99.9%. The mechanism involves enrichment in the absorption of light by surface plasma resonance of Au nanoparticles, the generation of active trapping oxygen species (HO·) through H$_2$O$_2$ decomposition, and interaction among CQDs and AuNPs under visible light [37].

*5.6. As a Catalyst for the Removal of Rhodamine B*

Preethi et al. prepared bluefluorescent CQDs from a natural carbon precursor (muskmelon peel) using a stirrer-assisted method. The synthesized CQDs were utilized as an excellent photocatalyst and a sonocatalyst for the degradation of Rhodamine B (RhB) dye. The efficiency of CQDs for the degradation of RhB is 99.11% in sunlight, with a degradation rate constant of 0.06943 min$^{-1}$ and 83.04% in ultrasonication. These results advocate that CQDs are an efficient catalyst for the breakdown of organic dyes in wastewater. The mechanism reveals the generation of •OH radicals during active species trapping experiment. •OH was confirmed by taking terephthalic acid (TA) as a scavenger. The dye molecules adsorbed on the surface of CQDs may be oxidized by these active species, ensuing in dye degradation [35].

*5.7. As a Catalyst in Azide-Alkyne Cycloadditions*

Liu and coworkers synthesized yellow light-emitting bio-friendly CQDs from Na$_2$[Cu(EDTA)] by thermolysis. Cu(I)-doped fluorescent CQDs were utilized for catalyzing the Huisgen 1,3-dipolar cycloaddition among azides and terminal alkynes, the classical example of "click chemistry". The possible mechanism behind this catalytic property using these CQDs was projected to be the UV-induced split of excitons. First of all, the escape of electrons from the CQDs occurs, resulting in the formation of holes to compete with Cu(I), and at last, Cu(I) is released from the CQDs. The high biocompatibility of this nanocatalyst was confirmed by Hep-2 cells, revealing intracellular detection [32].

*5.8. As H-Bond Catalysis in Aldol Condensations*

Han and coworkers synthesized CQDs by an electrochemical etching method and utilized them as efficient heterogeneous nanocatalysts for H-bond catalysis in aldol condensations. The catalytic activity was excellent (89% yields), with visible light irradiation. Highly efficient electron-accepting capabilities, novel photochemical properties, and functional hydroxyl and carboxylic groups on the surface are responsible for such soaring catalytic activities of CQDs [38]. The catalytic efficiency of CQDs was high in visible light irradiation, and almost no conversion was observed in the absence of light. The CQD-catalyzed aldol condensation was greatly influenced by solvents. Han et al. used different solvents such as ethanol, tetrahydrofuran (THF), acetone, chloroform (CHCl$_3$), and toluene. However, the highest yield (89%) was calculated when the solvent and reactant were acetone. These investigations exposed that CQDs acted as an outstanding catalyst for Aldol condensation. The mechanism revealed that the cationic or anionic intermediates were generated during catalytic reaction. The hydroxyl groups present on the CQDs edge act as extremely weak acids, which can form H-bonds with oxygenates [166,167]. Aldehydes and ketones, both reactants, were capable of forming H-bonds. They confirmed that the hydroxyl groups present on the surface of CQDs favor contact with aldehyde groups. When the reactions were carried out in the absence of a hydroxyl group, no product was obtained and free CQDs were unreactive. These results advocate that the capability of CQDs to intervene in reactions is through interfacial H-bond catalysis. In visible light irradiation, CQDs act as highly proficient electron acceptors and attract electrons from the O−H···O region, resulting in the development of a positive charge on hydrogen and oxygen, and the negative charge increases. This effect results in an increase in the s-character in the oxygen hybrid orbital, thereby leading to the strengthening of the O−H bond, which efficiently activates the C=O bond of the aldehyde group and accelerates the aldol condensation. Furthermore, the reaction-intermediate or transition-state species is stabilized by the enhanced O−H bonds, resulting in the highest yield of 89.4% [168,169].

*5.9. As a Catalyst for the Ring Opening of Epoxides*

In modern organic synthesis, acid catalytic reactions contribute a characteristic and imperative role [170]. Some carbon-based nanostructures such as sulfated-graphene/-tube/-active carbon materials have been utilized as acid catalysts in several catalytic applications [171]. However, they possess some limitations, such as the requirement of sufficient surface functionalization, low efficiency, and complex synthesis steps [172]. As a result, the development of carbon materials-based acid catalysts with high efficiency that are light-driven or light-enhanced are still required. Keeping these in mind, Li et al. described the synthesis of CQDs based on a novel, photoswitchable solid acid catalyst. The CQDs were synthesized from a graphite rod using an electrochemical method, doped with hydrogen sulfate groups (S-CQDs). They utilized S-CQDs as light-enhanced acid catalysts, which catalyze the ring opening of epoxides in the presence of nucleophiles and solvents (methanol and other primary alcohols). The mechanism revealed that the additional protons are released from the ionization of the -SO$_3$H group under visible light irradiation and, as a result, a stronger acid environment is offered for the opening reaction, and a higher yield as well as selectivity of the product is obtained compared to the process without light irradiation. The photoexcitation and charge separation in the CQDs create an electron-withdrawing effect from the acidic groups. The utilization of S-CQDs as visiblelight-responsive and convenient photocatalysts is a novel application of CQDs in green chemistry [40].

*5.10. As a Catalyst for the Degradation of Levofloxacin*

Levofloxacin (LEVO), also known as levaquin, is an important antibiotic medicine. Several bacterial infections such as pneumonia, acute bacterial sinusitis, urinary tract infection, *H. pypori*, and chronic prostatitis are treated by LEVO. It is also used to treat tuberculosis, pelvic inflammatory disease, or meningitis, along with other antibiotics [173,174]. However, the degradation of LEVO is typical. Although some techniques have been utilized for the degradation of LEVO, the degradation using CQD had not been discovered. Meng et al. synthesized CQDs@FeOOH nanoneedles for an efficient electro-catalytic degradation of LEVO. The CQDs were synthesized by a hydrothermal method from orange peels. With the help of a facile in situ growth method, the α-FeOOH was fabricated by using Fe$_2$(SO4)$_3$ and H$_2$O in 50 mL distilled water. Similarly, a CQDs@FeOOH electro-catalyst was prepared using the above method, except that 500 mL aqueous solutions of 0.5 g/L CQDs were used instead of 500 mL distilled water. By using CQDs@FeOOH, about 99.6% LEVO and 53.7% total organic carbon (TOC) could be competently removed after 60 min degradation. This high degradation performance for LEVO was due to the soaring mass transfer capability and the high % OH generation ability of the CQDs@FeOOH. Meng et al. proposed a possible LEVO degradation mechanism and also investigated the change in toxicity throughout LEVO degradation. The mechanism revealed the generation of both % OH and SO4% in LEVO degradation, but a dominant role was played by % OH. Liquid chromatography-mass spectrometry (LC-MS) results designated that the LEVO could be entirely decomposed by % OH under the de-piperazinylation, decarboxylation, and ring opening reaction. This novel work offers a proficient technique to reduce the quantity and toxicity of antibiotics in water [33].

*5.11. CQDs as Electrocatalyst*

CQDs are also utilized as electrocatalysts in hydrogen evolution reduction, oxygen evolution reaction, CO$_2$ reduction reaction and oxygen reduction reaction. The large surface area, good conductivity and fast charge transfer process of CQDs are responsible for the electrocatalytic applications [6].

**6. Conclusions and Future Perspectives**

The present review paper discusses the structures, synthetic methods, optical properties, and applications of CQDs as a catalyst. The structure of CQDs includes core–shell, either graphitic (sp$^2$) or amorphous (mixed sp$^2$/sp$^3$). CQDs are usually amorphous, having different functional groups such as amino, carboxyl, hydroxyl, etc. CQDs are synthesized by both the bottom-up and the top-down approach. The bottom-up method is better because it is ecofriendly and economically viable, but it has poor control over the size of CQDs.

In contrast, the top-down methods are expensive. For the synthesis of CQDs, chemical as well as biological precursors are used. CQDs possess admirable optical properties and have superior water solubility, low toxicity, biocompatibility, and ecofriendliness. The optical properties and QYs are essential parameters for the applications of CQDs in the field of nanomedicine, biosensing, chemical sensing, bioimaging, solar cells, drug delivery, and light-emitting diodes. In this review paper, we have focused on the applications of CQDs as a catalyst in the degradation of levofloxacin, the selective oxidation of amines and alcohols, azide-alkyne cycloadditions, the synthesis of multisubstituted 4H pyran, the selective oxidation of alcohols to aldehydes, the removal of Rhodamine B, cyclohexane oxidation, the ring opening of epoxides, and intrinsic peroxidase-mimetic enzyme activity. The mechanism suggests that the catalytic activity might be due to the presence of more active reaction sites, favorable charge transfer, improved structure stability, and enhanced electronic conductivity.

However, during the last fifteen years, several investigations have been carried out on CQDs, and numerous challenges require being resolved for the extensive adoption of CQDs. (1) It is difficult to synthesize CQDs of a desired structure and size because of the requirement of accurate control over different synthesis parameters. Therefore, to powerfully control the core structure, a manufacturing process could be developed which helps increase QYs and the large-scale production of CQDs. (2) In many research papers, it has not been reported why the fluorescence QY of doped and co-doped CQDs are high in contrast to the un-doped CQDs. Thus, in the future, it is possible to realize the basic fluorescence mechanism in doped and co-doped CQDs. (3) Most doped and co-doped CQDs emit blue fluorescence. Hence, it is challenging for the researcher to synthesize multicolor emission CQDs and utilize them in different applications in the future. (4) To broaden the spectrum of CQDs, efforts must be made, particularly in the near-IR region, so that the applications of CQDs can be widespread, such as in organic bioelectronics. (5) CQDs possess some limitations such as low reactivity, poor stability, short lifetime, etc., which prevents them from promising to be a good catalyst. Therefore, in the future, it will be possible to overcome these shortcomings.

Compared to other applications of CQDs, very few studies have been reported on the application of CQDs as a catalyst in organic synthesis. In detail, theoretical and experimental studies are required to carefully design CQD-based catalysts with attractive catalytic action and durable operation stability. The applications of CQDs as a catalyst in organic synthesis signify the flexibility of CQDs in the most unpredicted areas. It is inspiring to see the applications of CQDs in green chemistry and clean energy production. It looks obvious that the future of CQDs remains promising.

**Author Contributions:** Conceptualization, P.K.Y.; methodology, P.K.Y. and S.C.; literature investigation, P.K.Y., S.C., V.K., D.K. and S.H.H.; writing—original draft preparation, P.K.Y.; writing—review and editing, P.K.Y. and S.C.; visualization, P.K.Y., S.C., V.K. and D.K.; supervision, S.H.H. All authors have read and agreed to the published version of the manuscript.

**Funding:** This research received no external funding.

**Data Availability:** Not applicable.

**Acknowledgments:** The authors are thankful for the Indian Institute of Technology, BHU, India, for encouraging and facilitating us in pursuing this research. The authors also give thanks to Department of Chemistry, Jagatpur P.G. College, affiliated to MGKV University Varanasi, India, for providing a conductive atmosphere for research activities.

**Conflicts of Interest:** The authors declare no conflict of interest.

## References

1. Clancy, A.; Bayazit, M.K.; Hodge, S.A.; Skipper, N.T.; Howard, C.A.; Shaffer, M.S.P. Charged Carbon Nanomaterials: Redox Chemistries of Fullerenes, Carbon Nanotubes, and Graphenes. *Chem. Rev.* **2018**, *118*, 7363–7408. [CrossRef] [PubMed]
2. Lin, H.-S.; Jeon, I.; Xiang, R.; Seo, S.; Lee, J.-W.; Li, C.; Pal, A.; Manzhos, S.; Goorsky, M.S.; Yang, Y.; et al. Achieving high efficiency in solution-processed perovskite solar cells using C60/C70 mixed fullerenes. *ACS Appl. Mater. Interfaces* **2018**, *46*, 39590–39598. [CrossRef] [PubMed]
3. Georgakilas, V.; Perman, J.A.; Tucek, J.; Zboril, R. Broad Family of Carbon Nanoallotropes: Classification, Chemistry, and Applications of Fullerenes, Carbon Dots, Nanotubes, Graphene, Nanodiamonds, and Combined Superstructures. *Chem. Rev.* **2015**, *115*, 4744–4822. [CrossRef] [PubMed]
4. Rao, R.; Pint, C.L.; Islam, A.E.; Weatherup, R.S.; Hofmann, S.; Meshot, E.R.; Wu, F.; Zhou, C.; Dee, N.; Amama, P.B.; et al. Carbon nanotubes and related nanomaterials: Critical advances and challenges for synthesis toward mainstream commercial applica-tions. *ACS Nano* **2018**, *12*, 11756–11784. [CrossRef]
5. Patel, K.D.; Singh, R.K.; Kim, H.W. Carbon-based nanomaterials as an emerging platform for theranostics. *Materials Horizons* **2019**, *3*, 434–469. [CrossRef]
6. Wang, X.; Feng, Y.; Dong, P.; Huang, J. A Mini Review on Carbon Quantum Dots: Preparation, Properties, and Electrocatalytic Application. *Front. Chem.* **2019**, *7*, 671. [CrossRef]
7. Zuo, J.; Tao, J.; Zhao, X.; Xiong, X.; Xiao, S.; Zhu, Z. Preparation and application of fluorescent carbon dots. *J. Nanomater.* **2015**, *2015*, 787862. [CrossRef]
8. Xu, X.; Ray, R.; Gu, Y.; Ploehn, H.J.; Gearheart, L.; Raker, K.; Scrivens, W.A. Electrophoretic Analysis and Purification of Fluorescent Single-Walled Carbon Nanotube Fragments. *J. Am. Chem. Soc.* **2004**, *126*, 12736–12737. [CrossRef]
9. Sun, Y.-P.; Zhou, B.; Lin, Y.; Wang, W.; Fernando, K.S.; Pathak, P.; Meziani, M.J.; Harruff, B.A.; Wang, X.; Wang, H. Quantum-Sized Carbon Dots for Bright and Colorful Photoluminescence. *J. Am. Chem. Soc.* **2006**, *128*, 7756–7757. [CrossRef]
10. Ahmad, F.; Khan, A.M. Carbon quantum dots: Nanolights. *Int. J. Petrochem. Sci. Eng.* **2017**, *2*, 247–250. [CrossRef]
11. Yang, S.; Sun, J.; Li, X.; Zhou, W.; Wang, Z.; He, P.; Ding, G.; Xie, X.; Kang, Z.; Jiang, M. Large-scale fabrication of heavy doped carbon quantum dots with tunable-photoluminescence and sensitive fluorescence detection. *J. Mater. Chem. A* **2014**, *2*, 8660–8667. [CrossRef]
12. Guo, H.; Liu, Z.; Shen, X.; Wang, L. One-Pot Synthesis of Orange Emissive Carbon Quantum Dots for All-Type High Color Rendering Index White Light-Emitting Diodes. *ACS Sustain. Chem. Eng.* **2022**, *10*, 8289–8296. [CrossRef]
13. Toma, E.E.; Stoian, G.; Cojocaru, B.; Parvulescu, V.I.; Coman, S.M. ZnO/CQDs Nanocomposites for Visible Light Photodegradation of Organic Pollutants. *Catalysts* **2022**, *12*, 952. [CrossRef]
14. Subedi, S.; Rella, A.K.; Trung, L.G.; Kumar, V.; Kang, S.-W. Electrically Switchable Anisometric Carbon Quantum Dots Exhibiting Linearly Polarized Photoluminescence: Syntheses, Anisotropic Properties, and Facile Control of Uniaxial Orientation. *ACS Nano* **2022**, *16*, 6480–6492. [CrossRef]
15. Gu, L.; Zhang, J.; Yang, G.; Tang, Y.; Zhang, X.; Huang, X.; Zhai, W.; Fodjo, E.K.; Kong, C. Green preparation of carbon quantum dots with wolfberry as on-off-on nanosensors for the detection of Fe3+ and l-ascorbic acid. *Food Chem.* **2022**, *376*, 131898. [CrossRef]
16. Chen, Y.; Xue, B. A review on quantum dots modified g-C3N4-based photocatalysts with improved photocatalytic activity. *Catalysts* **2020**, *1*, 142. [CrossRef]
17. Parya, E.; Rhim, J.-W. Pectin/carbon quantum dots fluorescent film with ultraviolet blocking property through light conversion. *Colloids Surf. B Biointerfaces* **2022**, *219*, 112804.
18. Ajayan, P.M.; Zhou, O.Z. Applications of carbon nanotubes. *Carbon Nanotub.* **2001**, *80*, 391–425.
19. Baptista, F.R.; Belhout, S.A.; Giordani, S.; Quinn, S.J. Recent developments in carbon nanomaterial sensors. *Chem. Soc. Rev.* **2015**, *44*, 4433–4453. [CrossRef]
20. Chen, D.; Tang, L.; Li, J. Graphene-based materials in electrochemistry. *Chem. Soc. Rev.* **2010**, *39*, 3157–3180. [CrossRef]
21. Cayuela, A.; Benítez-Martínez, S.; Soriano, M.L. Carbon nanotools as sorbents and sensors of nanosized objects: The third way of analytical nanoscience and nanotechnology. *TrAC Trends Anal. Chem.* **2016**, *84*, 172–180. [CrossRef]
22. Pardo, J.; Peng, Z.; Leblanc, R.M. Cancer Targeting and Drug Delivery Using Carbon-Based Quantum Dots and Nanotubes. *Molecules* **2018**, *23*, 378. [CrossRef] [PubMed]
23. Sahar, T.; Abnous, K.; Taghdisi, S.M.; Ramezani, M.; Alibolandi, M. Hybrid car-bon-based materials for gene delivery in cancer therapy. *J. Control Release* **2020**, *318*, 158–175.
24. Mingjun, C.; Cao, Y.; Zhu, Y.; Peng, W.; Li, Y.; Zhang, F.; Xia, Q.; Fan, X. Oxidation-Modulated CQDs Derived from Covalent Organic Frameworks as Enhanced Fluorescence Sensors for the Detection of Chromium (VI) and Ascorbic Acid. *Ind. Eng. Chem. Res.* **2022**, *31*, 11484–11493.
25. Murali, G.; Kwon, B.; Kang, H.; Modigunta, J.K.R.; Park, S.; Lee, S.; Lee, H.; Park, Y.H.; Kim, J.; Park, S.Y.; et al. Hematoporphyrin Photosensitizer-Linked Carbon Quantum Dots for Photodynamic Therapy of Cancer Cells. *ACS Appl. Nano Mater.* **2022**, *5*, 4376–4385. [CrossRef]
26. Li, P.; Yu, M.; Ke, X.; Gong, X.; Li, Z.; Xing, X. Cytocompatible Amphipathic Carbon Quantum Dots as Potent Membrane-Active Antibacterial Agents with Low Drug Resistance and Effective Inhibition of Biofilm Formation. *ACS Appl. Bio Mater.* **2022**, *5*, 3290–3299. [CrossRef]

27. Wu, Y.; Qin, D.; Luo, Z.; Meng, S.; Mo, G.; Jiang, X.; Deng, B. High Quantum Yield Boron and Nitrogen Codoped Carbon Quantum Dots with Red/Purple Emissions for Ratiometric Fluorescent $IO_4^-$ Sensing and Cell Imaging. *ACS Sustain. Chem. Eng.* **2022**, *10*, 5195–5202. [CrossRef]
28. Kaur, A.; Pandey, K.; Kaur, R.; Vashishat, N.; Kaur, M. Nanocomposites of Carbon Quantum Dots and Graphene Quantum Dots: Environmental Applications as Sensors. *Chemosensors* **2022**, *10*, 367. [CrossRef]
29. Sharma, V.; Vishal, V.; Chandan, G.; Bhatia, A.; Chakrabarti, S.; Bera, M. Green, sustainable, and economical synthesis of fluorescent nitrogen-doped carbon quantum dots for applications in optical displays and light-emitting diodes. *Mater. Today Sustain.* **2022**, *19*, 100184. [CrossRef]
30. Wang, Y.; Chen, D.; Zhang, J.; Balogun, M.T.; Wang, P.; Tong, Y.; Huang, Y. Charge Relays via Dual Carbon-Actions on Nanostructured $BiVO_4$ for High Performance Photoelectrochemical Water Splitting. *Adv. Funct. Mater.* **2022**, *32*, 2112738. [CrossRef]
31. Rasooll, M.M.; Zarei, M.; Zolfigol, M.A.; Sepehrmansourie, H.; Omidi, A.; Hasani, M.; Gu, Y. Novel nano-architectured carbon quantum dots (CQDs) with phosphorous acid tags as an efficient catalyst for the synthesis of multisubstituted 4*H*-pyran with indole moieties under mild conditions. *RSC Adv.* **2021**, *11*, 25995–26007. [CrossRef]
32. Liu, Z.X.; Bin Chen, B.; Liu, M.L.; Zou, H.Y.; Huang, C.Z. Cu(i)-Doped carbon quantum dots with zigzag edge structures for highly efficient catalysis of azide–alkyne cycloadditions. *Green Chem.* **2017**, *19*, 1494–1498. [CrossRef]
33. Fanqing, M.; Wang, Y.; Chen, Z.; Hu, J.; Lu, G.; Ma, W. Synthesis of CQDs@ FeOOH nanoneedles with abundant active edges for efficient electro-catalytic degradation of levofloxacin: Degradation mechanism and toxicity assessment. *Appl. Catal. B Environ.* **2021**, *282*, 119597.
34. Rezaei, A.; Mohammadi, Y.; Ramazani, A.; Zheng, H. Ultrasound-assisted pseudohomogeneous tungstate catalyst for selective oxidation of alcohols to aldehydes. *Sci. Rep.* **2022**, *12*, 3367. [CrossRef]
35. Preethi, M.; Viswanathan, C.; Ponpandian, N. A metal-free, dual catalyst for the removal of Rhodamine B using novel carbon quantum dots from muskmelon peel under sunlight and ultrasonication: A green way to clean the environment. *J. Photochem. Photobiol. A Chem.* **2022**, *426*, 113765. [CrossRef]
36. Ye, J.; Ni, K.; Liu, J.; Chen, G.; Ikram, M.; Zhu, Y. Oxygen-Rich Carbon Quantum Dots as Catalysts for Selective Oxidation of Amines and Alcohols. *Chemcatchem* **2017**, *10*, 259–265. [CrossRef]
37. Ruihua, L.; Huang, H.; Li, H.; Liu, Y.; Zhong, J.; Li, Y.; Zhang, S.; Kang, Z. Metalnanopar-ticle/carbon quantum dot composite as a photocatalyst for high-efficiency cyclohexane oxidation. *ACS Catal.* **2014**, *1*, 328–336.
38. Han, Y.; Huang, H.; Zhang, H.; Liu, Y.; Han, X.; Liu, R.; Li, H.; Kang, Z. Carbon Quantum Dots with Photoenhanced Hydrogen-Bond Catalytic Activity in Aldol Condensations. *ACS Catal.* **2014**, *4*, 781–787. [CrossRef]
39. Pradeep Kumar, Y.; Singh, V.K.; Chandra, S.; Bano, D.; Kumar, V.; Talat, M.; Hasan, S.H. Green synthesis of fluorescent carbon quantum dots from azadirachtaindica leaves and their peroxidase-mimetic ac-tivity for the detection of H2O2 and ascorbic acid in common fresh fruits. *ACS Biomater. Sci. Eng.* **2018**, *2*, 623–632.
40. Li, H.; Sun, C.; Ali, M.; Zhou, F.; Zhang, X.; MacFarlane, D.R. Sulfated Carbon Quantum Dots as Efficient Visible-Light Switchable Acid Catalysts for Room-Temperature Ring-Opening Reactions. *Angew. Chem.* **2015**, *127*, 8540–8544. [CrossRef]
41. Wang, Y.; Hu, A. Carbon quantum dots: Synthesis, properties and applications. *J. Mater. Chem. C* **2014**, *34*, 6921–6939. [CrossRef]
42. Sofia, P.; Palomares, E.; Martinez-Ferrero, E. Graphene and carbon quantum dot-based materials in pho-tovoltaic devices: From synthesis to applications. *Nanomaterials* **2016**, *6*, 157.
43. Chae, A.Y.; Choi, S.J.; Paoprasert, N.P.; Park, S.Y.; In, I. Microwave-assisted synthesis of fuorescent carbon quantum dots from an A2/B3 monomer set. *RSC Adv.* **2017**, *7*, 12663–12669. [CrossRef]
44. Liu, Y.; Huang, H.; Cao, W.; Mao, B.; Liu, Y.; Kang, Z. Advances in carbon dots: From the perspective of traditional quantum dots. *Mater. Chem. Front.* **2020**, *4*, 1586–1613. [CrossRef]
45. Li, H.; Xu, Y.; Zhao, L.; Ding, J.; Chen, M.; Chen, G.; Li, Y.; Ding, L. Synthesis of tiny carbon dots with high quantum yield using multi-walled carbon nanotubes as support for selective "turn-off-on" detection of rutin and $Al^{3+}$. *Carbon* **2018**, *143*, 391–401. [CrossRef]
46. Zhao, M.; Zhang, J.; Xiao, H.; Hu, T.; Jia, J.; Wu, H. Facile *in situ* synthesis of a carbon quantum dot/graphene heterostructure as an efficient metal-free electrocatalyst for overall water splitting. *Chem. Commun.* **2019**, *55*, 1635–1638. [CrossRef]
47. Yuan, F.; Su, W.; Gao, F. Monolayer 2D polymeric fullerene: A new member of the carbon material family. *Chem* **2022**, *8*, 2079–2081. [CrossRef]
48. Zhu, H.; Wang, X.; Li, Y.; Wang, Z.; Yang, F.; Yang, X. Microwave synthesis of fluorescent carbon nanoparticles with electrochemi-luminescence properties. *Chem. Commun.* **2009**, *34*, 5118–5120. [CrossRef]
49. Li, H.; Zhang, Y.; Ding, J.; Wu, T.; Cai, S.; Zhang, W.; Cai, R.; Chen, C.; Yang, R. Synthesis of carbon quantum dots for application of alleviating amyloid-β mediated neurotoxicity. *Colloids Surfaces B Biointerfaces* **2022**, *212*, 112373. [CrossRef]
50. Cui, L.; Ren, X.; Wang, J.; Sun, M. Synthesis of homogeneous carbon quantum dots by ultrafast dual-beam pulsed laser ablation for bioimaging. *Mater. Today Nano* **2020**, *12*, 100091. [CrossRef]
51. Doñate-Buendía, C.; Fernández-Alonso, M.; Lancis, J.; Mínguez-Vega, G. Pulsed laser ablation in liq-uids for the production of gold nanoparticles and carbon quantum dots: From plasmonic to fluorescence and cell labelling. *J. Phys. Conf. Ser.* **2020**, *1537*, 012013. [CrossRef]

52. Jigang, Z.; Booker, C.; Li, R.; Zhou, X.; Sham, T.-K.; Sun, X.; Ding, Z. An electro-chemical avenue to blue luminescent nanocrystals from multiwalled carbon nanotubes (MWCNTs). *J. Am. Chem. Soc.* **2007**, *4*, 744–745.
53. Zhao, Q.-L.; Zhang, Z.-L.; Huang, B.-H.; Peng, J.; Zhang, M.; Pang, D.-W. Facile preparation of low cytotoxicity fluorescent carbon nanocrystals by electrooxidation of graphite. *Chem. Commun.* **2008**, *41*, 5116–5118. [CrossRef]
54. Zheng, L.; Chi, Y.; Dong, Y.; Lin, J.; Wang, B. Electrochemiluminescence of Water-Soluble Carbon Nanocrystals Released Electrochemically from Graphite. *J. Am. Chem. Soc.* **2009**, *131*, 4564–4565. [CrossRef]
55. Deng, J.; Lu, Q.; Mi, N.; Li, H.; Liu, M.; Xu, M.; Tan, L.; Xie, Q.; Zhang, Y.; Yao, S. Electrochemical Synthesis of Carbon Nanodots Directly from Alcohols. *Chem. A Eur. J.* **2014**, *20*, 4993–4999. [CrossRef]
56. Hou, Y.; Lu, Q.; Deng, J.; Li, H.; Zhang, Y. One-pot electrochemical synthesis of functionalized fluorescent carbon dots and their selective sensing for mercury ion. *Anal. Chim. Acta* **2015**, *866*, 6974. [CrossRef]
57. Bottini, M.; Tautz, L.; Huynh, H.; Monosov, E.; Bottini, N.; Dawson, M.I.; Bellucci, S.; Mustelin, T. Covalent decoration of multi-walled carbon nanotubes with silica nanoparticles. *Chem. Commun.* **2004**, *6*, 758–760. [CrossRef]
58. Michelsen, H.A.; Colket, M.B.; Bengtsson, P.-E.; D'anna, A.; Desgroux, P.; Haynes, B.S.; Miller, J.H.; Nathan, G.J.; Pitsch, H.; Wang, H. A review of terminology used to describe soot formation and evo-lution under combustion and pyrolytic conditions. *ACS Nano* **2020**, *10*, 12470–12490. [CrossRef]
59. Su, Y.; Xie, M.; Lu, X.; Wei, H.; Geng, H.; Yang, Z.; Zhang, Y. Facile synthesis and photoelectric properties of carbon dots with upconversion fluorescence using arc-synthesized carbon by-products. *RSC Adv.* **2013**, *4*, 4839–4842. [CrossRef]
60. Biazar, N.; Poursalehi, R.; Delavari, H. Optical and structural properties of carbon dots/TiO2 nanostructures prepared via DC arc discharge in liquid. *IP Conf. Proc.* **2018**, *1920*, 020033.
61. Wang, L.; Ruan, F.; Lv, T.; Liu, Y.; Deng, D.; Zhao, S.; Wang, H.; Xu, S. One step synthesis of Al/N co-doped carbon nanoparticles with enhanced photoluminescence. *J. Lumin.* **2015**, *158*, 1–5. [CrossRef]
62. Inderbir, S.; Arora, R.; Dhiman, H.; Pahwa, R. Carbon quantum dots: Synthesis, characterization and biomedical applications. *Turk. J. Pharm. Sci.* **2018**, *2*, 219–230.
63. Qu, Y.; Li, X.; Zhang, H.; Huang, R.; Qi, W.; Su, R.; He, Z. Controllable synthesis of a sponge-like Z-scheme N, S-CQDs/Bi2MoO6@TiO2 film with enhanced photocatalytic and antimicrobial activity under visi-ble/NIR light irradiation. *J. Hazard. Mater.* **2022**, *429*, 128310. [CrossRef] [PubMed]
64. Kaixin, C.; Zhu, Q.; Qi, L.; Guo, M.; Gao, W.; Gao, Q. Synthesis and Properties of Nitro-gen-Doped Carbon Quantum Dots Using Lactic Acid as Carbon Source. *Materials* **2022**, *15*, 466.
65. Henriquez, G.; Ahlawat, J.; Fairman, R.; Narayan, M. Citric Acid-Derived Carbon Quantum Dots Attenuate Paraquat-Induced Neuronal Compromise In Vitro and In Vivo. *ACS Chem. Neurosci.* **2022**, *13*, 2399–2409. [CrossRef]
66. Nammahachak, N.; Aup-Ngoen, K.K.; Asanithi, P.; Horpratum, M.; Chuangchote, S.; Ratanaphan, S.; Surareungchai, W. Hydrothermal synthesis of carbon quantum dots with size tunability via heterogeneous nucleation. *RSC Adv.* **2022**, *12*, 31729–31733. [CrossRef]
67. Jamila, G.S.; Sajjad, S.; Leghari, S.A.K.; Kallio, T.; Flox, C. Glucose derived carbon quantum dots on tungstate-titanate nanocomposite for hydrogen energy evolution and solar light catalysis. *J. Nanostruct. Chem.* **2021**, *12*, 611–623. [CrossRef]
68. Qiu, Y.; Li, D.; Li, Y.; Ma, X.; Li, J. Green carbon quantum dots from sustainable lignocellulosic biomass and its application in the detection of Fe3+. *Cellulose* **2021**, *29*, 367–378. [CrossRef]
69. Aayushi, K.; Maity, B.; Basu, S. Rice Husk-Derived Carbon Quantum Dots-Based Dual-Mode Nano-probe for Selective and Sensitive Detection of Fe3+ and Fluoroquinolones. *ACS Biomater. Sci. Eng.* **2022**, *11*, 4764–4776.
70. El-Brolsy, H.M.E.M.; Hanafy, N.A.N.; El-Kemary, M.A. Fighting Non-Small Lung Cancer Cells Using Optimal Functionalization of Targeted Carbon Quantum Dots Derived from Natural Sources Might Provide Potential Therapeutic and Cancer Bio Image Strategies. *Int. J. Mol. Sci.* **2022**, *23*, 13283. [CrossRef]
71. Kumari, M.; Chaudhary, G.R.; Chaudhary, S.; Umar, A.; Akbar, S.; Baskoutas, S. Bio-Derived Fluorescent Carbon Dots: Synthesis, Properties and Applications. *Molecules* **2022**, *27*, 5329. [CrossRef]
72. Yao, L.; Zhao, M.-M.; Luo, Q.-W.; Zhang, Y.-C.; Liu, T.-T.; Yang, Z.; Liao, M.; Tu, P.; Zeng, K.-W. Carbon Quantum Dots-Based Nanozyme from Coffee Induces Cancer Cell Ferroptosis to Activate Antitumor Immunity. *ACS Nano* **2022**, *16*, 9228–9239. [CrossRef]
73. Zhang, B.; Liu, C.; Liu, Y. A Novel One-Step Approach to Synthesize Fluorescent Carbon Nanoparticles. *Eur. J. Inorg. Chem.* **2010**, *2010*, 4411–4414. [CrossRef]
74. Castañeda-Serna, H.U.; Calderón-Domínguez, G.; García-Bórquez, A.; Salgado-Cruz, M.d.l.P.; Rebollo, R.R.F. Structural and luminescent properties of CQDs produced by microwave and conventional hy-drothermal methods using pelagic Sargassum as carbon source. *Opt. Mater.* **2022**, *126*, 112156. [CrossRef]
75. Hong, Y.; Chen, X.; Zhang, Y.; Zhu, Y.; Sun, J.; Swihart, M.T.; Tan, K.; Dong, L. One-pot hydrothermal synthesis of high quantum yield orange-emitting carbon quantum dots for sensitive detection of per-fluorinated compounds. *New J. Chem.* **2022**, *41*, 19658–19666. [CrossRef]
76. Ye, H.; Liu, B.; Wang, J.; Zhou, C.; Xiong, Z.; Zhao, L. A Hydrothermal Method to Generate Carbon Quantum Dots from Waste Bones and Their Detection of Laundry Powder. *Molecules* **2022**, *27*, 6479. [CrossRef]
77. Huo, X.; Liu, L.; Bai, Y.; Qin, J.; Yuan, L.; Feng, F. Facile synthesis of yellowish-green emitting carbon quantum dots and their applications for phoxim sensing and cellular imaging. *Anal. Chim. Acta* **2021**, *1206*, 338685. [CrossRef]

78. Liu, H.; Ye, T.; Mao, C. Fluorescent Carbon Nanoparticles Derived from Candle Soot. *Angew. Chem.* **2007**, *119*, 6593–6595. [CrossRef]
79. Liu, R.; Wu, D.; Liu, S.; Koynov, K.; Knoll, W.; Li, Q. An Aqueous Route to Multicolor Photoluminescent Carbon Dots Using Silica Spheres as Carriers. *Angew. Chem. Int. Ed.* **2009**, *48*, 4598–4601. [CrossRef]
80. Pan, D.; Zhang, J.; Li, Z.; Wu, C.; Yan, X.; Wu, M. Observation of pH-, solvent-, spin-, and excitation-dependent blue photoluminescence from carbon nanoparticles. *Chem. Commun.* **2010**, *46*, 3681–3683. [CrossRef]
81. Martindale, B.C.M.; Hutton, G.A.M.; Caputo, C.A.; Reisner, E. Solar Hydrogen Production Using Carbon Quantum Dots and a Molecular Nickel Catalyst. *J. Am. Chem. Soc.* **2015**, *137*, 6018–6025. [CrossRef] [PubMed]
82. Rong, M.; Feng, Y.; Wang, Y.; Chen, X. One-pot solid phase pyrolysis synthesis of nitrogen-doped carbon dots for $Fe^{3+}$ sensing and bioimaging. *Sens. Actuators B Chem.* **2017**, *245*, 868–874. [CrossRef]
83. Ma, C.; Zhou, Y.; Yan, W.; He, W.; Liu, Q.; Li, Z.; Wang, H.; Li, X. Predominant catalytic performance of nickel nanoparticles embedded into nitrogen-doped carbon quantum dot-based nanosheets for the nitrore-duction of halogenated nitrobenzene. *ACS Sustain. Chem. Eng.* **2022**, *25*, 8162–8171. [CrossRef]
84. Otten, M.; Hildebrandt, M.; Kühnemuth, R.; Karg, M. Pyrolysis and Solvothermal Synthesis for Carbon Dots: Role of Purification and Molecular Fluorophores. *Langmuir* **2022**, *38*, 6148–6157. [CrossRef] [PubMed]
85. Krishnamoorthy, K.; Veerapandian, M.; Mohan, R.; Kim, S.-J. Investigation of Raman and photoluminescence studies of reduced graphene oxide sheets. *Appl. Phys. A* **2011**, *106*, 501–506. [CrossRef]
86. Zong, J.; Zhu, Y.; Yang, X.; Shen, J.; Li, C. Synthesis of photoluminescent carbogenic dots using mesoporous silica spheres as nanoreactors. *Chem. Commun.* **2010**, *47*, 764–766. [CrossRef]
87. Ahlawat, A.; Rana, P.S.; Solanki, P.R. Studies of photocatalytic and optoelectronic properties of microwave synthesized and polyethyleneimine stabilized carbon quantum dots. *Mater. Lett.* **2021**, *305*, 130830. [CrossRef]
88. Architha, N.; Ragupathi, M.; Shobana, C.; Selvankumar, T.; Kumar, P.; Lee, Y.S.; Selvan, R.K. Microwave-assisted green synthesis of fluorescent carbon quantum dots from Mexican Mint extract for $Fe^{3+}$ detection and bio-imaging applications. *Environ. Res.* **2021**, *199*, 111263. [CrossRef]
89. Harshita, L.; Yadav, P.; Jain, Y.; Sharma, M.; Reza, M.; Agarwal, M.; Gupta, R. One-pot microwave-assisted synthesis of blue emissive multifunctional NSP co-doped carbon dots as a nanoprobe for sequential detection of Cr (VI) and ascorbic acid in real samples, fluorescent ink and logic gate operation. *J. Mol. Liq.* **2022**, *346*, 117088.
90. Larsson, M.A.; Ramachandran, P.; Jarujamrus, P.; Lee, H.L. Microwave Synthesis of Blue Emissive N-Doped Carbon Quantum Dots as a Fluorescent Probe for Free Chlorine Detection. *Sains Malays.* **2022**, *51*, 1197–1212. [CrossRef]
91. Jeong, G.; Lee, J.M.; Lee, J.a.; Praneerad, J.; Choi, C.A.; Supchocksoonthorn, P.; Roy, A.K.; In, I. Microwave-assisted synthesis of multifunctional fluorescent carbon quantum dots from A4/B2 polyamidation monomer sets. *Appl. Surf. Sci.* **2021**, *542*, 148471. [CrossRef]
92. Bourlinos, A.B.; Stassinopoulos, A.; Anglos, D.; Zboril, R.; Georgakilas, V.; Giannelis, E.P. Photoluminescent Carbogenic Dots. *Chem. Mater.* **2008**, *20*, 4539–4541. [CrossRef]
93. Tang, L.; Ji, R.; Cao, X.; Lin, J.; Jiang, H.; Li, X.; Teng, K.S.; Luk, C.M.; Zeng, S.; Hao, J.; et al. Deep Ultraviolet Photoluminescence of Water-Soluble Self-Passivated Graphene Quantum Dots. *ACS Nano* **2012**, *6*, 5102–5110. [CrossRef]
94. Hola, K.; Bourlinos, A.B.; Kozak, O.; Berka, K.; Siskova, K.M.; Havrdova, M.; Tucek, J.; Safarova, K.; Otyepka, M.; Giannelis, E.P.; et al. Photoluminescence effects of graphitic core size and surface functional groups in carbon dots: $COO^-$ induced red-shift emission. *Carbon* **2014**, *70*, 279–286. [CrossRef]
95. Sciortino, A.; Marino, E.; van Dam, B.; Schall, P.; Cannas, M.; Messina, F. Solvatochromism un-ravels the emission mechanism of carbon nanodots. *J. Phys. Chem. Lett.* **2016**, *17*, 3419–3423. [CrossRef]
96. Dager, A.; Uchida, T.; Maekawa, T.; Tachibana, M. Synthesis and characterization of Mono-disperse Carbon Quantum Dots from Fennel Seeds: Photoluminescence analysis using Machine Learning. *Sci. Rep.* **2019**, *9*, 14004. [CrossRef]
97. Zhang, W.; Yu, S.F.; Fei, L.; Jin, L.; Pan, S.; Lin, P. Large-area color controllable remote carbon white-light light-emitting diodes. *Carbon* **2015**, *85*, 344–350. [CrossRef]
98. Martindale, B.C.M.; Hutton, G.A.M.; Caputo, C.A.; Prantl, S.; Godin, R.; Durrant, J.R.; Reisner, E. Enhancing Light Absorption and Charge Transfer Efficiency in Carbon Dots through Graphitization and Core Nitrogen Doping. *Angew. Chem.* **2017**, *129*, 6559–6563. [CrossRef]
99. Tingting, Y.; Wang, H.; Guo, C.; Zhai, Y.; Yang, J.; Yuan, J. A rapid microwave synthesis of green-emissive carbon dots with solid-state fluorescence and pH-sensitive properties. *R. Soc. Open Sci.* **2018**, *7*, 180245.
100. Haitao, L.; Kang, Z.; Liu, Y.; Lee, S.-T. Carbon nanodots: Synthesis, properties and applications. *J. Mater. Chem.* **2012**, *46*, 24230–24253.
101. Zheng, Y.; Yang, D.; Wu, X.; Yan, H.; Zhao, Y.; Feng, B.; Duan, K.; Weng, J.; Wang, J. A facile approach for the synthesis of highly luminescent carbon dots using vitamin-based small organic molecules with benzene ring structure as precursors. *RSC Adv.* **2015**, *5*, 90245–90254. [CrossRef]
102. Hou, H.; Banks, C.E.; Jing, M.; Zhang, Y.; Ji, X. Carbon Quantum Dots and Their Derivative 3D Porous Carbon Frameworks for Sodium-Ion Batteries with Ultralong Cycle Life. *Adv. Mater.* **2015**, *27*, 7861–7866. [CrossRef]
103. Semeniuk, M.; Yi, Z.; Poursorkhabi, V.; Tjong, J.; Jaffer, S.; Lu, Z.-H.; Sain, M. Future per-spectives and review on organic carbon dots in electronic applications. *ACS Nano* **2019**, *6*, 6224–6255. [CrossRef] [PubMed]

104. Bomben, K.D.; Moulder, J.F.; Stickle, W.F.; Sobol, P.E. *Handbook of X-Ray Photoelectron Spectroscopy: A Reference Book of Standard Spectra for Identication and Interpretation of XPS, Physical Electronics, Eden Prairie*; Perkin-Elmer Corporation: Waltham, MA, USA, 1995.
105. Zhou, Y.; Sharma, S.K.; Peng, Z.; Leblanc, R.M. Polymers in Carbon Dots: A Review. *Polymers* **2017**, *9*, 67. [CrossRef] [PubMed]
106. Kolanowska, A.; Dzido, G.; Krzywiecki, M.; Tomczyk, M.M.; Łukowiec, D.; Ruczka, S.; Boncel, S. Carbon Quantum Dots from Amino Acids Revisited: Survey of Renewable Precursors toward High Quan-tum-Yield Blue and Green Fluorescence. *ACS Omega* **2022**, *45*, 41165–41176. [CrossRef]
107. Qiang, S.; Zhang, L.; Li, Z.; Liang, J.; Li, P.; Song, J.; Guo, K.; Wang, Z.; Fan, Q. New Insights into the Cellular Toxicity of Carbon Quantum Dots to *Escherichia coli*. *Antioxidants* **2022**, *11*, 2475. [CrossRef]
108. Gayen, B.; Palchoudhury, S.; Chowdhury, J. Carbon Dots: A Mystic Star in the World of Nanoscience. *J. Nanomater.* **2019**, 1–19. [CrossRef]
109. Hu, C.; Yu, C.; Li, M.; Wang, X.; Yang, J.; Zhao, Z.; Eychmüller, A.; Sun, Y.-P.; Qiu, J. Chemically Tailoring Coal to Fluorescent Carbon Dots with Tuned Size and Their Capacity for Cu(II) Detection. *Small* **2014**, *10*, 4926–4933. [CrossRef]
110. Wang, S.; Kirillova, K.; Lehto, X. Travelers' food experience sharing on social network sites. *J. Travel Tour. Mark.* **2016**, *34*, 680–693. [CrossRef]
111. Jiang, K.; Zhang, L.; Lu, J.; Xu, C.; Cai, C.; Lin, H. Triple-Mode Emission of Carbon Dots: Applications for Advanced Anti-Counterfeiting. *Angew. Chem.* **2016**, *128*, 7347–7351. [CrossRef]
112. Li, F.; Li, Y.; Yang, X.; Han, X.; Jiao, Y.; Wei, T.; Yang, D.; Xu, H.; Nie, G. Highly Fluorescent Chiral N-S-Doped Carbon Dots from Cysteine: Affecting Cellular Energy Metabolism. *Angew. Chem.* **2018**, *130*, 2401–2406. [CrossRef]
113. Anwar, S.; Ding, H.; Xu, M.; Hu, X.; Li, Z.; Wang, J.; Bi, H. Recent advances in synthesis, optical properties, and bio-medical applications of carbon dots. *ACS Appl. Bio Mater.* **2019**, *6*, 2317–2338. [CrossRef]
114. Jhonsi, M.A. Carbon Quantum Dots for Bioimaging. In *State of the Art in Nano-Bioimaging*; IntechOpen: London, UK, 2018; pp. 35–55.
115. Konstantinos, D. Carbon quantum dots: Surface passivation and functionalization. *Curr. Org. Chem.* **2016**, *6*, 682–695.
116. Li, L.; Dong, T. Photoluminescence tuning in carbon dots: Surface passivation or/and functionalization, heteroatom doping. *J. Mater. Chem. C* **2018**, *30*, 7944–7970. [CrossRef]
117. Peng, H.; Travas-Sejdic, J. Simple Aqueous Solution Route to Luminescent Carbogenic Dots from Carbohydrates. *Chem. Mater.* **2009**, *21*, 5563–5565. [CrossRef]
118. Wang, L.; Li, W.; Yin, L.; Liu, Y.; Guo, H.; Lai, J.; Han, Y.; Li, G.; Li, M.; Zhang, J.; et al. Full-color fluorescent carbon quantum dots. *Sci. Adv.* **2020**, *6*, eabb6772. [CrossRef]
119. Nisha, P.; Amrita, D. Theoretical study of Dependence of Wavelength on Size of Quantum Dot. *Int. J. Sci. Res. Dev.* **2016**, *4*, 126–130.
120. Kandasamy, G. Recent Advancements in Doped/Co-Doped Carbon Quantum Dots for Multi-Potential Applications. *C* **2019**, *5*, 24. [CrossRef]
121. Darragh, C.; Rocks, C.; Padmanaban, D.B.; Maguire, P.; Svrcek, V.; Mariotti, D. Environmentally friendly nitrogen-doped carbon quantum dots for next generation solar cells. *Sustain. Energy Fuels* **2017**, *7*, 1611–1619.
122. Gao, R.; Wu, Z.; Wang, L.; Liu, J.; Deng, Y.; Xiao, Z.; Fang, J.; Liang, Y. Green Preparation of Fluorescent Nitrogen-Doped Carbon Quantum Dots for Sensitive Detection of Oxytetracycline in Environmental Samples. *Nanomaterials* **2020**, *10*, 1561. [CrossRef]
123. Yu, J.; Liu, C.; Yuan, K.; Lu, Z.; Cheng, Y.; Li, L.; Zhang, X.; Jin, P.; Meng, F.; Liu, H. Luminescence Mechanism of Carbon Dots by Tailoring Functional Groups for Sensing Fe3+ Ions. *Nanomaterials* **2018**, *8*, 233. [CrossRef] [PubMed]
124. Zuo, G.; Xie, A.; Li, J.; Su, T.; Pan, X.; Dong, W. Large Emission Red-Shift of Carbon Dots by Fluorine Doping and Their Applications for Red Cell Imaging and Sensitive Intracellular $Ag^+$ Detection. *J. Phys. Chem. C* **2017**, *121*, 26558–26565. [CrossRef]
125. Baker, S.N.; Baker, G.A. Luminescent carbon nanodots: Emergent nanolights. *Angew. Chem. Int. Ed.* **2010**, *49*, 6726–6744. [CrossRef] [PubMed]
126. Gokus, T.; Nair, R.R.; Bonetti, A.; Böhmler, M.; Lombardo, A.; Novoselov, K.; Geim, A.K.; Ferrari, A.C.; Hartschuh, A. Making Graphene Luminescent by Oxygen Plasma Treatment. *ACS Nano* **2009**, *3*, 3963–3968. [CrossRef]
127. Demchenko, A.P.; Dekaliuk, M.O. Novel fluorescent carbonic nanomaterials for sensing and imaging. *Methods Appl. Fluoresc.* **2013**, *1*, 042001. [CrossRef]
128. Zhang, Q.; Wang, R.; Feng, B.; Zhong, X.; Ostrikov, K. Photoluminescence mechanism of carbon dots: Triggering high-color-purity red fluorescence emission through edge amino protonation. *Nat. Commun.* **2021**, *12*, 6856. [CrossRef]
129. Nguyen, H.A.; Srivastava, I.; Pan, D.; Gruebele, M. Unraveling the Fluorescence Mechanism of Carbon Dots with *Sub*-Single-Particle Resolution. *ACS Nano* **2020**, *14*, 6127–6137. [CrossRef]
130. An, Y.; Liu, C.; Li, Y.; Chen, M.; Zheng, Y.; Tian, H.; Shi, R.; He, X.; Lin, X. Preparation of Multicolour Solid Fluorescent Carbon Dots for Light-Emitting Diodes Using Phenylethylamine as a Co-Carbonization Agent. *Int. J. Mol. Sci.* **2022**, *23*, 11071. [CrossRef]
131. Cao, L.; Meziani, M.J.; Sahu, S.; Sun, Y.-P. Photoluminescence Properties of Graphene versus Other Carbon Nanomaterials. *Acc. Chem. Res.* **2012**, *46*, 171–180. [CrossRef]
132. Fang, Y.; Guo, S.; Li, D.; Zhu, C.; Ren, W.; Dong, S.; Wang, E. Easy Synthesis and Imaging Applications of Cross-Linked Green Fluorescent Hollow Carbon Nanoparticles. *ACS Nano* **2011**, *6*, 400–409. [CrossRef]
133. Shen, J.; Zhu, Y.; Chen, C.; Yang, X.; Li, C. Facile preparation and upconversionlumi-nescence of graphene quantum dots. *Chem. Commun.* **2011**, *9*, 2580–2582. [CrossRef]

134. Li, X.; Wang, H.; Shimizu, Y.; Pyatenko, A.; Kawaguchi, K.; Koshizaki, N. Preparation of carbon quantum dots with tunable photoluminescence by rapid laser passivation in ordinary organic solvents. *Chem. Commun.* **2010**, *47*, 932–934. [CrossRef]
135. Nourbakhsh, A.; Cantoro, M.; Vosch, T.; Pourtois, G.; Clemente, F.; van der Veen, M.; Hofkens, J.; Heyns, M.M.; De Gendt, S.; Sels, B.F. Bandgap opening in oxygen plasma-treated graphene. *Nanotechnology* **2010**, *21*, 435203. [CrossRef]
136. Dekaliuk, M.O.; Viagin, O.; Malyukin, Y.V.; Demchenko, A.P. Fluorescent carbon nanomateri-als:"Quantum dots" or nanoclusters? *Phys. Chem. Chem. Phys.* **2014**, *30*, 16075–16084. [CrossRef]
137. Bibekananda, D.; Karak, N. A green and facile approach for the synthesis of water soluble fluorescent carbon dots from banana juice. *RSC Adv.* **2013**, *22*, 8286–8290.
138. Dong, Y.; Shao, J.; Chen, C.; Li, H.; Wang, R.; Chi, Y.; Lin, X.; Chen, G. Blue luminescent graphene quantum dots and graphene oxide prepared by tuning the carbonization degree of citric acid. *Carbon* **2012**, *50*, 4738–4743. [CrossRef]
139. Zhi, B.; Yao, X.; Cui, Y.; Orr, G.; Haynes, C.L. Synthesis, applications and potential photoluminescence mechanism of spectrally tunable carbon dots. *Nanoscale* **2019**, *11*, 20411–20428. [CrossRef]
140. De Caluwé, E.; Halamouá, K.; Van Damme, P. Adansoniadigitata, L. A review of traditional uses, phytochemistry and pharmacology. *Afr. Focus* **2010**, *1*, 11–51. [CrossRef]
141. Lu, C.; Su, Q.; Yang, X. Ultra-long room-temperature phosphorescent carbon dots: pH sensing and dual-channel detection of tetracyclines. *Nanoscale* **2019**, *11*, 16036–16042. [CrossRef]
142. Gao, W.; Zhang, S.; Wang, G.; Cui, J.; Lu, Y.; Rong, X.; Gao, C. A review on mechanism, applications and influencing factors of carbon quantum dots based photocatalysis. *Ceram. Int.* **2022**, *48*, 35986–35999. [CrossRef]
143. Das, R.; Bandyopadhyay, R.; Pramanik, P. Carbon quantum dots from natural resource: A review. *Mater. Today Chem.* **2018**, *8*, 96–109. [CrossRef]
144. Xu, J.; Tao, J.; Su, L.; Wang, J.; Jiao, T. A Critical Review of Carbon Quantum Dots: From Synthesis toward Applications in Electrochemical Biosensors for the Determination of a Depression-Related Neurotransmitter. *Materials* **2021**, *14*, 3987. [CrossRef]
145. Dong, Y.-L.; Zhang, H.-G.; Rahman, Z.U.; Su, L.; Chen, X.-J.; Hu, J.; Chen, X.-G. Graphene oxide−Fe3O4 magnetic nanocomposites with peroxidase-like activity for colorimetric detection of glucose. *Nanoscale* **2012**, *4*, 3969–3976. [CrossRef]
146. Ryan, B.J.; Carolan, N.; Ó'Fágáin, C. Horseradish and soybean peroxidases: Comparable tools for alternative niches? *Trends Biotechnol.* **2006**, *24*, 355–363. [CrossRef] [PubMed]
147. Gao, L.; Zhuang, J.; Nie, L.; Zhang, J.; Zhang, Y.; Gu, N.; Wang, T.; Feng, J.; Yang, D.; Perrett, S.; et al. Intrinsic peroxidase-like activity of ferromagnetic nanoparticles. *Nat. Nanotechnol.* **2007**, *2*, 577–583. [CrossRef] [PubMed]
148. Floss, M.A.; Fink, T.; Maurer, F.; Volk, T.; Kreuer, S.; Müller-Wirtz, L.M. Exhaled Aldehydes as Biomarkers for Lung Diseases: A Narrative Review. *Molecules* **2022**, *27*, 5258. [CrossRef]
149. Li, J.; Yao, S.-L.; Liu, S.-J.; Chen, Y.-Q. Fluorescent sensors for aldehydes based on luminescent metal–organic frameworks. *Dalton Trans.* **2021**, *50*, 7166–7175. [CrossRef]
150. Ibáñez, D.; González-García, M.B.; Hernández-Santos, D.; Fanjul-Bolado, P. Spectroelectrochemical Enzyme Sensor System for Acetaldehyde Detection in Wine. *Biosensors* **2022**, *12*, 1032. [CrossRef]
151. Mohammadi, M.; Khazaei, A.; Rezaei, A.; Huajun, Z.; Xuwei, S. Ionic-Liquid-Modified Carbon Quantum Dots as a Support for the Immobilization of Tungstate Ions ($WO_4^{2-}$): Heterogeneous Nanocatalysts for the Oxidation of Alcohols in Water. *ACS Sustain. Chem. Eng.* **2019**, *7*, 5283–5291. [CrossRef]
152. Mahamuni, N.N.; Gogate, P.R.; Pandit, A.B. Selective synthesis of sulfoxides from sulfides using ultrasound. *Ultrason. Sonochem.* **2007**, *2*, 135–142. [CrossRef] [PubMed]
153. An, H.; Luo, H.; Xu, T.; Chang, S.; Chen, Y.; Zhu, Q.; Huang, Y.; Tan, H.; Li, Y.-G. Visible-Light-Driven Oxidation of Amines to Imines in Air Catalyzed by Polyoxometalate–Tris(bipyridine)ruthenium Hybrid Compounds. *Inorg. Chem.* **2022**, *61*, 10442–10453. [CrossRef]
154. Chen, W.; Li, H.; Song, J.; Zhao, Y.; Ma, P.; Niu, J.; Wang, J. Binuclear Ru(III)-Containing Polyoxometalate with Efficient Photocatalytic Activity for Oxidative Coupling of Amines to Imines. *Inorg. Chem.* **2022**, *61*, 2076–2085. [CrossRef] [PubMed]
155. Kawahara, N.; Palasin, K.; Asano, Y. Novel Enzymatic Method for Imine Synthesis via the Oxi-dation of Primary Amines Using D-Amino Acid Oxidase from Porcine Kidney. *Catalysts* **2022**, *5*, 511. [CrossRef]
156. Pyun, J. Graphene oxide as catalyst: Application of carbon materials beyond nanotechnology. *Angew. Chem. Int. Ed.* **2011**, *1*, 46–48. [CrossRef]
157. Su, D.S.; Zhang, J.; Frank, B.; Thomas, A.; Wang, X.; Paraknowitsch, J.; Schlögl, R. Metal-Free Heterogeneous Catalysis for Sustainable Chemistry. *Chemsuschem* **2010**, *3*, 169–180. [CrossRef]
158. Yu, H.; Peng, F.; Tan, J.; Hu, X.; Wang, H.; Yang, J.; Zheng, W. Selective Catalysis of the Aerobic Oxidation of Cyclohexane in the Liquid Phase by Carbon Nanotubes. *Angew. Chem.* **2011**, *123*, 4064–4068. [CrossRef]
159. Shiri, M. Indoles in Multicomponent Processes (MCPs). *Chem. Rev.* **2012**, *112*, 3508–3549. [CrossRef]
160. Vidhya, L.N.; Thirumurugan, P.; Noorulla, K.M.; Perumal, P.T. InCl3 mediated one-pot multicomponent synthesis, anti-microbial, antioxidant and anticancer evaluation of 3-pyranyl indole derivatives. *Bioorganic Med. Chem. Lett.* **2010**, *17*, 5054–5061.
161. Gomha, S.M.; Abdel-Aziz, H.A. Synthesis of new heterocycles derived from 3-(3-methyl-1H-indol-2-yl)-3-oxopropanenitrile as potent antifungal agents. *Bull. Korean Chem. Soc.* **2012**, *9*, 2985–2990. [CrossRef]
162. Fadda, A.A.; El-Mekabaty, A.; Mousa, I.A.; Elattar, K.M. Chemistry of 3-(1H-Indol-3-yl)-3-oxopropanenitrile. *Synth. Commun.* **2014**, *44*, 1579–1599. [CrossRef]

163. Thirumurugan, P.; Nandakumar, A.; Muralidharan, D.; Perumal, P.T. Simple and convenient approach to the Kröhnke pyridine type synthesis of functionalized indol-3-yl pyridine derivatives using 3-cyanoacetyl indole. *J. Comb. Chem.* **2010**, *1*, 161–167. [CrossRef] [PubMed]
164. Irfan, S.M.; Khan, P.; Abid, M.; Khan, M.M. Design, synthesis, and biological evaluation of novel fused spiro-4 H-pyran derivatives as bacterial biofilm disruptor. *ACS Omega* **2019**, *16*, 16794–16807.
165. Kumar, S.P.; Silakari, O. The current status of O-heterocycles: A synthetic and medicinal overview. *ChemMedChem* **2018**, *11*, 1071–1087.
166. Zhao, L.; Sun, Z.; Ma, J. Novel Relationship between Hydroxyl Radical Initiation and Surface Group of Ceramic Honeycomb Supported Metals for the Catalytic Ozonation of Nitrobenzene in Aqueous Solution. *Environ. Sci. Technol.* **2009**, *43*, 4157–4163. [CrossRef]
167. McKeen, J.C.; Yan, Y.S.; Davis, M.E. Proton Conductivity in Sulfonic Acid-Functionalized Zeolite Beta: Effect of Hydroxyl Group. *Chem. Mater.* **2008**, *20*, 3791–3793. [CrossRef]
168. Doyle, A.G.; Jacobsen, E.N. Small-molecule H-bond donors in asymmetric catalysis. *Chem. Rev.* **2007**, *12*, 5713–5743. [CrossRef]
169. Chen, X.; Brauman, J.I. Hydrogen bonding lowers intrinsic nucleophilicity of solvated nucleophiles. *J. Am. Chem. Soc.* **2008**, *45*, 15038–15046. [CrossRef]
170. Kitano, M.; Nakajima, K.; Kondo, J.N.; Hayashi, S.; Hara, M. Protonated Titanate Nanotubes as Solid Acid Catalyst. *J. Am. Chem. Soc.* **2010**, *132*, 6622–6623. [CrossRef]
171. Navalon, S.; Dhakshinamoorthy, A.; Alvaro, M.; Garcia, H. Carbocatalysis by graphene-based materials. *Chem. Rev.* **2014**, *12*, 6179–6212. [CrossRef]
172. Chang, B.; Fu, J.; Tian, Y.; Dong, X. Multifunctionalized Ordered Mesoporous Carbon as an Efficient and Stable Solid Acid Catalyst for Biodiesel Preparation. *J. Phys. Chem. C* **2013**, *117*, 6252–6258. [CrossRef]
173. Fekry, A.M. An Innovative Simple Electrochemical Levofloxacin Sensor Assembled from Carbon Paste Enhanced with Nano-Sized Fumed Silica. *Biosensors* **2022**, *12*, 906. [CrossRef] [PubMed]
174. Sitara, E.; Ehsan, M.F.; Nasir, H.; Iram, M.; Bukhari, A.B. Synthesis, characterization and photocatalytic activity of MoS2/ZnSe heterostructures for the degradation of levofloxacin. *Catalysts* **2020**, *12*, 1380. [CrossRef]

**Disclaimer/Publisher's Note:** The statements, opinions and data contained in all publications are solely those of the individual author(s) and contributor(s) and not of MDPI and/or the editor(s). MDPI and/or the editor(s) disclaim responsibility for any injury to people or property resulting from any ideas, methods, instructions or products referred to in the content.

*Review*

# The Behavior of Carbon Dots in Catalytic Reactions

**Lerato L. Mokoloko [1,2,\*], Roy P. Forbes [2] and Neil J. Coville [2,\*]**

[1] Institute for Nanotechnology and Water Sustainability, College of Science, Engineering and Technology, University of South Africa, Florida, Johannesburg 1709, South Africa

[2] DSI-NRF Centre of Excellence in Catalysis and the Molecular Sciences Institute, School of Chemistry, University of the Witwatersrand, Johannesburg 2050, South Africa; roy.forbes@wits.ac.za

\* Correspondence: mokolll@unisa.ac.za (L.L.M.); neil.coville@wits.ac.za (N.J.C.); Tel.: +27-11-717-6738 (N.J.C.)

**Abstract:** Since their discovery in 2004, carbon dots (CDs), with particle sizes < 10 nm, have found use in various applications, mainly based on the material's fluorescent properties. However, other potential uses of CDs remain relatively unexplored when compared to other carbon-based nanomaterials. In particular, the use of CDs as catalysts and as supports for use in catalytic reactions, is still in its infancy. Many studies have indicated the advantages of using CDs in catalysis, but there are difficulties associated with their stability, separation, and aggregation due to their small size. This small size does however allow for studying the interaction of small catalyst particles with small dimensional supports, including the inverse support interaction. However, recent studies have indicated that CDs are not stable under high temperature conditions (especially >250 °C; with and without a catalyst) suggesting that the CDs may agglomerate and transform under some reaction conditions. The agglomeration of the metal in a CD/metal catalyst, especially because of the CDs agglomeration and transformation at high temperature, is not always considered in studies using CDs as catalysts, as post-reaction analysis of a catalyst is not always undertaken. Further, it appears that under modest thermal reaction conditions, CDs can react with some metal ions to change their morphology, a reaction that relates to the metal reducibility. This review has thus been undertaken to indicate the advantages, as well as the limitations, of using CDs in catalytic studies. The various techniques that have been used to evaluate these issues is given, and some examples from the literature that highlight the use of CDs in catalysis are described.

**Keywords:** carbon dots; thermal stability; metal support; heterogeneous catalysis

## 1. Introduction

Carbon allotropes are multifunctional materials, due to their unique physical and chemical properties. Carbon allotropes can be chemically modified by other elements via functionalization or doping, and they can also be used in combination with other materials to form carbon–carbon or metal–carbon composite materials [1]. The modification of carbon allotropes helps to enhance their properties, and also widens their spectrum of applications. Hence, carbon allotropes such as graphite, graphene, fullerene, carbon nanotubes (CNTs), carbon nanofibers (CNFs), carbon black (CB), carbon nano-onions (CNOs), carbon spheres (CSs) and carbon dots (CDs) have been successfully incorporated in fields such as nanomedicine, electronics, sensor fabrication and catalysis [1–4].

Carbon's many allotropes have shown great potential when used as a catalyst support, and the carbon can even act as a catalyst in its own right [1,5]. The enormous interest in carbon-based support materials is due to their surface chemistry, variable surface area and the porosity of the carbon [1,5]. The use of carbon allotropes is also enhanced by their electronic properties, which are influenced by their structure and the carbon atom valence. The electronic effects can promote a high dispersion of the supported (metal) catalyst, and also of surface defects, and this can enhance their capability for gas storage, adsorption and/or separation processes [5,6].

Heterogeneous metal catalysts usually comprise a metallic catalyst that is anchored or supported on the surface of a material that serves to enhance the surface area of the metal while also improving the metal stability during chemical reactions [7]. Support materials are important for immobilizing and anchoring the active metal catalyst, improving its dispersion and durability, and aiding it in avoiding deactivation during a catalytic reaction [8]. Many studies have shown that a catalytic reaction rate is influenced by the catalyst dispersibility [9]. Metal–support interactions (MSI) are an important factor to consider when selecting a catalyst support, as the support directly influences the dispersibility, sintering, reducibility, and overall performance of the catalyst. MSIs are effects measured by the physical and chemical interactions between the metal catalyst and the support [10–12]. The stronger the MSI, the more resistant the catalyst is to deactivation via sintering during a reaction [12,13]. Carbon, in its various allotropes, generally forms a weak MSI with a metal catalyst. However, the MSI can be improved by means of functionalizing or doping of the carbon surfaces with heteroatoms. These effects create defects in the carbon nanostructure, and also alter the electronic properties of the carbon allotrope. Consequently, the dispersibility of the metal catalyst is increased and metal sintering/deactivation is reduced by doping/functionalization [12].

Most carbon allotropes are relatively 'stable' under harsh reaction conditions; this makes them less susceptible to structural and/or chemical changes during catalysis [12]. This is important, because the structural disintegration of a catalyst support can lead to deactivation of a catalyst. The properties of carbon supports have been studied under different conditions using varying pH, solvents, temperatures, time, etc. For example, carbon allotropes have successfully been used as supports for titania in photocatalysis [14], platinum in electrocatalysis [15], and cobalt, iron and ruthenium in catalysis for fuel production under a wide range of reaction conditions [12].

Carbon dots (CDs) are a fairly new type of carbon nanomaterial. CDs are defined as zero-dimensional quasi-spherical carbon nanomaterials with particle sizes below 10 nm [16]. Their general structure consists of $sp^2$- and $sp^3$-type carbons, with a large number of functional groups or polymer chains attached to their surfaces [17–19]. Numerous reviews have summarized the potential of CDs in many traditional and emerging areas, such as photoluminescence (PL) and photoelectrochemical-driven sensing, catalysis, imaging, and biomedicine applications, where they have been shown to be superior to other carbon allotropes [16,20]. Further, CDs, unlike metal-based quantum dots, have a high water dispersibility, low toxicity, and good biocompatibility [21]. Thousands of papers have been written about CDs in terms of their synthesis, characterization and uses. The aim of this review is not to reproduce the information on the synthesis, applications, and the properties of CDs, which has been summarized in scores of reviews, but to indicate the transformation of CDs into other shaped materials, which is associated with temperature and the presence of catalysts.

CDs, like many carbon allotropes, have been used as both catalysts and catalyst 'supports,' and reviews on the use of CDs both as carbocatalysts and as composites with metal and metal oxide supports have been reported [22–24]. As is known, there are many advantages to using CDs and CD/metal complexes in catalysis, but there are also limitations on the use of these carbons. For example, in many of the studies it is not clear if the integrity of the carbon has been retained, and the degree to which metal sintering has taken place during their synthesis or in their catalytic reactions. Indeed, some studies have revealed that CDs and CD/metal materials, lose their shape/size during reaction, especially at high temperatures (>100 °C) and in the presence of easy-to-reduce metals. This has been observed in carbon–carbon coupling reactions [25], hydrogen evolution reactions [26] and hydrogenation reactions, where the CDs either decomposed or were transformed into other carbon structures [27,28].

In this review we have evaluated data on the use of CDs and CD/metal catalysts in catalysis. One of the advantages of using these two types of catalysts, relative to more typical catalyst systems, relates to the easy access to CDs which can be produced at low cost from any carbon source, including waste materials. Since the metal particles are either the same size or larger than the CDs, the terminology CD/metal has been used to describe the catalysts made from CDs and metal particles. We have not comprehensively reviewed all reports where CDs and CD/metal materials have been used in catalysis, but have rather chosen some typical literature reactions to highlight the challenges associated with the use of CDs in catalytic reactions. In particular, we wish to highlight the structural changes or destruction of the CDs that can occur during a reaction that can ultimately lead to metal agglomeration. The techniques that can be used to evaluate the stability of CDs, especially post reaction, are discussed with special reference to metal and carbon agglomeration.

## 2. The Structure of Carbon Dots (CDs)

### 2.1. Types of CDs

CDs are usually divided into four distinct subgroups. These are graphene quantum dots (GQDs), carbon quantum dots (CQDs), carbon nanodots (CNDs) and carbonized polymer dots (CPDs) [29–32] (see Figure 1). In the literature, these four types are usually called carbon dots (CDs), and generally they are all regarded as being similar in their catalytic (and many other) properties. However, this may not always be true. To date, this issue has not been explored, and there is evidence to suggest that the different CDs will react differently as a function of temperature and when in the presence of metals that can aid their decomposition [27]. In this review, the term CDs will be used for all carbon dots shown in Figure 1.

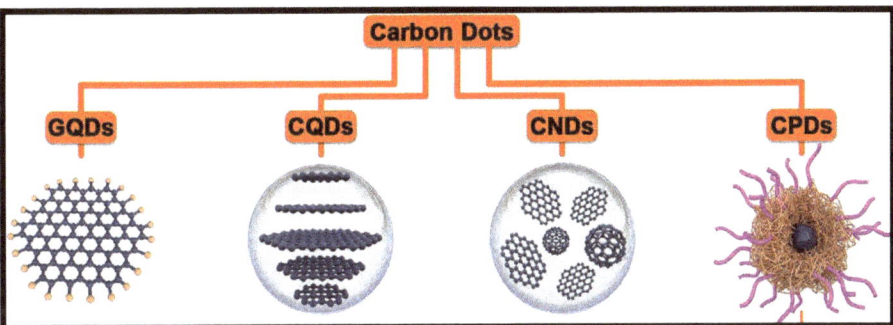

**Figure 1.** The proposed representative structures of a graphene quantum dot (GQD), carbon quantum dot (CQD), carbon nanodot (CND), and carbonized polymer dot (CPD). Adapted with permission from Wiley-VCH Verlag [33].

CDs must be differentiated from carbon nano-onions (CNOs) and onion-like nanocarbons (OLNCs), which are also in the nanometer range [34]. CNOs and OLNCs have structures with carbon layers similar to those found in a concentric-like onion structure (see Figure 2), while CDs (Figure 1) tend to have the carbon layers arranged as parallel graphene sheets (Figure 1). The CNOs/OLNCs are prepared at high temperatures (>500 °C), while CDs are typically prepared at T < 200 °C. This results in differences in their thermal stabilities that impact on their chemical properties.

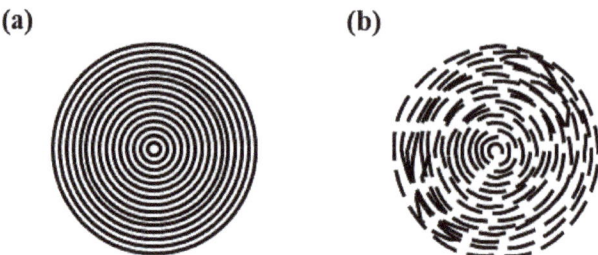

**Figure 2.** Graphical representation showing the differences between (**a**) a carbon nano-onion (CNOs), and (**b**) an onion-like nanocarbon (OLCN).

There are two methods used to modify the structure/surface of CDs. One method is by functionalization, and this is described in more detail below. The other is by making the CDs with precursors in which reactants provide non-carbon atoms to the CD. This is referred to as doping. The most typical dopant is nitrogen, and the addition thereof in quantities ranging from 1 to 10% can lead to substantial changes in the surface chemistry of a carbon material, including CDs. Many CDs have also been doped with metal ions, and this modification can also impact their chemistry and catalytic reactions [35].

*2.2. Surface Properties of CDs*

CDs have many useful properties associated with their small size and functional groups. While the surface of the CD is important in correlating with their physical and chemical properties, the role of the core is still poorly understood [36]. Thus, most studies on the use of CDs that have been performed have related to their surface chemistry [37]. The surface of a CD contains many functional groups, and these groups are responsible for the catalytic activity of the CD. CDs are typically synthesized with both oxidizing and reducing groups that are used to bring about organic redox transformation reactions. These functional groups are also used to modify their PL spectra. Their surface chemistry can be modified by classical procedures associated with modifying any carbon surface. The surface modification can be achieved by covalent and non-covalent bonding of reactants with the CD surface, as reported in work by Yan et al. [38]. This type of modification is carried out to improve the properties of the CDs for a specific application such as biosensing, or metal and molecule detection [38]. Examples are shown in Figure 3.

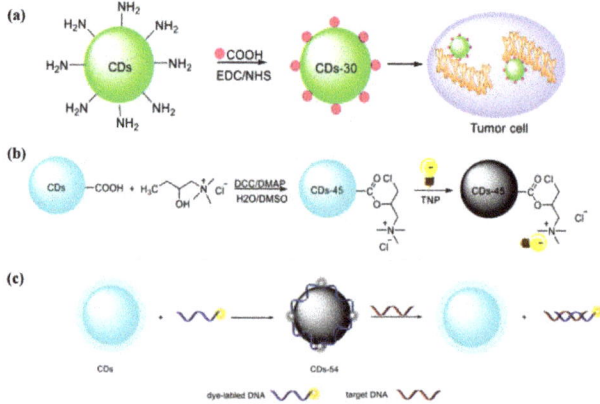

**Figure 3.** Post-surface functionalization of different CDs via (**a**) amide coupling-type reaction (covalent bonding), (**b**) esterification (covalent bonding) and (**c**) pi–pi interactions (non-covalent bonding). Adapted with permission from [38], Copyright 2018, Springer Nature.

The surface charge on the CD can be modified by changing the functional groups. The starting precursors can be manipulated in order to obtain CDs that are either hydrophilic or hydrophobic [39]. Typically, CDs are synthesized using multifunctional organic chemicals, and this produces hydrophilic CDs. Similarly, hydrophobic CDs can be produced using aliphatic chemicals such as dodecylamine [39,40]. It is further noted that the surface charge on the CD can be modified by changing the functional groups, and this can be done without post modification of the CDs [41]. Thus, CDs with different surface charges [41,42] and polarities [40,43] have been reported. For example, most CDs are made with a negative charge (associated with $COO^-$ groups). An important approach to generate positively charged CDs is by modifying CD surfaces with ionic liquids (ILs) and then annealing the material at *ca.* 240 °C. The IL-covered CDs were then used to detect metal ions in solution [41] or to make inks [42].

CDs can also react with themselves. For example, CD–CD linkages have been achieved from CDs that were made from $C_{60}$ fullerene (by a base reaction). The CDs self-assembled when they were freeze dried [44]. After annealing (600–800 °C), the materials were used in capacitor studies. The self-assembly was proposed to be achieved with ice crystals acting as a template [44]. Studies have also been carried out in which a chemical reaction between two different CDs produced CD assemblies with surface properties associated with the different CDs used. Zhou et al. made three different types of CDs, and under room temperature conditions and with different CD combinations, they were linked together via the functional groups on the CDs. In one instance, the CDs, after reaction with each other, gave nanostructures that were used as drug nanocarriers [45,46]. Many other examples have also been reported [27].

Self-assembly of CDs has been achieved by the use of tannic acid. Modification of the functional groups on the CDs with ionic liquids provided for a good interaction of the CDs, with the negative charges on the tannic acid allowing for formation of assembled species (in some cases, $d_{CD} > 100$ nm), which were readily detected by Tyndall cone measurements and transmission electron microscopy (TEM) measurements [47]. Supramolecular organization of CDs after alkylation of amine-functionalized CDs [39] has led to a series of alkyl-functionalized CDs, which could be separated by chromatography and that formed an organized structure in the solid state. The new assemblies were used in nonlinear optical studies [40]. A recent extension of this work described the self-assembly of CDs made from chiral cholesteryl to make thermotropic liquid crystals with a range of architectures [43]. These self-assembled CDs could provide an entry into novel structures for use as both carbocatalysts and to make CD-metal catalysts.

A key property associated with a CD surface is its hydrophilicity, which has led to the extensive use of CDs in medicinal chemistry [48]. Another important property is their photoluminescence (PL), which has allowed them to be used in sensing devices [49]. Typically, addition of a metal reactant to the CD functional groups results in a decrease in their PL spectrum, and hence these types of experiments are usually conducted to detect metal ions within various media [50]. The changes in the PL spectra of CDs will be influenced, in a catalytic reaction, by the varying concentrations of products and reactants that could bind to the CD surface, thus allowing for the exploration of a reaction mechanism. The addition of polymers to a CD surface has been found to lead to improved PL properties [29].

Due to their small size (<10 nm), the separation and purification of CDs is not simple. Further, yields of the purified CDs are not always reported, and it is thus difficult to assess the usefulness of many synthetic strategies. While CDs have a small size, their surface area tends to vary, and can be lower than expected. The surface area can vary, for example, between 16.4 $m^2\,g^{-1}$ [51] and 1690 $m^2\,g^{-1}$ [52].Thus, interaction with a metal ion or particle will be limited by this property. However, the many surface groups can be used to reduce metal ions to metal particles and in so doing lead to the formation of small metal particles, by limiting the site of the reduction. As expected, when CD surfaces, as in all carbons, are doped with N atoms, metal particle agglomeration is reduced [53].

The role of carbons as supports is limited by their reactions under oxygen, hydrogen or inert gases. In the presence of oxygen, most carbons will oxidize (to CO, $CO_2$) below 600 °C [54,55], while under $H_2$, the carbon can react to form $CH_4$, typically at temperatures above 500 °C [56]. Under an inert atmosphere, surface groups on the carbon can be removed at temperatures dependent on the carbon-to-element bond. In the absence of a catalyst, and under an inert atmosphere, the carbon core can be stable to temperatures above 600 °C. CDs, because of their size, have a high surface-to-bulk (core) carbon atom ratio [57]. Thus, all the reactions listed above can be expected to be modified when CDs are used, in relation to reactions with larger carbon molecules.

### 2.3. Synthesis of Carbon Dots and Their Application as Reducing Agents

Many papers and review articles have been written on the synthesis of CDs and this topic will not be discussed in detail here. CDs can be synthesized by "top-down" procedures, by cutting down larger carbon allotropes such as graphene, fullerene and CNTs using strong oxidizing agents like sulphuric and nitric acid [58]. The CDs can also be prepared by the "bottom-up" process, and are generally made from precursors that contain functional groups that are typically retained from their synthesis precursors [29]. For example, CDs can be prepared from highly oxygenated starting materials such as ascorbic acid, sucrose, and citric acid [59,60], using a "bottom-up" synthesis approach (Figure 4). Further functionalization using a variety of methods can be carried out to advance the surface chemistry and other properties of the CDs [29,38]. The CD surface groups affect their overall chemical behavior, such as their electronic properties. These electronic properties have been exploited for oxidizing and reducing metals, and in this way generate CD/metal catalysts [24,61].

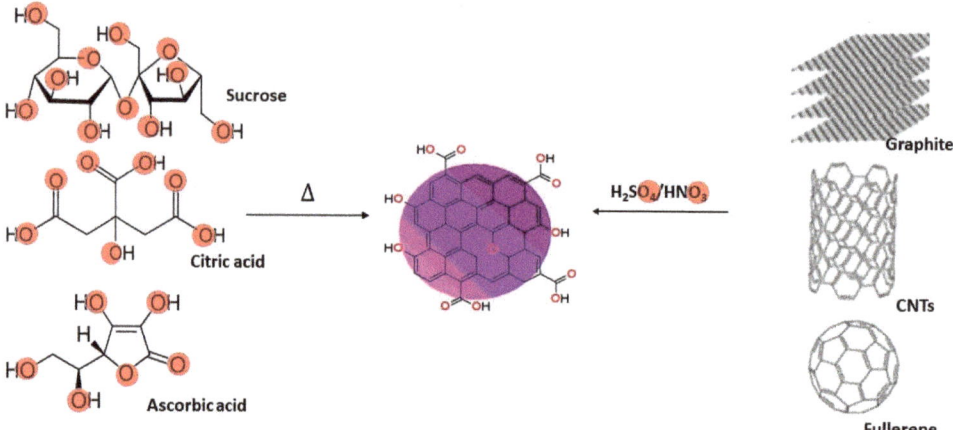

**Figure 4.** Schematic presentation of the "bottom-up" and "top-down" CDs synthesis procedures from oxygen-rich starting materials (highlighted in red) or by reacting 'large' carbon allotropes with oxidizing acids.

There are numerous studies that have been reported in which CD surface oxygen groups have been modified/reduced after the CDs have been synthesized. These are referred to as reduced CDs (r-CDs). Reducing agents used include $NaBH_4$, ascorbic acid, sodium citrate, and hydrazine hydrate, with $NaBH_4$ being the most effective reducing agent [62]. The r-CDs have been reported to have better luminescence properties than their pristine (more highly oxidized) CD counterparts [37]. These r-CDs have been used to reduce strong oxidizing agents such as $KMnO_4$, $KIO_4$ and $K_2Cr_2O_7$. These reagents in turn have been used to selectively oxidize the O-H groups in r-CDs to C=O [63]. They have also been used to reduce metal ions, to generate metal catalysts.

## 3. Carbon Dots in Catalysis

The main role of any carbon in the field of catalysis is that of the carbon acting either as a catalyst (called carbocatalysis) or as a support for a heterogenous catalyst [22]. Because of their small size, CDs add this extra feature of support dimension to their use in catalysis, relative to other carbon supports.

### 3.1. Applications of Carbon Dots-Based Catalysts

Excellent reviews exist on the use of CDs as carbocatalysts [22–24,64,65] in thermal, photocatalytic and electrocatalytic applications. Many simple and classic chemical transformation have occurred in the presence of CDs (Figure 5) [24].

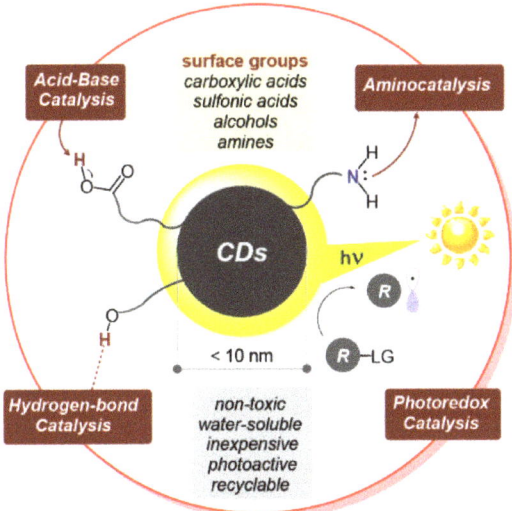

**Figure 5.** CD application in nano-organocatalysis and photocatalysis. (Adapted with permission from [24] Copyright 2020, American Chemical Society).

As will be seen in these reviews, much focus has been on their surface functionalization and use, rather than on morphology changes of the CDs observed after or during reactions. This is understandable, as in most instances very small amounts of catalyst are used, and re-use studies have indicated that the carbon 'catalyst' is stable after many reaction cycles. However, some studies have shown that the CDs can be converted into other morphologies, typically when reactions occur at high temperatures or in the presence of easily reduced metal ions. The difficulty in using smaller carbon particles (CDs) was hinted at in an early review on the use of nanocarbons in catalysis, viz. "However, it is useful to comment that it is not always proven that functionalized nanocarbons act as real catalysts; e.g., they are not consumed during the reaction" [66]. Some examples where post-reaction studies of CDs have been reported and have revealed CD conversion to other morphologies are given below.

- A limitation of using carbon dots in catalysis

The conversion of CDs into carbons with a different framework was noted in early studies on the synthesis of CNTs when the CDs were decomposed at high temperature in porous anodic aluminas [67]. Later studies showed that CDs, made from acetone, could be easily converted into carbons called porous carbon frameworks (PCFs) at high temperatures of between 400 °C and 800 °C [68]. Transmission electron microscopy (TEM), scanning electron microscopy (SEM) and Fourier-transform infrared spectroscopy (FT-IR) studies showed that conversion of the CD morphology occurred. A mechanism involving Na ions interacting with the CDs was suggested, in which carbon nanosheets were formed by CD decomposition to carbon atoms which then self-assembled to form sheets; these materials were used in electrochemical studies [68]. The synthesis of P- and N-doped porous materials (NPCNs) was also achieved by adding the above CDs to amino trimethylene phosphonic acid, and heating to 800–1000 °C (Figure 6). The NPCNs and porous carbon frameworks (PCFs) were used in electrochemical studies that gave exceptional behavior as metal-free ORR catalysts [69].

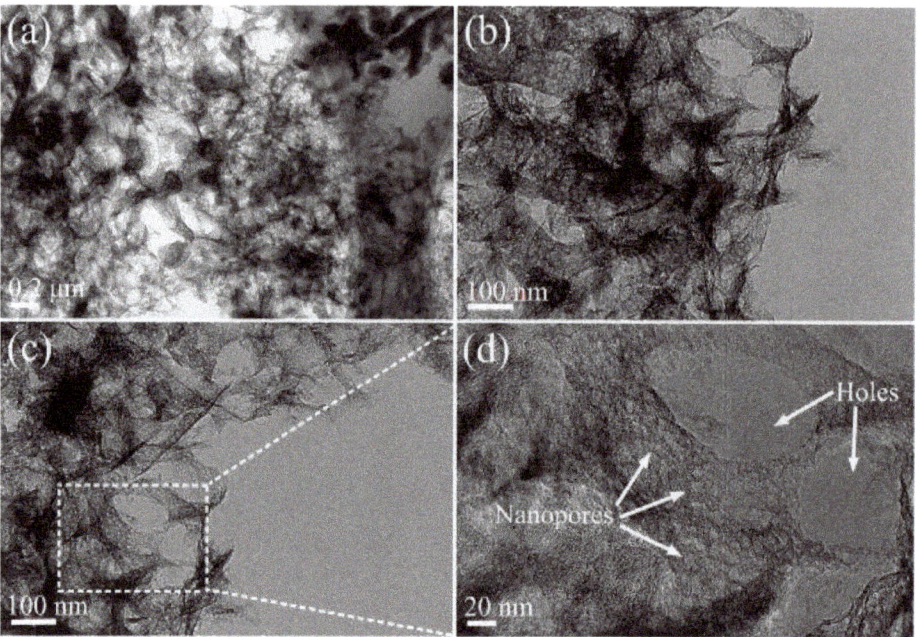

**Figure 6.** TEM images of NPCN-900 (i.e., carbon heated to 900 °C) at (**a**) 0.2μm scale, (**b**,**c**) 100 nm scale, highlighting the nanoporosity and holes in their structure (**c**), as well as an (**d**) HRTEM (20 nm scale) image of the highlighted nanopores and holes (reproduced with permission from [69], 2017 Elsevier Ltd.).

The reaction of CDs made from acetaldehyde mixed with NaHPO$_4$ at temperatures ranging from 400 °C to 900 °C gave P-doped carbon nanosheets (P-CNSs) [70]. TEM and SEM studies of the CDs (made without the P addition) and the P-CNSs showed that sheet-like materials had been made. These materials were studied for their electrochemical behavior. They showed good sodium ion storage when used as an anode material for sodium-ion batteries [70]. The thermal conversion of CDs made from sucrose or glucose [71,72] has been monitored, and the data clearly indicated a simple morphology change to a layered carbon material with graphene-like structure. These conversion reactions are described in Section 6. In summary, numerous examples have shown that CDs can readily convert to sheet-like materials under thermal conditions.

### 3.2. Metals Doped into CDs and Metals Supported on CDs as Catalysts

Metals salts can be added to CDs in two different ways: (i) during the CD synthesis to give metal-doped CDs [35] and (ii) after reaction, to give metal-supported CD materials. In the literature, the catalysts in which the metal is (i) on/in the surface of a CD, (ii) covered by a carbon layer or, (iii) on a carbon layered material made from a CD, have been given various names. In this review all will be referred to as a CD/metal material, to indicate that the CD is generally smaller than the metal particle.

Different types of CD/metal composites have been studied as catalysts in a variety of organic reactions, including carbon–carbon bond formation, oxidation, reduction, hydrogenation, heterocyclic synthesis, multi-component synthesis, and simple organic conversions under light- or mild-temperature conditions ($\leq$100 °C). This data has been reviewed [22,23]. Furthermore, CDs have also been studied as catalysts in a variety of nano-organocatalytic and nano-photocatalytic reactions [24]. For example, Li et al. have reported on the use of novel nanocomposites made by doping CDs with a variety of metals (Cu, Zn, Co, Fe, etc.) to improve the optical and electronic properties of the CDs for use in photo-/electrocatalysis [35].

When considering data from the literature, it is clear that the mixture of metal ions and CDs at low temperatures leads to a range of possible chemical interactions. Thus, reports have shown that addition of metal ions leads to (i) coordination compounds with the CD surface [73], (ii) metal reduction by the CD surface groups [74], and even (iii) reduction of the metal by the carbon core [25]. These reactions are discussed below.

The ability of the CD core to reduce a metal is determined by the reducibility of the metal. Ellingham diagrams have been used to indicate the role of carbon in reducing bulk metal oxides to a metal, and some date indicating this are shown in Figure 7 [75]. Any metal oxide above the free energy of carbon in the diagram can be reduced by carbon. While it is expected that nano-sized metal oxides will have different phase diagrams and hence different free energy values from those in the Ellingham diagram, the differences will be small, and will allow for similar generalizations to be made. As can be noted, a reduction reaction can occur at T < 100 °C for some metals. Further, the addition of metal ions to carbon, where the metal appears below the carbon free energy line, will not be expected to bring about a reaction with carbon. Post analysis of reactions to support the above, where reported, suggest no reaction with the carbon CD core has occurred and that the CDs have retained their size/shape [35].

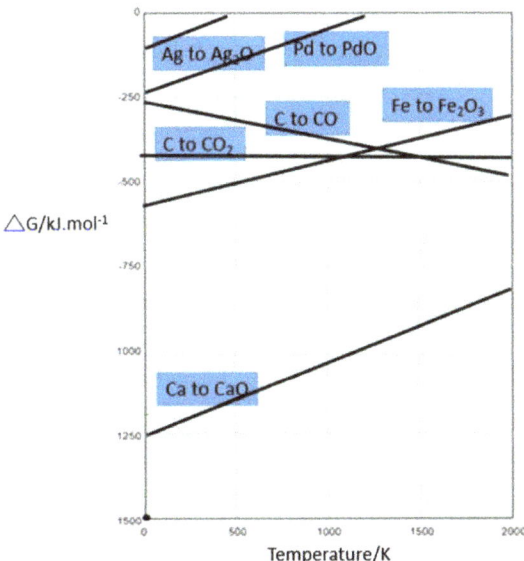

**Figure 7.** Ellingham diagram showing some metal oxides and carbon.

### 3.2.1. High-Temperature Reactions of Metals Supported on CDs

Studies have shown that when high-temperature reactions (T > 400 °C) are used to react metal ions with large spherical carbon materials, the carbon itself acts as a reducing agent and in so doing gives a surface with the metal particle 'embedded' in the carbon. For example, the reaction of Co ions with carbon spheres (d = *ca*. 450 nm) under an inert gas, at T *ca*. 450 °C, produced Co/C catalysts with small reduced Co particles for use in the Fischer–Tropsch (FT) reaction (Figure 8). These particle sizes compared with the sizes of Co particles produced under $H_2$ gas, but with a better ability to prevent Co agglomeration during the FT reaction [76]. The above suggests that the reduction of metal ions on carbon should be possible on smaller carbon spheres, the CDs. However, as the size of the carbon sphere is reduced, there comes a point at which the carbon and the metal particle will have similar sizes, and this could influence the resulting metal–carbon interaction.

**Figure 8.** An illustration showing the reaction of $CoO_x$ with a carbon sphere (cobalt oxide =  ; cobalt metal =  ).

Indeed, the study of reactions in which the support is comparable in size to the metal is referred to as inverse support catalysis (Figure 9) [77–79]. While this concept is well known when metal oxide supports are used ($SiO_2/TiO_2/Al_2O_3$), this concept has rarely been exploited with carbon as the support.

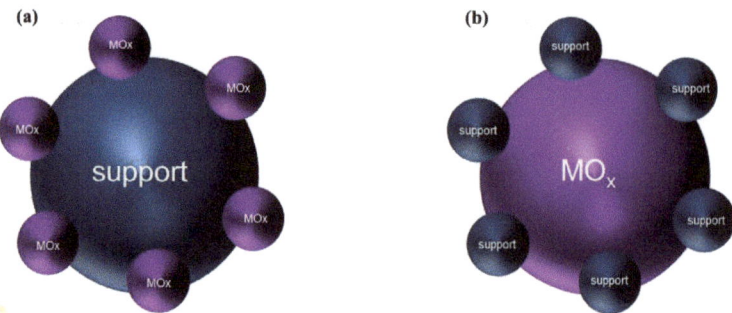

**Figure 9.** An illustration of (**a**) a conventional support; where small metal catalyst particles are dispersed on a large surface area support, and (**b**) an inverse support; small amounts of a support material are dispersed on the surface of a metal catalyst (forming "nano islands" around the metal).

A previous attempt was made to study catalysts made with Co and CDs (with similar small dimensions) in the Fischer–Tropsch reaction [28]. In the reaction (220 °C/10 bar pressure) the CDs were found to decompose, and this led to Co agglomeration. It is clear that both surface groups and the CD core were altered/removed in the reaction. The residual carbon support showed no CDs, and the CDs were completely transformed into a layered carbon material, as shown in Figure 10. The changes to the CDs were accompanied by the simultaneous reduction of the Co active metal phase. This data clearly illustrated how the support material changed alongside the active metal during the reduction treatment. It is unclear if the Co metal dictates the changes to the CDs, or vice versa.

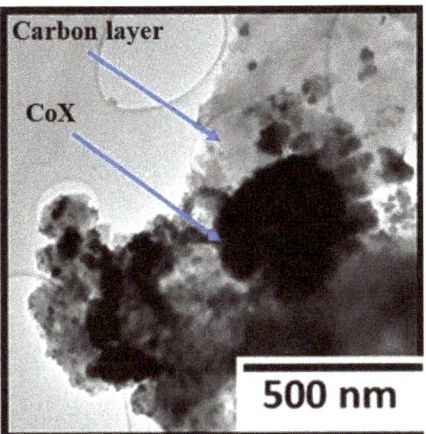

Figure 10. TEM image of CoX/CDs after reaction [28].

The data are consistent with the high-temperature studies of CDs in the absence of a catalyst (see [71]). This could limit the potential use of CDs as metal supports under reducing conditions, but may open up new ways of making metals supported on carbons with unexpected morphologies.

Other studies have also appeared in the literature in which similar observations have been made. CDs were prepared from citric acid and ethylenediamine. To the CDs was added nickel nitrate, and the mixture annealed under a nitrogen atmosphere at 300, 400, 500, and 600 °C for 3 h. The catalysts were used for the nitro-reduction of halogenated nitrobenzenes. The Ni@NCDs (NCD = nitrogen-doped CD) exhibited a nanosheet structure with Ni nanoparticles (6.88 nm) embedded in the NCDs, as observed in high-resolution TEM (HRTEM) images (Figure 11). Ni metal particles could be seen forming from NiO, even at 300 °C (detected by XRD studies) and the NiO had disappeared by 600 °C [80].

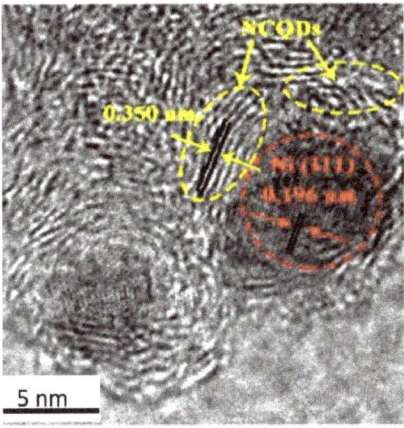

Figure 11. HRTEM image of Ni encompassed by carbon layers made from CDs (adapted with permission from American Chemical Society [80]).

Studies have also shown that CDs can be added to metal complexes to produce carbon covered metal oxide composites. Thus, the addition of CDs (made from acetone) added to TiCl$_3$ (and CTAB) after annealing at 800 °C gave carbon-covered TiO$_2$ (Figure 12). This provides an excellent method of producing total coverage of the TiO$_2$ by carbon. The petal-like structures were used to 'store' sodium ions for use in battery studies [81].

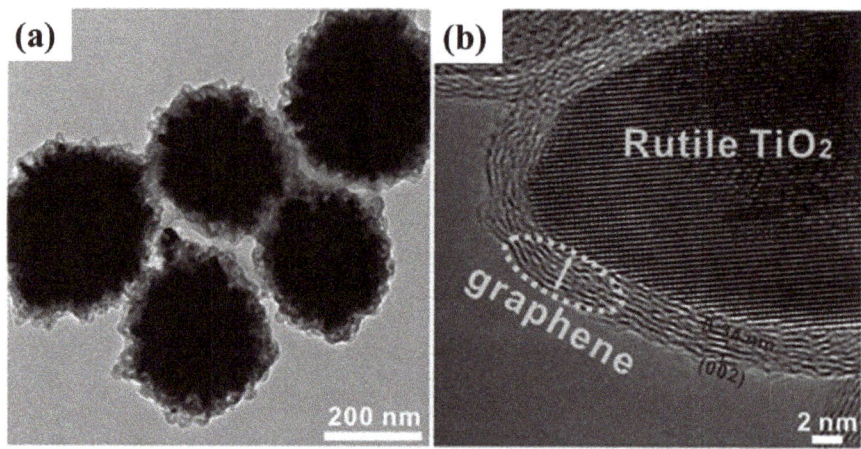

**Figure 12.** (a) TEM and (b) HRTEM images of titania covered by carbon layers made from CDs. Adapted with permission from [81], 2016 WILEY-VCH Verlag GmbH and Co. KGaA, Weinheim, Germany).

These results are consistent with the data from Ellingham diagrams discussed above.

3.2.2. Low-Temperature Reactions of Metals Supported on/in CDs

Many reactions have been studied on metal-CD composites in catalysis at temperatures lower than 150 °C. In these studies, the CDs act as a reducing agent for the reduction of a metal salt (and as a capping agent) that is to be used as a catalyst. Any pre-existing O-H groups on the CDs are sufficient to reduce metal salts and metal oxides into a lower oxidation state, and even to the metallic state. It is less likely that the carbon core itself acts as a reducing agent at these low temperatures, but this will be influenced by the metal under study. In general, very few post-reaction studies have been reported to evaluate any changes that may have occurred at the surface or to the carbon after the use of the CD in chemical reactions.

- Examples where the CD structure is retained

Since CDs can reduce metals, it is important to use appropriate reaction conditions and metals (or metal oxides) to make CD/metal composites. Examples where CDs have been loaded onto metals (or metal oxides) and retained their morphology are given below. For example, simple coordination of metal ions onto the CD surface has been shown by studies on single-atom (Fe) catalysts that were added to CDs [82]. The coordination of Fe$^{3+}$ with the carboxyl groups on the CD was studied by the extended X-ray absorption fine structure (EXAFS) technique, and results clearly indicated that the metal had not been reduced and that the Fe was directly linked to the CD [73]. The 'coordination' of CDs with metal oxide particles has also been observed, e.g., with Fe$_3$O$_4$ particles [83].

Studies in which the surface groups on CDs have been used to reduce metal ions and then bind metal particles to the CDs are well known. Ferrocene, acetone and hydrogen peroxide were used to make a CDs/Fe$_3$O$_4$@CS product in a solvothermal reaction that showed CDs (3.9–9.8 nm) and Fe$_3$O$_4$ embedded in a carbon sphere (CS). The average diameter of the nanocomposite, which resembled a pomegranate fruit, was 451.9 nm

(Figure 13). The HRTEM studies clearly showed the co-existence of the CDs and the FeOx particles. The CDs/Fe$_3$O$_4$@CS were used in peroxymonosulfate, persulfate and H$_2$O$_2$ studies, with and without visible-light illumination, and in ibuprofen degradation studies [84]. Interestingly, when commercial Fe$_3$O$_4$ particles and glucose/acetic acid (to make CDs) were heated at 140 °C for various time periods (4 h–18 h), the products formed showed CDs attached to the Fe$_3$O$_4$ particles [85]. In contrast, glutaric acid-functionalized Fe$_3$O$_4$ particles (14–20 nm) when reacted with CDs (2.5 nm) made from polyacrylamide at 270 °C gave Fe$_3$O$_4$ particles encapsulated by carbon layers [86].

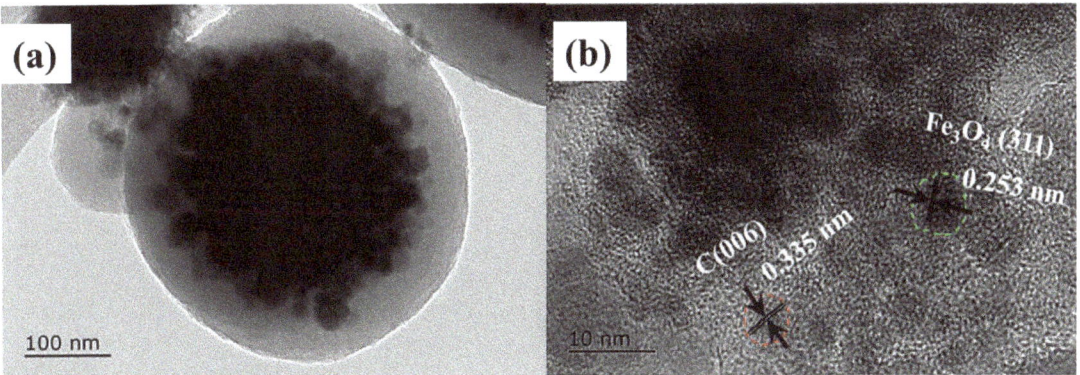

**Figure 13.** (a) TEM and (b) HRTEM images of a CDs/Fe$_3$O$_4$@CS catalyst at different magnifications. Adapted with permission from [84], Copyright 2020 Elsevier.

CDs were produced from ethylenediamine and citric acid, and these were added to copper acetate solutions in different CD/Cu ratios. The photocatalytic reaction of copper acetate and CDs produced Cu/CDs (4–6 nm), as detected by HRTEM studies (Figure 14). The interaction of the Cu with the CD can be clearly seen in the image. The larger the CD/Cu ratio, the larger the particles that were formed. The Cu/CD catalysts were used for the photocatalytic hydrogen evolution from lactic acid solutions [87].

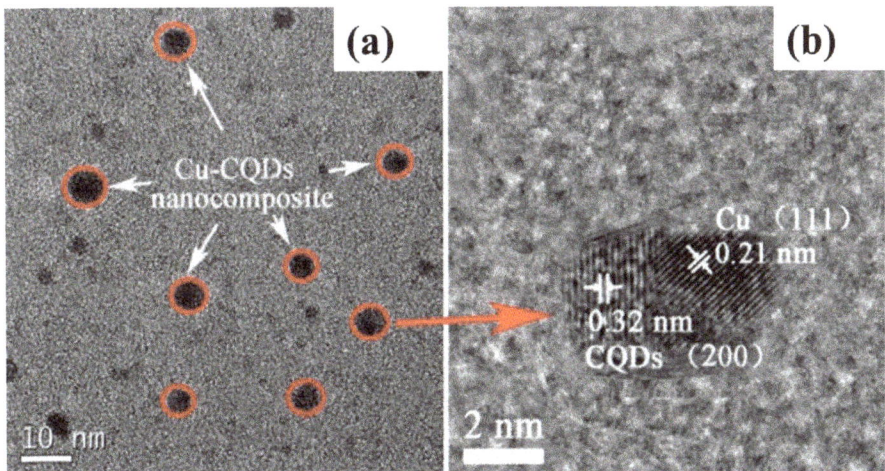

**Figure 14.** (a) TEM and (b) HRTEM images of Cu/CDs NPs (reproduced with permission from [87], Copyright 2017 Elsevier.

In another study, the polyoxometalate (POM) clusters, $Na_6[H_2PtW_6O_{24}]$ and $Na_6[H_2PtMo_6O_{24}]$, and carbon dots (CDs) were added together, and were used in the hydrogen oxidation reaction (HOR) in acid [88]. The CDs (*ca.* 5 nm) were made by electrolyzing graphite rods. The CDs improved the catalytic performance of the POM by enhancing the electron acquisition ability of Pt. The HRTEM image (Figure 15) clearly shows well-dispersed CDs on the POM.

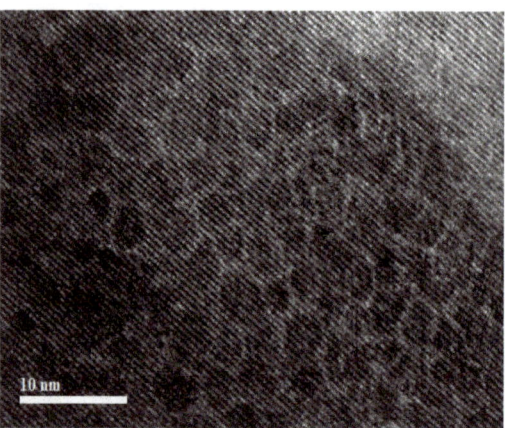

**Figure 15.** HRTEM image of $Na_6[H_2PtW_6O_{24}]$ with CDs observed on the surface. Adapted with permission from [88], Copyright 2021 Elsevier.

Lu et al. synthesized CDs from ethylene glycol by electrolyzing an electrolyte solution. The as-prepared CDs (particle size range: 2–4 nm) were then used to reduce $HAuCl_4$ and $AgNO_3$ to AuNPs and AgNPs, respectively, using a facile room temperature method [61]. HRTEM images showed the presence of metallic Au particles (12–14 nm). The particle size range for the Ag nanoparticles was 6 to 8 nm. In the study, analysis of the samples using FT-IR spectroscopy and X-ray photoelectron spectroscopy (XPS), before and after the reduction reaction, showed that the O-H group concentration on the CDs surface were significantly reduced, while the carbonyl group concentration increased post reduction. Additionally, the as-prepared metal nanoparticles showed good dispersibility, and this was associated with possible hydrogen bonding between the residual CD hydroxyl groups and the metal nanoparticles [61]. The CD-reduced metal catalysts were active in the colorimetric detection of $H_2O_2$ and glucose. Later, Yang et al. developed a AuPd bimetallic catalyst [89] using CDs prepared from ethylene glycol following the procedure reported by Lu et al. [61]. The reduction of the metals by CDs was also associated with the presence of the hydroxyl functional groups on the surface of the CDs. The prepared AuPd nanoparticles were tested for the catalytic reduction of 4-nitrophenol (4-NP) to 4-aminophenol (4-AP).

Another interesting study was conducted by Jin et al., who prepared four different types of CDs from sucrose, sucrose and acetylcholine chloride, sucrose and mercaptosuccinic acid, and sucrose and N-acetyl-L-cysteine as precursors to produce pristine CDs (with C and O functionalities only), N-doped CDs, S-doped CDs, and N-S doped CDs. The TEM results showed that doping the CDs, especially with S-functional groups, produced small-sized Ag nanoparticles [90]. The prepared AgNPs showed good antibacterial properties. CDs prepared from polyethyleneimine [91] and chitosan [92] have also been used in the preparation of stable AgNPs from $AgNO_3$. XPS data of the Ag nanoparticles indicated the presence of C, possibly acting as a capping agent.

In another study, Pd nanoparticles (4.7 nm) supported on activated carbon were added to CDs (1.7–3.6 nm) made from citric acid and L-cysteine, to give N−S (nitrogen and sulfur)-doped carbon quantum dots (N,S-CDs). The mixture was dried at 100 °C and the product used for the liquid-phase selective hydrogenation of ρ-chloronitrobenzene. HRTEM data revealed that the Pd interacts with the CDs. (Figure 16) [93], and that coverage does not appear to be complete.

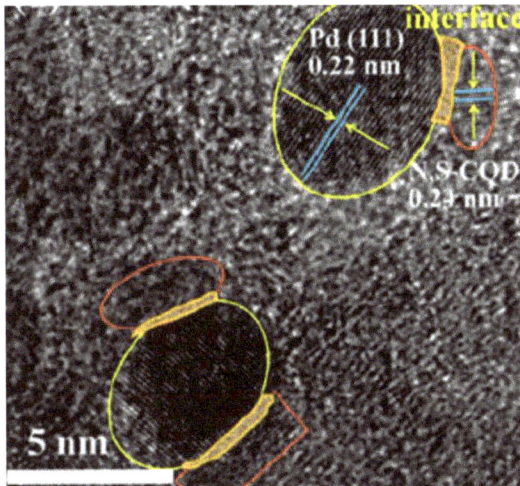

**Figure 16.** HRTEM image for Pd/N,S-CDs catalysts. Adapted with permission from [93], copyright 2019 American Chemical Society).

Liu et al. produced CDs made by the electrochemical ablation of graphite [94]. To a solution containing the CDs was added a $HAuCl_4$ solution to give an Au nanoparticles/CDs catalyst used in the selective photocatalytic oxidation of cyclohexane. Similar studies were performed on Cu/CD and Ag/CD catalysts made from Ag or Cu ions, and CD mixtures. HRTEM studies showed an interaction between the CDs and the Au (and Cu and Ag) (Figure 17). This was confirmed by X-ray absorption spectroscopy experiments on the Au/CDs composites when the Au/CDs composites were exposed to visible light; EXAFS data indicated Au–C interactions [94].

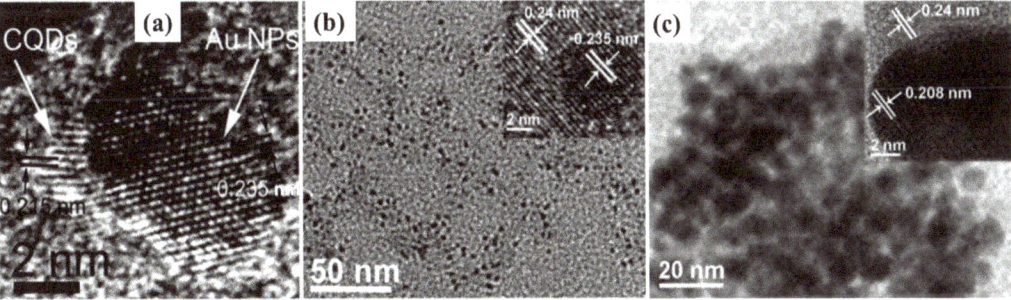

**Figure 17.** HRTEM images of (**a**) Au/CD, (**b**) Ag/CD and (**c**) Cu/CD catalysts. Adapted with permission from [94], copyright 2014, American Chemical Society.

The above examples show that CDs make good reducing and stabilizing agents for different metals. In summary, it appears that in the examples described above, the CD-metal interaction involves (i) simple coordination chemistry and (ii) reduction of the metal by CD surface functional groups. Additionally, the metal nanoparticles produced usually have uniform size, are well dispersed and have good stability (stable for months, post-synthesis).

- Examples where the CD structure is lost

The examples below show studies in which the CDs have lost their morphology during reaction between a metal salt and the CD. In some studies, the layered carbon takes the shape of the formed metal nanoparticles. These CD-transformed carbons, containing metals, were used in further catalytic studies.

Ag nanoparticles with estimated particle sizes of 40 nm were produced by reducing AgNO$_3$ in a solution containing CDs, at 50 °C for 5 min [95]. The CDs (~2–6 nm sizes), contained −OH and −NH$_2$ surface groups, and it was believed that these groups were responsible for the reduction of the metal salt into metallic Ag. Post analysis of the CDs after reduction was reported. The resulting CDs-Ag nanocomposite had a core-shell structure (called Ag@CD), with the Ag nanoparticles encapsulated inside the CD layers. The thickness of this carbon shell was found to be about 2 nm (see Figure 18a). HRTEM data analysis confirmed the conversion of the Ag salt to metallic Ag. The Ag, surrounded by C, was successfully used for the catalytic oxidation of TMB in the presence of H$_2$O$_2$, the plasmon-enhanced-driven photocatalytic reaction of p-nitrothiophenol (PNTP) into 4,4′-dimercaptoazobenzene, and the catalytic-driven reduction of PNTP to PATP in the presence of NaBH$_4$ [95].

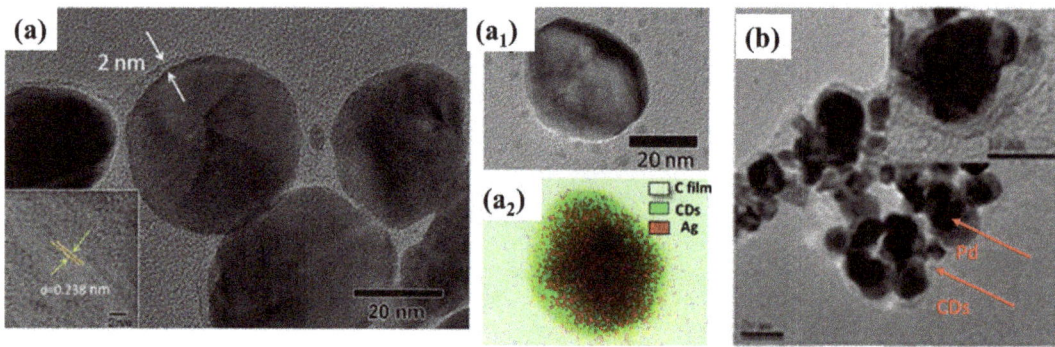

Figure 18. (a) TEM images (a$_1$) HRTEM image and (a$_2$) C and Ag element mapping of Ag@CDs (copyright 2016, American Chemical Society [95]), and (b) Pd@CDs nanocomposite (copyright 2013, Royal Society of Chemistry [25]), showing that the CDs are transformed into thin shells that encapsulate the metal catalysts.

Interestingly, Dey at el. produced CDs of estimated particle sizes of 6.6 nm from clotted cream. The CDs were then mixed with H$_2$PdCl$_4$ and refluxed at 100 °C for 6 h. From this, a nanocomposite of Pd@CDs was formed, and the HRTEM revealed that the reduced Pd nanoparticles were encapsulated inside a ~3.8 nm carbon layer (Figure 18b). Reduction of the Pd salt to metallic Pd was confirmed by XRD data [25]. The Pd@CDs was tested for activity in the Heck and Suzuki-based coupling reactions, e.g., in the reaction of phenylboronic acid and bromobenzene to give biphenyl. Beyond 4 h reaction time, the catalyst deactivated. PVP was then added to the Pd@CDs to further enhance the dispersibility of the catalyst. This led to an improved conversion of biphenyl from 45% to 95% and a >4 h reaction time, presumably due to a limited agglomeration of the Pd [25].

A study by Zhang et al. showed that CDs produced from chitosan could reduce Rh$^{3+}$ to its active catalytic form, Rh$^0$ [96]. The CDs were synthesized using a microwave-assisted hydrothermal reaction of chitosan, and the obtained CD particle size was *ca.* 9.6 nm. The prepared CDs were then mixed with RhCl$_3$·3H$_2$O and allowed to react for 1 h at 120 °C. Post reaction, the sample was analyzed using XRD, and the data showed the presence of metallic Rh$^0$. TEM results showed that the resulting Rh nanoparticles and CDs formed clustered structures, and their average particle sizes after synthesis at two different CD:Rh salt concentrations (4:1 and 6:1), were 23.4 nm and 27.8 nm, respectively. The TEM images recorded in the study are shown in Figure 19, and were reported to show a close interaction of the Rh particles with the CDs. The prepared catalysts were used to hydrogenate polybutadiene (HTPB) and hydroxy-terminated butadiene-acrylonitrile (HTBN), and showed a high degree of hydrogenation at 80 °C [96]. The CDs also acted as a stabilizing agent during the reactions.

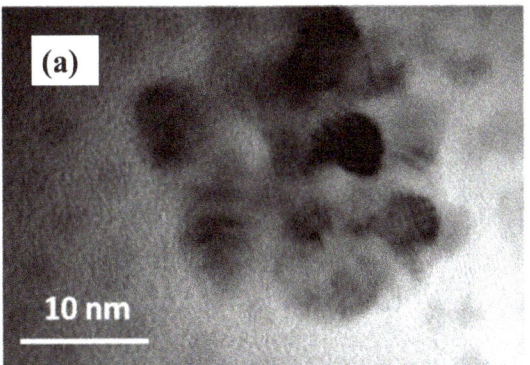
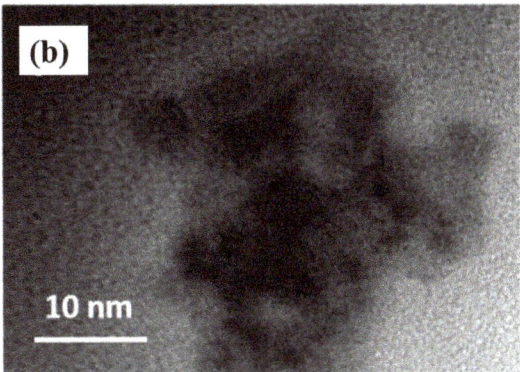

**Figure 19.** HRTEM images of Rh-CD composites prepared using (**a**) 4:1 and (**b**) 6:1 CDs:Rh salt (reproduced with permission from [96], copyright 2017 Elsevier).

CD precursors, when mixed with $ZnCl_2$, led to the formation of graphitic sheet layered materials in which the Zn is said to link the CDs together. In the absence of Zn, CDs are formed (from citric acid/urea/autoclave at 180 °C). The sheet size was affected by the zinc/carbon ratio, where an increase in Zn produced larger graphene sheets [97].

In summary, it is still not clear as to how the conversion of CDs to the carbon sheets takes place. Loss of the functional groups must lead to a change in the CD morphology, but this in itself would not lead to an obvious stitching of the carbon layers. Also, at high temperatures, a competition will exist between carbon oxidation (by the metal oxide) and carbon stitching.

### 3.2.3. Post Reduction of CDs as Metal Supports

Post reduction of a CD to give CDs that have been used to support metals have been reported. In the post-reduction process, C=O groups are converted to $CH_2OH$ groups (Figure 20) [62,63,98]. For example, Wang et al. obtained CDs by treating carbon black ("lampblack") in acid under reflux. The obtained CDs were further treated with $NaBH_4$ to produce r-CDs with approximate particle sizes of 3.4 [98]. It was observed that the C=O groups found in the pristine CDs were reduced to C-OH groups. The r-CDs were used for the synthesis of Au metal nanoparticles. When $HAuCl_4$ was mixed with r-CDs and heated at different temperatures (40, 60 and 80 °C) for 24 h [98], the obtained Au nanoparticles had averages sizes of 7 ± 2.1 nm, 16.4 ± 3.8 nm, and 15.9 ± 4.2 nm, respectively. Interestingly, it was observed that the r-CDs were oxidized back to CDs after reaction with the metal salt; the resulting CDs showed an increase in C=O peaks, as detected in the photoluminescence emission spectrum. No post synthesis of the CD-derived Au (e.g., by TEM analysis) was performed. The obtained Au nanoparticles were kept for 6 months without any aggregation. It was believed that the CDs acted as capping agents for the resulting Au nanoparticles, leading to their stability. However, there was no TEM evidence seen for a carbon layer. These nanoparticles were used for the catalytic oxidation of 3,3′,5,5′-tetramethylbenzidine (TMB) by $H_2O_2$, and the CD-reduced Au nanoparticles showed superior performance to Au nanoparticles obtained after treating $HAuCl_4$ with citrate.

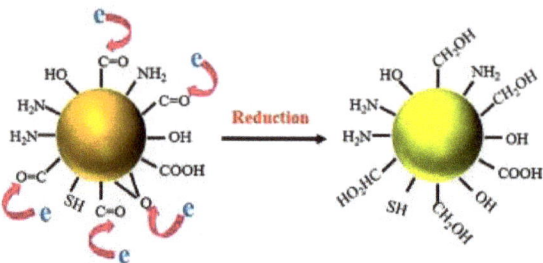

**Figure 20.** Conversion of CDs to r-CDs by C=O reduction to $CH_2OH$ groups (adapted with permission from [62], copyright 2015 Royal Society of Chemistry).

Zhuo et al. synthesized CDs from cysteine using a microwave-assisted hydrothermal reaction. The CDs were then reduced using sodium borohydride [62]. The average particle sizes of the CDs and r-CDs were 2.0 and 2.3 nm, respectively. The FT-IR spectra of CDs showed a C=O peak at 1639 cm$^{-1}$, which shifted to 1651 cm$^{-1}$ in the r-CDs, signifying a reduction of the C=O bonds. Additionally, the C=O peak at ~287.9 eV in the XPS was reduced significantly in size for the r-CDs sample, while the O-H peaks increased, further confirming the reduction of the CDs [62]. The reduced CDs (r-CDs) were then used to reduce $AgNO_3$ and $HAuCl_4$ to metallic Ag and Au [62]. The obtained r-CDs-Ag and r-CDs-Au materials had average particle sizes ranging between 6–10 nm and 2–3 nm, respectively.

In summary, the ability to pre-reduce CDs should permit an enhanced ability to reduce metal ions for catalytic reactions.

## 4. Techniques to Evaluate CD and CD/Metal Transformations

Many different techniques can be used to establish the structure and composition of CDs and CD/metal composites. However, of importance in CD studies is the post analysis of the carbons. As noted in the sections above, the CDs can be modified during reaction. Electron microscopy studies (TEM/SEM) are the most useful techniques to evaluate changes in the CDs, as the CDs typically have dimensions <10 nm. TEM analysis in particular provides a means of establishing if the spherical CDs change into layered materials or core/shell structures. However, it has been noted that CDs may sometimes not be seen in microscope images because of their low contrast, relative to the substrate used [92].

HRTEM has been used extensively to indicate these changes. Various spectroscopic techniques have also been used to determine the bulk and surface structure of CDs and the changes in the CDs with temperature after chemical reactions. These include infrared, photoluminescence, solid state NMR and Raman studies, as well as XPS studies. However, the changes that occur do not necessarily provide definitive data on the difference between CDs and their conversion to layered materials. Careful analysis does show changes in the Raman D and G band intensity ratios, as well as changes in NMR and IR data that relate to C=O/C-O/COOH ratios. XRD studies can follow the changes in the carbon by monitoring the carbon–carbon layers in the structure. The 002 peak in the PXRD data tends to be broad, but does shift as the structure is modified. The use of total X-ray scattering experiments enables the collection of data on the local structural order that exists within the CDs. This means that useful data can be extracted from nano-sized CDs, since a total scattering experiment is insensitive to structural disorder and insensitive to their small size [71,72].

Traditional laboratory-based XRD studies can follow the changes in the carbon by monitoring the carbon–carbon layers in the structure. The 002 peak in the XRD data tends to be broad, but does shift along the x-axis, as the structure is modified. Similarly, total X-ray scattering experiments provide data on the local structural order that exists within the CDs. This type of data contains both Bragg and diffuse X-ray scattering, which allows for a comprehensive analysis of the CD atomic structure. This means that useful information can be extracted via the pair distribution function (PDF) analysis of the data collected on nano-sized CDs, which produce primarily diffuse scattering. This is due to the technique being insensitive to structural disorder and the small size inherent to CDs [71,72]. Accordingly, PDF analysis of high-energy synchrotron X-ray data typically produces information that is more accurate and provides a more representative description of the nanoscale structure of the CDs than what is currently achievable with standard laboratory-based X-ray equipment. Consider the radial functions that were extracted from synchrotron X-ray data that were collected on a set of CDs, each of which was calcined at different temperatures (Figure 21).

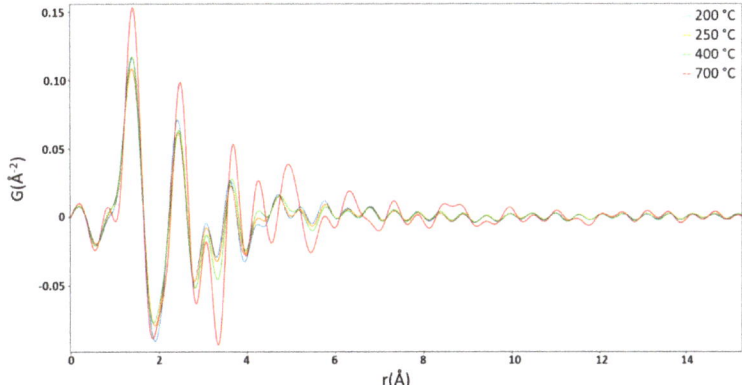

**Figure 21.** A comparison of the radial distribution function extracted from total scattering data collected on a set of CDs prepared as a function of calcination temperature.

The data in Figure 21 shows the general similarities of the CDs in this sample set up to r = 3Å. The peak at r = 1.38 Å is due to the carbon–carbon double bonds in the structure of the CDs. Beyond this point in the data, departures are seen, as the sample calcined at the highest temperature (700 °C) has better long-range order, with perturbations in the data that extend further afield than when compared to that of the sample calcined at lower temperatures (200, 250 and 400 °C). From this qualitative analysis of the data it is possible to show that an increased calcination temperature produced increasingly more crystalline CDs. By comparison, similar data collected on these samples using ordinary XRD data would typically have shown some evidence of these features with long data acquisitions. However, owing to its brilliance, the sensitivity of the data and speed with which they are generated at a synchrotron are unmatched.

- Summary of CD conversion reactions

A consideration of papers published in the area leads to some generalizations on metal-CD mixtures that can be made, and the role this will have in catalysis.

(i) CDs can react with each other, through physical or chemical bonds, to create CD assemblies. These assemblies could then be used as carbocatalysts.

(ii) CDs can react at temperatures of *ca.* 200 °C to convert to carbon sheet-like structures and the graphicity of the sheets has been shown to increase with temperature.

(iii) CDs can lose many of their surface functional groups at temperatures between 200 and 450 °C, and this will impact on their chemical properties.

(iv) In the presence of easily reducible metal oxide catalysts, CDs can be converted to carbon sheet-like materials at temperatures < 100 °C. The ability to achieve this will be dependent on the metal reducibility, with directions given by Ellingham diagrams.

(v) Ellingham diagrams have been determined for bulk metals, but metal nanoparticles are expected to have similar temperature versus free energy values.

(vi) The chemical process by which a CD is converted to a sheet-like carbon is not known.

## 5. Future Directions

It is hoped that the review has presented some useful thoughts on the use of CDs as a metal support in catalysis. It is clear from the many reports and reviews in the literature that the area is a fruitful one for further studies.

There are a number of key issues that need to be addressed to provide an understanding of the interaction between CDs and metal, i.e., when the CD-metal is retained and when the CD converts to another morphology in the presence/absence of a metal.

(i)     Is the CD conversion dependent on the type of CD?
(ii)    Is the conversion dependent on the type of functional group?
(iii)   How does the conversion from a CD to other dimensional carbon structures take place?
(iv)   The morphology conversion reaction is temperature dependent. What are the temperature ranges in which the CD/metal acts as a catalyst with/without a morphology change?
(v)    Is there a carbon layer thickness in which a carbon will only be oxidized by a metal but not undergo a change in morphology? For example, in the presence of Co it is clear that CSs (d = 400 nm) only undergo carbon removal while CDs (d < 10 nm) undergo a morphology change.
(vi)   In situ studies could provide valuable information on the CD conversion process.

The review also indicates that post analysis of CD/metal catalysts is needed; the current stability repeat studies do not necessarily give information on the morphology of the active catalyst.

## 6. Conclusions

The use of CDs and metal-CDs has been extensively reported in the literature. The data suggest that the CDs (as carbocatalysts), when studied at low temperatures (<150 °C) appear to retain their morphology in the reactions. It is possible that morphology changes could occur at the higher temperature, but lack of post-catalysis data has generally not allowed for this to be confirmed. Further, while the surface groups can be modified in the reaction, this does not appear to influence the carbon core. At higher temperatures, the CDs are converted to sheet-like carbons. These new carbons have also been studied as catalysts, e.g., in electrochemical reactions.

Metal-doped CDs also appear to act as classic carbocatalysts/metal catalysts in low-temperature catalytic reactions. High-temperature studies could provide information on CD morphology changes influenced by the metal dopants.

The addition of metals to CDs to make different metal-CD catalysts appears to be dependent on the metal reducibility and the CD reduction ability. If high temperatures are used, the metal/metal salt can react with the carbon core and remove or change the morphology of the carbon.

The use of temperature/pressure/carbon functional groups to produce retention/transformation of the CD structure thus provides exciting possibilities for further studies in this area of catalysis using CDs.

**Author Contributions:** Conceptualization: N.J.C. and L.L.M.; Data curation and formal analysis: N.J.C., L.L.M. and R.P.F.; Writing—first draft: L.L.M.; Funding—N.J.C.; Writing—review & editing: N.J.C., L.L.M. and R.P.F. All authors have read and agreed to the published version of the manuscript.

**Funding:** We wish to thank the University of the Witwatersrand, the DSI-NRF Centre of Excellence in Catalysis (c*change) and the NRF for financial support.

**Acknowledgments:** We wish to also acknowledge the European Synchrotron Radiation Facility (ESRF) for provision of synchrotron radiation facilities under proposal number MA-5435, and we would like to thank Jonathan Wright for assistance and support in using beamline ID11 (DOI: 10.15151/ESRF-ES-1028463059).

**Conflicts of Interest:** The authors declare no conflict of interest.

## References

1. Coville, N.J.; Mhlanga, S.D.; Nxumalo, E.N.; Shaikjee, A. A review of shaped carbon nanomaterials. *S. Afr. J. Sci.* **2011**, *107*, 44–58. [CrossRef]
2. Nasir, S.; Hussein, M.Z.; Zainal, Z.; Yusof, N.A. Carbon-Based Nanomaterials/Allotropes: A Glimpse of Their Synthesis, Properties and Some Applications. *Materials* **2018**, *11*, 295. [CrossRef] [PubMed]
3. Karthik, P.S.; Himaja, A.L.; Singh, S.P. Carbon-allotropes: Synthesis methods, applications and future perspectives. *Carbon Lett.* **2014**, *15*, 219–237. [CrossRef]
4. Tiwari, S.K.; Kumar, V.; Huczko, A.; Oraon, R.; De Adhikari, A.; Nayak, G.C. Magical Allotropes of Carbon: Prospects and Applications. *Crit. Rev. Solid State Mater. Sci.* **2016**, *41*, 257–317. [CrossRef]
5. Rodriguez-reinoso, F. The role of carbon materials in heterogeneous catalysis. *Carbon* **1998**, *36*, 159–175. [CrossRef]
6. Berseth, P.A.; Harter, A.G.; Zidan, R.; Blomqvist, A.; Araújo, C.M.; Scheicher, R.H.; Ahuja, R.; Jena, P. Carbon nanomaterials as catalysts for hydrogen uptake and release in $NaAlH_4$. *Nano Lett* **2009**, *9*, 1501–1505. [CrossRef]
7. Védrine, J.C. Heterogeneous catalysis on metal oxides. *Catalysts* **2017**, *7*, 341. [CrossRef]
8. Tauster, S.J.; Fung, S.C.; Baker, R.T.K.; Horsley, J.A. Strong Interactions in Supported-Metal Catalysts. *Science* **1981**, *211*, 1121–1125. [CrossRef]
9. Iglesia, E. Design, synthesis, and use of cobalt-based Fischer-Tropsch synthesis catalysts. *Appl. Catal. A Gen.* **1997**, *161*, 59–78. [CrossRef]
10. Macheli, L.; Carleschi, E.; Doyle, B.P.; Leteba, G.; van Steen, E. Tuning catalytic performance in Fischer-Tropsch synthesis by metal-support interactions. *J. Catal.* **2021**, *395*, 70–79. [CrossRef]
11. Tsakoumis, N.E.; Rønning, M.; Borg, Ø.; Rytter, E.; Holmen, A. Deactivation of cobalt based Fischer—Tropsch catalysts: A review. *Catal. Today* **2010**, *154*, 162–182. [CrossRef]
12. Xiong, H.; Jewell, L.L.; Coville, N.J. Shaped Carbons As Supports for the Catalytic Conversion of Syngas to Clean Fuels. *ACS Catal.* **2015**, *5*, 2640–2658. [CrossRef]
13. Macheli, L.; Roy, A.; Carleschi, E.; Doyle, B.P.; van Steen, E. Surface modification of $Co_3O_4$ nanocubes with TEOS for an improved performance in the Fischer-Tropsch synthesis. *Catal. Today* **2020**, *343*, 176–182. [CrossRef]
14. Nguyen, B.H.; Nguyen, V.H.; Vu, D.L. Photocatalytic composites based on titania nanoparticles and carbon nanomaterials. *Adv. Nat. Sci. Nanosci. Nanotechnol.* **2015**, *6*, 033001. [CrossRef]
15. Li, L.; Hu, L.; Li, J.; Wei, Z. Enhanced stability of Pt nanoparticle electrocatalysts for fuel cells. *Nano Res.* **2015**, *8*, 418–440. [CrossRef]
16. Baker, S.N.; Baker, G.A. Luminescent Carbon Nanodots: Emergent Nanolights. *Angew. Chem. Int. Ed.* **2010**, *49*, 6726–6744. [CrossRef]
17. Kang, Z.; Lee, S.T. Carbon dots: Advances in nanocarbon applications. *Nanoscale* **2019**, *11*, 19214–19224. [CrossRef]
18. Kurian, M.; Paul, A. Recent trends in the use of green sources for carbon dot synthesis—A short review. *Carbon Trends* **2021**, *3*, 100032. [CrossRef]
19. Li, Z.; Wang, L.; Li, Y.; Feng, Y.; Feng, W. Frontiers in carbon dots: Design, properties and applications. *Mater. Chem. Front.* **2019**, *3*, 2571–2601. [CrossRef]
20. Sharma, A.; Das, J. Small molecules derived carbon dots: Synthesis and applications in sensing, catalysis, imaging, and biomedicine. *J. Nanobiotechnol.* **2019**, *17*, 92. [CrossRef]
21. Yuan, F.; Yuan, T.; Sui, L.; Wang, Z.; Xi, Z.; Li, Y.; Li, X.; Fan, L.; Chen, A.; Jin, M. Engineering triangular carbon quantum dots with unprecedented narrow bandwidth emission for multicolored LEDs. *Nat. Commun.* **2018**, *9*, 2249. [CrossRef] [PubMed]
22. Chen, B.B.; Liu, M.L.; Huang, C.Z. Carbon dot-based composites for catalytic applications. *Green Chem.* **2020**, *22*, 4034–4054. [CrossRef]
23. Manjupriya, R.; Roopan, S.M. Carbon dots-based catalyst for various organic transformations. *J. Mater. Sci.* **2021**, *56*, 17369–17410. [CrossRef]
24. Rosso, C.; Filippini, G.; Prato, M. Carbon Dots as Nano-Organocatalysts for Synthetic Applications. *ACS Catal.* **2020**, *10*, 8090–8105. [CrossRef]
25. Dey, D.; Bhattacharya, T.; Majumdar, B.; Mandani, S.; Sharma, B.; Sarma, T.K. Carbon dot reduced palladium nanoparticles as active catalysts for carbon–carbon bond formation. *Dalton Trans.* **2013**, *42*, 13821. [CrossRef]
26. Jana, J.; Chung, J.S.; Hur, S.H. Carbon dot supported bimetallic nanocomposite for the hydrogen evolution reaction. *J. Alloys Compd.* **2021**, *859*, 157895. [CrossRef]
27. Mokoloko, L.L.; Forbes, R.P.; Coville, N.J. The Transformation of 0-D Carbon Dots into 1-, 2- and 3-D Carbon Allotropes: A Minireview. *Nanomaterials* **2022**, *12*, 2515. [CrossRef]
28. Mokoloko, L.L.; Forbes, R.P.; Coville, N.J. Use of carbon dots-derived graphene-like sheets on the autoreduction of cobalt nanoparticles for Fischer-Tropsch synthesis: A limitation on the use of carbon supports. Presented at the Gauteng Catalysis Seminar, University of the Witwatersrand, Gauteng, South Africa, 2022. *To be published*.
29. Xia, C.; Zhu, S.; Feng, T.; Yang, M.; Yang, B. Evolution and Synthesis of Carbon Dots: From Carbon Dots to Carbonized Polymer Dots. *Adv. Sci.* **2019**, *6*, 1901316. [CrossRef]
30. Sarkar, S.; Banerjee, D.; Ghorai, U.; Das, N.; Chattopadhyay, K. Size dependent photoluminescence property of hydrothermally synthesized crystalline carbon quantum dots. *J. Lumin.* **2016**, *178*, 314–323. [CrossRef]

31. Li, L.; Dong, T. Photoluminescence tuning in carbon dots: Surface passivation or/and functionalization, heteroatom doping. *J. Mater. Chem. C* **2018**, *6*, 7944–7970. [CrossRef]
32. Guo, R.; Li, L.; Wang, B.; Xiang, Y.; Zou, G.; Zhu, Y.; Hou, H.; Ji, X. Functionalized carbon dots for advanced batteries. *Energy Storage Mater.* **2021**, *37*, 8–39. [CrossRef]
33. Ai, L.; Shi, R.; Yang, J.; Zhang, K.; Zhang, T.; Lu, S. Efficient Combination of G-$C_3N_4$ and CDs for Enhanced Photocatalytic Performance: A Review of Synthesis, Strategies, and Applications. *Small* **2021**, *17*, 2007523. [CrossRef] [PubMed]
34. Mongwe, T.H.; Coville, N.J.; Maubane-Nkadimeng, M.S. Synthesis of onion-like carbon nanoparticles by flame pyrolysis. In *Nanoscience*; SPR-Nanoscience, Royal Society of Chemistry: London, UK, 2022; pp. 198–220.
35. Li, X.; Fu, Y.; Zhao, S.; Xiao, J.; Lan, M.; Wang, B.; Zhang, K.; Song, X.; Zeng, L. Metal ions-doped carbon dots: Synthesis, properties, and applications. *Chem. Eng. J.* **2022**, *430*, 133101. [CrossRef]
36. Zhou, Y.; Zhang, W.; Leblanc, R.M. Structure-Property-Activity Relationships in Carbon Dots. *J. Phys. Chem. B* **2022**, *126*, 10777–10796. [CrossRef]
37. Ding, H.; Li, X.H.; Chen, X.B.; Wei, J.S.; Li, X.B.; Xiong, H.M. Surface states of carbon dots and their influences on luminescence. *J. Appl. Phys.* **2020**, *127*, 231101. [CrossRef]
38. Yan, F.; Jiang, Y.; Sun, X.; Bai, Z.; Zhang, Y.; Zhou, X. Surface modification and chemical functionalization of carbon dots: A review. *Microchim. Acta* **2018**, *185*, 424. [CrossRef] [PubMed]
39. Yin, K.; Lu, D.; Wang, L.-P.; Zhang, Q.; Hao, J.; Li, G.; Li, H. Hydrophobic Carbon Dots from Aliphatic Compounds with One Terminal Functional Group. *J. Phys. Chem. C* **2019**, *123*, 22447–22456. [CrossRef]
40. Yin, K.; Lu, D.; Tian, W.; Zhang, R.; Yu, H.; Gorecka, E.; Pociecha, D.; Godbert, N.; Hao, J.; Li, H. Ordered Structures of Alkylated Carbon Dots and Their Applications in Nonlinear Optics. *J. Mater. Chem. C* **2020**, *8*, 8980–8991. [CrossRef]
41. Xie, Z.; Sun, X.; Jiao, J.; Xin, X. Ionic liquid-functionalized carbon quantum dots as fluorescent probes for sensitive and selective detection of iron ion and ascorbic acid. *Colloids Surf. A Physicochem. Eng. Asp.* **2017**, *529*, 38–44. [CrossRef]
42. Sun, X.; Yin, K.; Liu, B.; Zhou, S.; Cao, J.; Zhang, G.; Li, H. Carbon quantum dots in ionic liquids: A new generation of environmentally benign photoluminescent inks. *J. Mater. Chem. C Mater.* **2017**, *5*, 4951–4958. [CrossRef]
43. Yin, K.; Feng, N.; Godbert, N.; Xing, P.; Li, H. Self-Assembly of Cholesteryl Carbon Dots with Circularly Polarized Luminescence in Solution and Solvent-Free Phases. *J. Phys. Chem. Lett.* **2023**, *14*, 1088–1095. [CrossRef] [PubMed]
44. Chen, G.; Wu, S.; Hui, L.; Zhao, Y.; Ye, J.; Tan, Z.; Zeng, W.; Tao, Z.; Yang, L.; Zhu, Y. Assembling carbon quantum dots to a layered carbon for high-density supercapacitor electrodes. *Sci. Rep.* **2016**, *6*, 19028. [CrossRef] [PubMed]
45. Zhou, Y.; Mintz, K.J.; Cheng, L.; Chen, J.; Ferreira, B.C.; Hettiarachchi, S.D.; Liyanage, P.Y.; Seven, E.S.; Miloserdov, N.; Pandey, R.R.; et al. Direct conjugation of distinct carbon dots as Lego-like building blocks for the assembly of versatile drug nanocarriers. *J. Colloid Interface Sci.* **2020**, *576*, 412–425. [CrossRef]
46. Zhou, Y.; Chen, J.; Miloserdov, N.; Zhang, W.; Mintz, K.J.; Ferreira, B.C.L.B.; Micic, M.; Li, S.; Peng, Z.; Leblanc, R.M. Versatile drug nanocarrier assembly via conjugation of distinct carbon dots. *Mor. J. Chem.* **2020**, *8*, 994–1007.
47. Sun, X.; Wang, H.; Qi, J.; Zhou, S.; Li, H. Supramolecular self-assemblies formed by co-assembly of carbon dots and tannic acid. *Dyes Pigments* **2021**, *190*, 109287. [CrossRef]
48. Jana, P.; Dev, A. Carbon quantum dots: A promising nanocarrier for bioimaging and drug delivery in cancer. *Mater. Today Commun.* **2022**, *32*, 104068. [CrossRef]
49. Roy, P.; Chen, P.; Periasamy, A.P.; Chen, Y.; Chang, H. Photoluminescent carbon nanodots: Synthesis, physicochemical properties and analytical applications. *Mater. Today* **2015**, *18*, 447–458. [CrossRef]
50. Batool, M.; Junaid, H.M.; Tabassumb, S.; Kanwala, F.; Abid, K.; Fatima, Z.; Shah, A.T. Metal Ion Detection by Carbon Dots—A Review. *Crit. Rev. Anal. Chem.* **2022**, *52*, 756–767. [CrossRef]
51. Wang, Q.; Huang, X.; Long, Y.; Wang, X.; Zhang, H.; Zhu, R.; Liang, L.; Teng, P.; Zheng, H. Hollow luminescent carbon dots for drug delivery. *Carbon* **2013**, *59*, 192–199. [CrossRef]
52. Ren, X.; Zhang, F.; Guo, B.; Gao, N.; Zhang, X. Synthesis of N-doped Micropore carbon quantum dots with high quantum yield and dual-wavelength photoluminescence emission from biomass for cellular imaging. *Nanomaterials* **2019**, *9*, 495. [CrossRef]
53. Li, B.; Zhang, L.; Zhang, J.; Su, Y. Recent Insight in Transition Metal Anchored on Nitrogen-Doped Carbon Catalysts: Preparation and Catalysis Application. *Electrochem* **2022**, *3*, 520–537. [CrossRef]
54. Xiong, H.; Motchelaho, M.A.M.; Moyo, M.; Jewell, L.L.; Coville, N.J. Correlating the preparation and performance of cobalt catalysts supported on carbon nanotubes and carbon spheres in the Fischer-Tropsch synthesis. *J. Catal.* **2011**, *278*, 26–40. [CrossRef]
55. Kumar, K.S.; Pittala, S.; Sanyadanam, S.; Paik, P. A new single/few-layered graphene oxide with a high dielectric constant of 106: Contribution of defects and functional groups. *RSC Adv.* **2015**, *5*, 14768–14779. [CrossRef]
56. Moyo, M.; Motchelaho, M.A.M.; Xiong, H.; Jewell, L.L.; Coville, N.J. Promotion of Co/carbon sphere Fischer-Tropsch catalysts by residual K and Mn from carbon oxidation by $KMnO_4$. *Appl. Catal. A Gen.* **2012**, *413–414*, 223–229. [CrossRef]
57. Mansuriya, B.D.; Altintas, Z. Carbon Dots: Classification, Properties, Synthesis, Characterization, and Applications in Health Care-An Updated Review (2018–2021). *Nanomaterials* **2018**, *11*, 2525. [CrossRef] [PubMed]
58. Xu, Y.; Wu, M.; Feng, X.Z.; Yin, X.B.; He, X.W.; Zhang, Y.K. Reduced carbon dots versus oxidized carbon dots: Photo- and electrochemiluminescence investigations for selected applications. *Chem. Eur. J.* **2013**, *19*, 6282–6288. [CrossRef]
59. De, B.; Karak, N. A green and facile approach for the synthesis of water soluble fluorescent carbon dots from banana juice. *RSC Adv.* **2013**, *3*, 8286–8290. [CrossRef]

60. Chu, K.W.; Lee, S.L.; Chang, C.J.; Liu, L. Recent progress of carbon dot precursors and photocatalysis applications. *Polymers* **2019**, *11*, 689. [CrossRef] [PubMed]
61. Lu, Q.; Deng, J.; Hou, Y.; Wang, H.; Li, H.; Zhang, Y.; Yao, S. Hydroxyl-rich C-dots synthesized by a one-pot method and their application in the preparation of noble metal nanoparticles. *Chem. Commun.* **2015**, *51*, 7164–7167. [CrossRef]
62. Zhuo, Y.; Zhong, D.; Miao, H.; Yang, X. Reduced carbon dots employed for synthesizing metal nanoclusters and nanoparticles. *RSC Adv.* **2015**, *5*, 32669–32674. [CrossRef]
63. Zheng, H.; Wang, Q.; Long, Y.; Zhang, H.; Huang, X.; Zhu, R. Enhancing the luminescence of carbon dots with a reduction pathway. *Chem. Commun.* **2011**, *47*, 10650–10652. [CrossRef] [PubMed]
64. Dandia, A.; Saini, P.; Sethi, M.; Kumar, K.; Saini, S.; Meena, S.; Meena, S.; Parewa, V. Nanocarbons in quantum regime: An emerging sustainable catalytic platform for organic synthesis. *Catal. Rev. Sci. Eng.* **2021**, *65*, 874–928. [CrossRef]
65. Corti, V.; Bartolomei, B.; Mamone, M.; Gentile, G.; Prato, M.; Filippini, G. Amine-Rich Carbon Dots as Novel Nano-Aminocatalytic Platforms in Organic Synthesis. *Eur. J. Org. Chem.* **2022**, *2022*, e202200879. [CrossRef] [PubMed]
66. Su, D.S.; Perathoner, S.; Centi, G. Nanocarbons for the development of advanced catalysts. *Chem. Rev.* **2013**, *113*, 5782–5816. [CrossRef]
67. Cheng, H.; Zhao, Y.; Fan, Y.; Xie, X.; Qu, L.; Shi, G. Graphene-quantum-dot assembled nanotubes: A new platform for efficient Raman enhancement. *ACS Nano* **2012**, *6*, 2237–2244. [CrossRef]
68. Hou, H.; Banks, C.E.; Jing, M.; Zhang, Y.; Ji, X. Carbon Quantum Dots and Their Derivative 3D Porous Carbon Frameworks for Sodium-Ion Batteries with Ultralong Cycle Life. *Adv. Mater.* **2015**, *27*, 7861–7866. [CrossRef]
69. Jiang, H.; Wang, Y.; Hao, J.; Liu, Y.; Li, W.; Li, J. N and P co-functionalized three-dimensional porous carbon networks as efficient metal-free electrocatalysts for oxygen reduction reaction. *Carbon* **2017**, *122*, 64–73. [CrossRef]
70. Hou, H.; Shao, L.; Zhang, Y.; Zou, G.; Chen, J.; Ji, X. Large-Area Carbon Nanosheets Doped with Phosphorus: A High-Performance Anode Material for Sodium-Ion Batteries. *Adv. Sci.* **2016**, *4*, 1600243. [CrossRef]
71. Mokoloko, L.L.; Matsoso, B.J.; Forbes, R.P.; Barrett, D.H.; Moreno, B.D.; Coville, N.J. Evolution of large-area reduced graphene oxide nanosheets from carbon dots via thermal treatment. *Carbon Trends* **2021**, *4*, 100074. [CrossRef]
72. Luo, H.; Lari, L.; Kim, H.; Hérou, S.; Tanase, L.C.; Lazarov, V.K.; Titirici, M.-M. Structural evolution of carbon dots during low temperature pyrolysis. *Nanoscale* **2022**, *14*, 910–918. [CrossRef]
73. Wu, X.; Yu, F.; Han, Y.; Jiang, L.; Li, Z.; Zhu, J.; Xu, Q.; Tedesco, A.C.; Zhang, J.; Bi, H. Enhanced chemodynamic and photoluminescence efficiencies of Fe-O$_4$ coordinated carbon dots via the core-shell synergistic effect. *Nanoscale* **2022**, *15*, 376–386. [CrossRef] [PubMed]
74. Gao, J.; Zhao, S.; Guo, S.; Wang, H.; Sun, Y.; Yao, B.; Liu, Y.; Haung, H.; Kang, Z. Carbon quantum dot-covered porous Ag with enhanced activity for selective electroreduction of $CO_2$ to CO. *Inorg. Chem. Front.* **2019**, *6*, 1453–1460. [CrossRef]
75. Atkins, P.W.; Overton, T.L.; Rourke, J.P.; Weller, M.T.; Armstrong, F.A. *Shriver and Atkins' Inorganic Chemistry*, 5th ed.; Oxford University Press: Oxford, UK, 2010; p. 171, ISBN 978-1-42-921820-7.
76. Xiong, H.; Moyo, M.; Rayner, M.K.; Jewell, L.L.; Billing, D.G.; Coville, N.J. Autoreduction and Catalytic Performance of a Cobalt Fischer-Tropsch Synthesis Catalyst Supported on Nitrogen-Doped Carbon Spheres. *ChemCatChem* **2010**, *2*, 514–518. [CrossRef]
77. Petersen, A.P.; Claeys, M.; Kooyman, P.J.; van Steen, E. Cobalt-Based Fischer—Tropsch Synthesis: A Kinetic Inverse Model System. *Catalysts* **2019**, *9*, 794. [CrossRef]
78. Zhang, J.; Medlin, J.W. Catalyst design using an inverse strategy: From mechanistic studies on inverted model catalysts to applications of oxide-coated metal nanoparticles. *Surf. Sci. Rep.* **2018**, *73*, 117–152. [CrossRef]
79. Nabaho, D.; Niemantsverdriet, J.W.; Claeys, M.; van Steen, E. Hydrogen spillover in the Fischer-Tropsch synthesis: An analysis of platinum as a promoter for cobalt-alumina catalysts. *Catal. Today* **2016**, *261*, 17–27. [CrossRef]
80. Ma, C.; Zhou, Y.; Yan, W.; He, W.; Liu, Q.; Li, Z.; Wang, H.; Li, G.; Yang, Y.; Han, W.; et al. Predominant Catalytic Performance of Nickel Nanoparticles Embedded into Nitrogen-Doped Carbon Quantum Dot-Based Nanosheets for the Nitroreduction of Halogenated Nitrobenzene. *ACS Sustain. Chem. Eng.* **2022**, *10*, 8162–8171. [CrossRef]
81. Zhang, Y.; Foster, C.W.; Banks, C.E.; Shao, L.; Hou, H.; Zou, G.; Chen, J.; Huang, Z.; Ji, X. Graphene-Rich Wrapped Petal-Like Rutile $TiO_2$ tuned by Carbon Dots for High-Performance Sodium Storage. *Adv. Mater.* **2016**, *28*, 9391–9399. [CrossRef]
82. Li, X.; Ding, S.; Lyu, Z.; Tieu, P.; Wang, M.; Feng, Z.; Pan, X.; Zhou, Y.; Niu, X.; Du, D.; et al. Single-Atomic Iron Doped Carbon Dots with Both Photoluminescence and Oxidase-Like Activity. *Small* **2022**, *18*, 2203001. [CrossRef]
83. Zeng, Y.; Zhao, B.; Luo, X.; Wu, F. Glucose-sensitive colorimetric sensor based on peroxidase mimics activity of carbon dots-functionalized $Fe_3O_4$ nanocomposites. *Diam. Relat. Mater.* **2023**, *136*, 109914. [CrossRef]
84. Zhang, B.T.; Wang, Q.; Zhang, Y.; Teng, Y.; Fan, M. Degradation of ibuprofen in the carbon dots/$Fe_3O_4$@carbon sphere pomegranate-like composites activated persulfate system. *Sep. Purif. Technol.* **2020**, *242*, 116820. [CrossRef]
85. Juang, R.S.; Ju, Y.-C.; Liao, C.-S.; Lin, K.-S.; Lu, H.-C.; Wang, S.-F.; Sun, A.-C. Synthesis of Carbon Dots on $Fe_3O_4$ Nanoparticles as Recyclable Visible-Light Photocatalysts. *IEEE Trans. Magn.* **2017**, *53*, 2710541. [CrossRef]
86. Abbas, M.W.; Soomro, R.A.; Kalwar, N.H.; Zahoor, M.; Avci, A.; Pehlivan, E.; Hallam, K.R.; Willander, M. Carbon quantum dot coated $Fe_3O_4$ hybrid composites for sensitive electrochemical detection of uric acid. *Microchem. J.* **2019**, *146*, 517–524. [CrossRef]
87. Zhang, P.; Song, T.; Wang, T.; Zeng, H. In-situ synthesis of Cu nanoparticles hybridized with carbon quantum dots as a broad spectrum photocatalyst for improvement of photocatalytic $H_2$ evolution. *Appl. Catal. B* **2017**, *206*, 328–335. [CrossRef]

88. Zhou, Y.; Yu, F.; Lang, Z.; Nie, H.; Wang, Z.; Shao, M.; Liu, Y.; Tan, H.; Li, Y.; Kang, Z. Carbon dots/PtW$_6$O$_{24}$ composite as efficient and stable electrocatalyst for hydrogen oxidation reaction in PEMFCs. *Chem. Eng. J.* **2021**, *426*, 130709. [CrossRef]
89. Yang, L.; Liu, X.; Lu, Q.; Huang, N.; Liu, M.; Zhang, Y.; Yao, S. Catalytic and peroxidase-like activity of carbon based-AuPd bimetallic nanocomposite produced using carbon dots as the reductant. *Anal. Chim. Acta* **2016**, *930*, 23–30. [CrossRef] [PubMed]
90. Jin, J.C.; Xu, X.Q.; Dong, P.; Lai, L.; Lan, J.Y.; Jiang, F.L.; Liu, Y. One-step synthesis of silver nanoparticles using carbon dots as reducing and stabilizing agents and their antibacterial mechanisms. *Carbon N. Y.* **2015**, *94*, 129–141. [CrossRef]
91. Liu, T.; Dong, J.X.; Liu, S.G.; Li, N.; Lin, S.M.; Fan, Y.Z.; Lei, J.L.; Luo, H.Q.; Li, N.B. Carbon quantum dots prepared with polyethyleneimine as both reducing agent and stabilizer for synthesis of Ag/CQDs composite for Hg$^{2+}$ ions detection. *J. Hazard. Mater.* **2017**, *322*, 430–436. [CrossRef]
92. Shen, L.; Chen, M.; Hu, L.; Chen, X.; Wang, J. Growth and stabilization of silver nanoparticles on carbon dots and sensing application. *Langmuir* **2013**, *29*, 16135–16140. [CrossRef]
93. Lu, C.; Zhu, Q.; Zhang, X.; Ji, H.; Zhou, Y.; Wang, H.; Liu, Q.; Nie, J.; Han, W.; Li, X. Decoration of Pd Nanoparticles with N and S Doped Carbon Quantum Dots as a Robust Catalyst for the Chemoselective Hydrogenation Reaction. *ACS Sustain. Chem. Eng.* **2019**, *7*, 8542–8553. [CrossRef]
94. Liu, R.; Huang, H.; Li, H.; Liu, Y.; Zhong, J.; Li, Y.; Zhang, S.; Kang, Z. Metal nanoparticle/carbon quantum dot composite as a photocatalyst for high-efficiency cyclohexane oxidation. *ACS Catal.* **2014**, *4*, 328–336. [CrossRef]
95. Jin, J.; Zhu, S.; Song, Y.; Zhao, H.; Zhang, Z.; Guo, Y.; Li, J.; Song, W.; Yang, B.; Zhao, B. Precisely Controllable Core-Shell Ag@Carbon Dots Nanoparticles: Application to in Situ Super-Sensitive Monitoring of Catalytic Reactions. *ACS Appl. Mater. Interfaces* **2016**, *8*, 27965. [CrossRef]
96. Zhang, J.; Chen, Y.; Tan, J.; Sang, H.; Zhang, L.; Yue, D. The synthesis of rhodium/carbon dots nanoparticles and its hydrogenation application. *Appl. Surf. Sci.* **2017**, *396*, 1138–1145. [CrossRef]
97. Bharathi, G.; Nataraj, D.; Premkumar, S.; Sowmiya, M.; Senthilkumar, K.; Thangadurai, T.D.; Khyzhun, O.Y.; Gupta, M.; Phase, D.; Patra, N.; et al. Graphene Quantum Dot Solid Sheets: Strong blue-light-emitting & photocurrent-producing band-gap-opened nanostructures. *Sci. Rep.* **2017**, *7*, 1085.
98. Wang, X.; Long, Y.; Wang, Q.; Zhang, H.; Huang, X.; Zhu, R.; Teng, P.; Liang, L.; Zheng, H. Reduced state carbon dots as both reductant and stabilizer for the synthesis of gold nanoparticles. *Carbon* **2013**, *64*, 499–506. [CrossRef]

**Disclaimer/Publisher's Note:** The statements, opinions and data contained in all publications are solely those of the individual author(s) and contributor(s) and not of MDPI and/or the editor(s). MDPI and/or the editor(s) disclaim responsibility for any injury to people or property resulting from any ideas, methods, instructions or products referred to in the content.

Article

# Anodic Catalyst Support via Titanium Dioxide-Graphene Aerogel (TiO$_2$-GA) for A Direct Methanol Fuel Cell: Response Surface Approach

Siti Hasanah Osman [1], Siti Kartom Kamarudin [1,2,*], Sahriah Basri [1] and Nabila A. Karim [1]

[1] Fuel Cell Institute, Universiti Kebangsaan Malaysia, Bangi 43600, Selangor, Malaysia
[2] Department of Chemical and Process Engineering, Faculty of Engineering and Built Environment, Universiti Kebangsaan Malaysia, Bangi 43600, Selangor, Malaysia
* Correspondence: ctie@ukm.edu.my

**Abstract:** The direct methanol fuel cell (DMFC) has the potential for portable applications. However, it has some drawbacks that make commercialisation difficult owing to its poor kinetic oxidation efficiency and non-economic cost. To enhance the performance of direct methanol fuel cells, various aspects should be explored, and operational parameters must be tuned. This research was carried out using an experimental setup that generated the best results to evaluate the effectiveness of these variables on electrocatalysis performance in a fuel cell system. Titanium dioxide-graphene aerogel (TiO$_2$-GA) has not yet been applied to the electrocatalysis area for fuel cell application. As a consequence, this research is an attempt to boost the effectiveness of direct methanol fuel cell electrocatalysts by incorporating bifunctional PtRu and TiO$_2$-GA. The response surface methodology (RSM) was used to regulate the best combination of operational parameters, which include the temperature of composite TiO$_2$-GA, the ratio of Pt to Ru (Pt:Ru), and the PtRu catalyst composition (wt%) as factors (input) and the current density (output) as a response for the optimisation investigation. The mass activity is determined using cyclic voltammetry (CV). The best-operating conditions were determined by RSM-based performance tests at a composition temperature of 202 °C, a Pt/Ru ratio of (1.1:1), and a catalyst composition of 22%. The best response is expected to be 564.87 mA/mg$_{PtRu}$. The verification test is performed, and the average current density is found to be 568.15 mA/mg$_{PtRu}$. It is observed that, after optimisation, the PtRu/TiO$_2$-GA had a 7.1 times higher current density as compared to commercial PtRu. As a result, a titanium dioxide-graphene aerogel has potential as an anode electrocatalyst in direct methanol fuel cells.

**Keywords:** titanium dioxide-graphene aerogel; methanol oxidation; anodic electrocatalyst; response surface methodology

**Citation:** Osman, S.H.; Kamarudin, S.K.; Basri, S.; Karim, N.A. Anodic Catalyst Support via Titanium Dioxide-Graphene Aerogel (TiO$_2$-GA) for A Direct Methanol Fuel Cell: Response Surface Approach. *Catalysts* **2023**, *13*, 1001. https://doi.org/10.3390/catal13061001

Academic Editors: Indra Neel Pulidindi, Archana Deokar and Aharon Gedanken

Received: 21 March 2023
Revised: 13 April 2023
Accepted: 23 April 2023
Published: 14 June 2023

**Copyright:** © 2023 by the authors. Licensee MDPI, Basel, Switzerland. This article is an open access article distributed under the terms and conditions of the Creative Commons Attribution (CC BY) license (https://creativecommons.org/licenses/by/4.0/).

## 1. Introduction

The interaction of the fuel with methanol generates energy, which is used to power an electrochemical cell, i.e., a direct methanol fuel cell, thus directly converting electrical energy without using combustion. Consequently, DMFC is an innovative application that provides facilities and resources to power portable technologies such as digital electronic equipment, notebook computers [1,2], mobile phones, and other popular devices among the general public [3,4]. It has several advantages, such as a modest system design, the possibility of using instant refueling [5], and low-volume and lightweight packaging. A DMFC could produce a significant amount of specific energy [5–7] using methanol with a low operating temperature, a high energy density, and an easy start-up [8]. However, apart from the technical advantages, DMFC has to find the best ways to eliminate There is no requirement for fuel reformation, yet it is classified as a zero-emission power system [9,10]. Nevertheless, its application is hindered by a few significant barriers that keep it from being commercialised: high material charges [11], low efficiency, methanol crossover from the anode to the cathode [12], and catalyst poisoning during operation [9].

Generally, transition metal oxides, such as $TiO_2$ [13], $WO_3$ [14], and $SnO_2$ [15], have been revealed as carbon support alternatives for electrocatalyst stability and activity enhancement in direct fuel methanol [16]. $TiO_2$ is one of these metal oxides that researchers are particularly interested in because of its cost-effectiveness and corrosion resistance. However, the low electrical conductivity of $TiO_2$ limits its use in fuel cells. $TiO_2$ is required to overcome low electrical conductivity in support of fuel cells, which coincides with challenges in the form of catalyst structure. Furthermore, $TiO_2$ can improve MOR performance for two reasons. The first is that $TiO_2$ empowers methanol oxidation via photopic excitation due to its 3.2 eV band gap [17], and the second is that it can reduce the CO toxicity of the catalyst via its cleaning ability, by which ordinarily metal oxides like $TiO_2$ adsorb OH and, as a result, convert the catalyst-adsorbed CO to $CO_2$ to restore the active sites of Pt NPs for continuous electrocatalytic activity and effectively mitigate the CO poisoning of Pt. Shahid et al. [18] reported that $TiO_2$/Pt significantly enhanced the electrocatalytic performance of methanol oxidation. Furthermore, Zhao and team [19] reported that a Pt/graphene-$TiO_2$ catalyst with $TiO_2$ and graphene as the mixed support exhibits high activity in comparison with Pt/graphene prepared by the same process. Other than that, Lou et al. [20] identified a three-component Pt/$TiO_2$-rGO electrocatalyst with significantly improved methanol oxidation electrocatalytic performance. Interestingly, a more convenient, simple, and quick method of preparing catalysts is required. In addition to the structural design, the rate capability can be enhanced with highly conductive materials, such as $TiO_2$, for which the use of this combination has been widely adopted. Highly conductive materials include metal oxides [21], carbon [22,23], and carbon-based materials (e.g., graphene [24,25], nanofibers [26,27], and carbon nanotubes [28,29]). Graphene has had a positive effect on $TiO_2$ due to its superior conductivity, chemical stability, and high specific surface area. However, three-dimensional (3D) graphene aerogel (GA) can effectively inhibit graphene rearrangement and provide graphene-based composites with a large specific surface area, fast electron transport kinetics, and more active sites.

A substantial amount of research is indeed required before this technology can be widely used in the years ahead. Remarkable progress has been made in the essential components, such as the system, membrane, and catalyst. The majority of the use of a multi-component catalyst has been the focus of catalyst investigation. Studies have demonstrated that when Pt and Ru's catalytic activity is compared on $TiO_2$ and graphene supports, both metals show comparable activity towards methanol oxidation on both supports. The size and distribution of the metal nanoparticles on the support, the support material's surface structure, and the reaction circumstances are only a few examples of the variables that might affect the performance of the catalysts [18,30]. Basri et al. [30] recommended PtRuFeNi/MWCNT, a novel multi-component anode catalyst. Kim et al. [31] proposed the PtRu/C-Au/$TiO_2$ electrocatalyst. Abdullah et al. studied PtRu/$TiO_2$-CNF [32] in 2018 and PtRu/Mxene [33] in 2020. All studies resulted in improved reaction kinetics and DMFC performance. The membrane was also used in several boosting studies: Thiam et al. [12] established a Nafion/Pd-$SiO_2$ composite membrane; Ahmad et al. [34] introduced a Nafion-PBI-ZP hybrid membrane; Shaari and Kamarudin [35] proposed a crosslinked sodium alginate/sulfonated graphene oxide as a polymer electrolyte membrane; and You et al. [36] studied a sulfonated polyimides/sulfonated rice husk ash (SPI/sRHA) composite membrane. All studies were created to identify the methanol crossover issue in a DMFC.

The optimisation process is critical to improving the performance of DMFCs. Previously, optimisation was done using the one-factor-at-a-time method. Conversely, this approach did not take into account factor interaction and did not reflect the actual effect of the factor on the response [37]. The method of resolving the matter using mathematically analysed issues is referred to as optimisation [38]. RSM appears to be the most reassuring optimisation method that has recently been seen by researchers in a variety of fields [39–43]. This method optimises, establishes, and improves processes by combining mathematical and statistical techniques [44]. RSM could indeed investigate the effect of independent variables in a system, either independently or in groups, and it can also reduce the number of experimental tests requisite to statistically analyse the process due to various factors [45]. The issue that our assessment is attempting to address is the 'optimal' value of the factors influencing electrocatalytic activity for methanol oxidation. These investigations discovered that the mathematical model of RSM may be employed for exact estimates and optimisation. Although DMFC optimisation has been studied extensively, electrocatalyst optimisation has received little attention.

Throughout this investigation, the bifunctional catalyst PtRu combined with $TiO_2$-GA is developed for the first time for use in DMFCs. As a consequence, the purpose of this study is to develop and optimise the $TiO_2$-GA integrated electrocatalyst performance for MOR using the RSM idea. To gain a better understanding of RSM, three parameters are manipulated: composite temperature, Pt/Ru ratio, and PtRu catalyst composition as well as the response of the current density to electrocatalyst activity in DMFC performance. Following the $TiO_2$-GA synthesis two methods are combined to make a fine composite, namely the hydrothermal and freeze-drying methods. PtRu is then coated onto $TiO_2$-GA nanoparticles. An X-ray diffraction analysis (XRD), Brunauer-Emmett-Teller (BET), field emission scanning electron microscope (FESEM), and transmission electron microscopy (TEM) were used to examine the PtRu/$TiO_2$-GA electrocatalyst for physical. The study is made more fascinating by the use of cyclic voltammetry to examine the RSM optimiser depending on three manipulation parameters and one response. The RSM was used to create a model that corresponded to the parameters analysed, and the findings of that analysis could be used in subsequent design space analyses. This research has become more intriguing with the demonstration that MOR performance is superior to commercial DMFC electrocatalysts in which the strong bond between the PtRu of the catalyst itself and the supporter, i.e., $TiO_2$-GA, in a unique 3D structure encompasses a broad surface-active site on the electrocatalyst to react.

## 2. Results and Discussion

### 2.1. Structure of Synthesised PtRu/TiO₂-GA Electrocatalyst and Electrochemical Testing

Throughout this research, FESEM was utilised to investigate the morphology of the bifunctional Pt and Ru distributions on the $TiO_2$-GA structure as well as the elemental and external mapping of the electrocatalyst. According to this, it is advantageous to boost the activity of the catalyst during the CV electrochemical test. Figure S1 (supplementary document) depicts the $TiO_2$-GA and electrocatalyst morphologies. $TiO_2$-GA and PtRu/$TiO_2$-GA FESEM images are acquired at magnifications of 500X and 5kX, respectively. Figure 1a of $TiO_2$-GA shows that the 3D $TiO_2$-GA structure was successfully developed, with a better-defined porosity structure within the ultrathin layer of the aerogel matrix. Figure 1b is the FESEM image of the PtRu/$TiO_2$-GA electrocatalyst, showing large particles dispersed and covering the 3D $TiO_2$-GA structure. Instead of that, Figure 1c described the original figure of the EDX and mapping for overall elements' particles. EDX and mapping analysis are used to further determine the existence of particles in $TiO_2$-GA, as shown in Figure 1d–h. The findings demonstrate that five elements, namely Pt, Ru, Ti, C, and O, are present in the electrocatalyst. Most of these elements are required in the electrocatalyst, and no impurities are present in the sample. Pt and Ru particles are similarly dispersed over the $TiO_2$-GA structure, according to the electrocatalyst mapping study. It would be useful to establish active response regions throughout the catalytic activity, potentially improving MOR. Nevertheless, there are a few accumulations of Pt and Ru on the sample as a result of the influence of NaOH overuse during the adjusting of the pH in the deposition process [46].

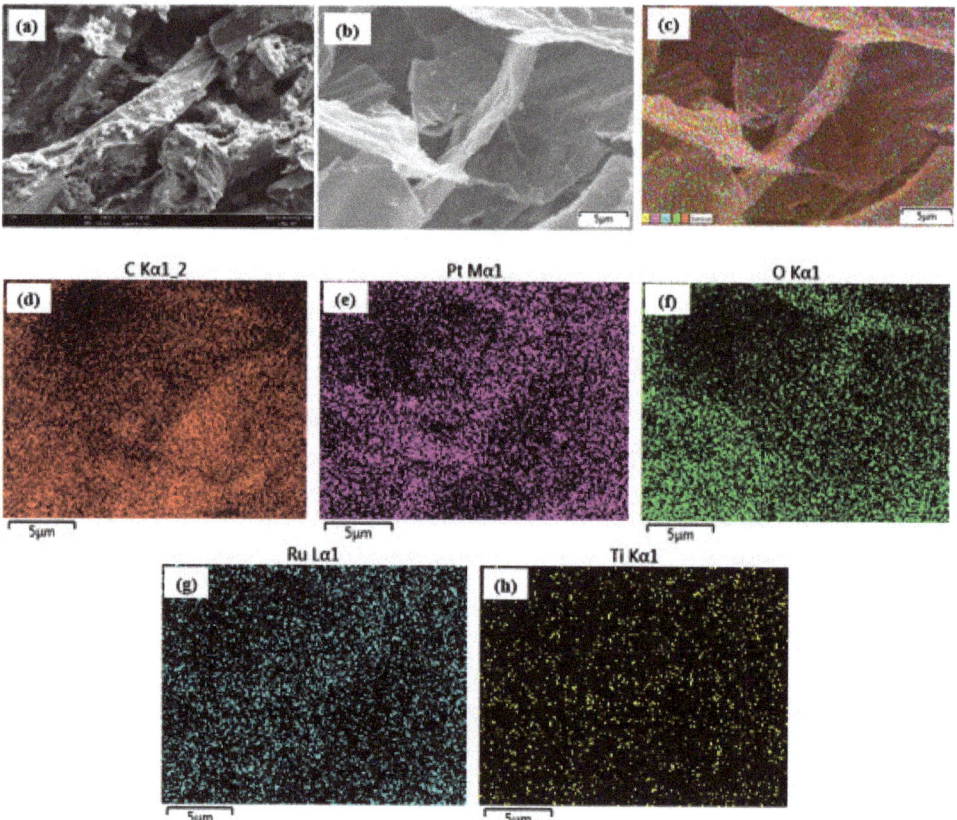

**Figure 1.** Surface morphology for (**a**) SEM of TiO$_2$-GA, (**b**) SEM of PtRu/TiO$_2$-GA, (**c**) EDX analysis of PtRu/TiO$_2$-GA, and (**d–h**) Mapping analysis for PtRu/TiO$_2$-GA.

The "electron relay effect" is a mechanism through which the support material (carbon) can deliver electrons to the Pt catalyst. During a sequence of electron transfers, the carbon support material serves as a source of electrons, which are then transported to the Pt catalyst. Several stages can be taken to move the electrons from the carbon support material to the Pt catalyst:

1. Whenever the electrons from the carbon support material interact with the Pt catalyst, they are first moved to the Pt-C bond contact.
2. After being transmitted from the Pt-C interface to the Pt surface, these electrons can subsequently be employed to speed up the methanol oxidation process.
3. Once the Pt catalyst uses these electrons to speed up the process, it may eventually lose its effectiveness. Yet, by supplying a source of electrons, the electron transfer from the support material contributes to maintaining the catalytic activity of the Pt catalyst.

In general, this process of electron transfer from the support to the catalyst promotes the MOR by offering a source of electrons that the Pt catalyst can exploit to speed up the reaction. This can increase the overall effectiveness of the reaction and enable the catalyst to sustain its activity over time.

Furthermore, the bifunctional mechanism that shifts the onset potential to a low potential region also improves catalytic performance during the MOR process. Other second metals, such as Ru, lower the potential region and increase catalytic activity, which is consistent with the bifunctional mechanism.

The electro-oxidation of methanol using a PtRu catalyst can be determined by the following equations, depending on the bifunctional mechanism:

$$Pt + CH_3OH \rightarrow Pt - CH_3OH_{ads} \rightarrow Pt - COH_{ads} \rightarrow 3H + 3e^- \rightarrow Pt - CO_{ads} + H^+ + e^-, \quad (1)$$

$$Ru + H_2O \rightarrow Ru - OH_{ads} + H^+ + e^-, \quad (2)$$

$$Pt - COH_{ads} + Ru - Oh_{ads} \rightarrow Pt + Ru + CO_2 + 2H^+ + 2e^-, \quad (3)$$

$$Pt - CO_{ads} + Ru - OH_{ads} \rightarrow Pt + Ru + CO_2 + H^+ + e^- \quad (4)$$

The proposed mechanism of PtRu/TiO$_2$-GA for the enhanced catalytic MOR activity as shown in Figure 2 can be defined as follows: (1) The unusual electronic reaction of the metal support between PtRu and TiO$_2$ results in electron transfer from the support to the metal, resulting in an increase in electron density at PtRu downwards, corresponding to the PtRu d-band centre. In general, the centre of the d-band is driven by the ability of surface d-electrons to participate in adsorption bonding. Based on the downward displacement of the d-band centre, reducing the metal back bond to CO can reduce the bonding energy between CO and PtRu atoms [47–49]. (2) In the catalyst, OH$_{ads}$, or oxygen-containing species, can be adsorbed. TiO$_2$ has the unique ability to convert CO-poisoning species (CO$_{ads}$) on PtRu bifunctional to CO$_2$, releasing the active sites of PtRu bifunctional by promoting electrooxidation activity towards methanol and CO [50]. (3) Pt, Ru, graphene aerogel, and TiO$_2$ have a good synergy effect, resulting in a four-junction structure that effectively improves electrocatalytic performance [20].

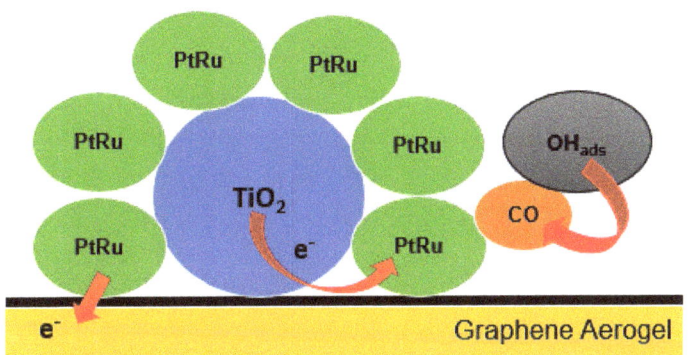

**Figure 2.** The proposed PtRu/TiO$_2$–GA pathway for enhanced catalytic MOR activity.

*2.2. One-Factor-At-A-Time*

The anode electrocatalyst, mainly composed of PtRu/TiO$_2$-GA for a DMFC, is the first time it has been accomplished in this investigation. This electrocatalyst is made up of four major components: the primary catalysts are Pt and Ru, with TiO$_2$ and graphene performing as support catalysts in an aerogel framework. Pt is widely regarded as the most effective catalyst for reactions where the carbon monoxide (CO) bond intensity can be reduced using reactions, such as hydrogen oxidation, oxygen reduction, and a combination of Pt and Ru, thus boosting catalytic performance. The inclusion of TiO$_2$ provides high thermal and electrochemical stability, whereas the electrocatalyst reaction surface area is increased by using an aerogel structure. As compared to conventional catalysts, all of these components work together to dramatically improve performance. Nevertheless, modifications of several of the elements might have an impact on the total performance of the catalyst, so identifying the optimal quantities of the catalysts is crucial to enhancing performance. The composition of this material versus TiO$_2$-GA must be developed to minimise production costs because the PtRu catalyst is a critical component that comes at a significant expense.

There are three parameters selected by respondents in this investigation, of which the major parameter is the temperature of the composition, °C (TiO$_2$-GA), the ratio of Pt to Ru, and catalyst composition interaction; PtRu is the main catalyst, while TiO$_2$-GA is the supporting catalyst. Most variables measured are optimised for half-cell performance, and the parameters were preferred for this study because they have a powerful influence on the DMFC performance based on electrocatalyst, which is comparable to what other scholars have found [22,32,51–54]. This value has been chosen based on the level of investigation from prior studies, which is a mix of PtRu and metal oxide support and graphene aerogel as an electrocatalyst [22,32,51–54]. The variety selected in the screening process was included in this research for the temperature of composition TiO$_2$-GA, which is varied in the range of 180–220 °C, the ratio of Pt to Ru in the range of 0.5–2, and the catalyst composition in the range of 10–30%.

Figure 3a–c present graphs illustrating the determined current density for the temperature of composition TiO$_2$-GA, the ratio of Pt to Ru, and catalyst composition over the experimental range. The maximum current density acquired using a 200 °C composite temperature and a 1:1 ratio is shown in the half-cell DMFC performance results (Pt:Ru). In addition, in the screening process, a catalyst composition of 20% by weight is used. The curve of the graph, on the other hand, shows that the most effective performance of DMFC can be achieved at temperature compositions of 180–220 °C. The second factor, the ratio of Pt to Ru in the range of 0.5–1.5, has a graph pattern that shows the ratio range increasing and decreasing. The catalyst composition in the range is the final factor in this study (15–25%). The results showed that the maximum current density could be achieved between 107.96 and 554.16 mA/mg$_{PtRu}$, with the greatest result from the screening process occurring at 608.17 mA/mg$_{PtRu}$. In the RSM optimisation process, the range of values for the variables is used.

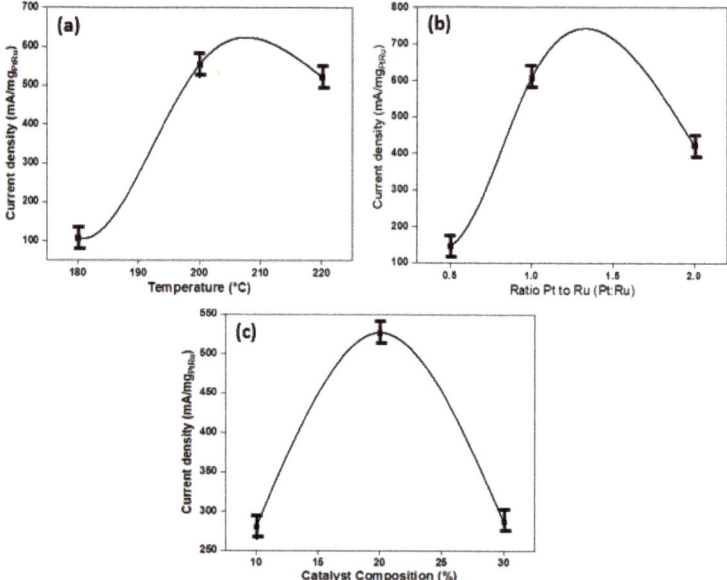

**Figure 3.** (a) The effect of temperature composite on the current density. (b) The effect of the ratio Pt to Ru (Pt:Ru) on the current density. (c) The effect of catalyst composition on the current density.

## 2.3. Optimisation Using RSM

The RSM CCD method optimisation process used 20 sessions for three study components. The experimental results for current density, including all assessments, are shown in Table 1. The regression technique was used to fit a quadratic approach to the results obtained. The current density response is modelled via Equation (1).

$$Y = 555.54 + 45.05 \times A + 22.64 \times B + 27.41 \times C + 4.16 \times A \times B + 1.76 \times A \times C \\ + 28.35 \times B \times C + \\ (-235.21) \times A2 + (-164.13) \times B2 + (-35.57) \times C2 \quad (5)$$

where Y denotes the current density (mA/mg$_{PtRu}$) and A, B, and C denote the temperature of the composition (°C), the ratio of Pt to Ru, and catalyst composition, respectively. ANOVA is a statistical analysis that compares modifications in variable-level combinations to adjustments due to random errors in response measurement [37]. The $p$-value and F-value are calculated using ANOVA, and a smaller $p$-value and a larger F-value indicate that interpreting design factor variability throughout mean data has more confidence.

**Table 1.** Summary results of the 20 experiments involved.

| Run | Factor A | Factor B | Factor C | Response 1 | |
|---|---|---|---|---|---|
| | | | | Predicted Value | Actual Value |
| 1 | 200.00 | 1.00 | 20.00 | 555.53 | 617.67 |
| 2 | 200.00 | 1.00 | 15.00 | 492.56 | 460.39 |
| 3 | 200.00 | 1.00 | 20.00 | 555.53 | 537.40 |
| 4 | 200.00 | 1.00 | 25.00 | 547.38 | 522.85 |
| 5 | 180.00 | 1.00 | 20.00 | 275.27 | 237.45 |
| 6 | 220.00 | 1.50 | 25.00 | 249.99 | 249.02 |
| 7 | 180.00 | 0.50 | 25.00 | 54.39 | 60.5 |
| 8 | 220.00 | 0.50 | 25.00 | 139.71 | 148.20 |
| 9 | 220.00 | 1.50 | 15.00 | 134.95 | 143.01 |
| 10 | 200.00 | 1.00 | 20.00 | 555.53 | 523.22 |
| 11 | 200.00 | 1.50 | 20.00 | 414.05 | 390.39 |
| 12 | 200.00 | 0.50 | 20.00 | 368.77 | 335.74 |
| 13 | 180.00 | 1.50 | 15.00 | 40.05 | 45.73 |
| 14 | 220.00 | 0.50 | 15.00 | 138.07 | 141.35 |
| 15 | 200.00 | 1.00 | 20.00 | 555.53 | 608.41 |
| 16 | 180.00 | 0.50 | 15.00 | 59.79 | 74.94 |
| 17 | 200.00 | 1.00 | 20.00 | 555.53 | 607.98 |
| 18 | 220.00 | 1.00 | 20.00 | 365.37 | 346.51 |
| 19 | 180.00 | 1.50 | 25.00 | 148.06 | 158.94 |
| 20 | 200.00 | 1.00 | 20.00 | 555.53 | 551.9 |

Table 2 shows the outcomes of an ANOVA for a quadratic response surface model as well as the significance of each coefficient. This model's F-value and $p$-value > F are 52.34 and 0.05, respectively, indicating its significance. There was only a 0.01% probability that the results were due to noise. Besides, the 0.90 lack of fit verifies its significance; the result demonstrates that this fitted model is relevant. These findings indicate that such a model could be used to predict the outcome of research in this area. The standard deviation and the determination coefficient, $R^2$, may be used to assess the appropriateness of this model. The standard deviation and $R^2$ for this model were 40.54 and 0.9792, respectively. This means that the model can account for 97.92% of the total variation in the reaction. In the meantime, the 'Pred $R^2$' of 0.9409 agrees reasonably with the 'Adj $R^2$' of 0.9605. The signal-to-noise ratio can be measured with greater precision, with a value greater than 4 being favoured. The adequate precision for the current density model is 17.984, indicating an adequate signal. To recap, the model accurately predicts future responses by fitting the experimental data. A plot of the residuals is included as part of the diagnostic model.

Table 2. Results of ANOVA analysis for current density model.

| Source | Sum of Squares | DF | Mean Square | F-Value | p-Value Prob > F |
|---|---|---|---|---|---|
| Model | $7.74 \times 10^5$ | 9 | 86,002.39 | 52.34 | <0.0001 significant |
| A: Temperature | 20,298.36 | 1 | 20,298.36 | 12.35 | 0.0056 |
| B: Ratio Pt to Ru (Pt:Ru) | 5124.56 | 1 | 5124.56 | 3.12 | 0.1078 |
| C: Catalyst Composition (wt%) | 7512.59 | 1 | 7512.59 | 4.57 | 0.0582 |
| $A^2$ | $1.521 \times 10^5$ | 1 | $1.521 \times 10^5$ | 92.60 | <0.0001 |
| $B^2$ | 74,077.24 | 1 | 74,077.24 | 45.08 | <0.0001 |
| $C^2$ | 3479.08 | 1 | 3479.08 | 2.12 | 0.1763 |
| AB | 138.13 | 1 | 138.13 | 0.084 | 0.7778 |
| AC | 24.79 | 1 | 24.79 | 0.015 | 0.9047 |
| BC | 6430.57 | 1 | 6430.57 | 3.91 | 0.0761 |
| Residual | 16,431.12 | 10 | 1643.11 | | |
| Lack of Fit | 7780.73 | 5 | 1556.15 | 0.90 | 0.5449 not significant |
| Pure Error | 8650.39 | 5 | 1730.08 | | |
| Correlation Total | $7.905 \times 10^5$ | 19 | | | |
| Standard Deviation | 40.54 | | $R^2$ | 0.9792 | |
| Mean | 338.08 | | Adj $R^2$ | 0.9605 | |
| | | | Pred $R^2$ | 0.9409 | |
| | | | Adeq $R^2$ | 17.984 | |

The diagnostic portion of RSM analysis is another process. In this section, graphs will be used to properly assess the model fit and transformation preference. Figure 4 depicts the model fit error, which is commonly referred to as a residual plot. Figure 4a shows a straight line for the residual normal probability plot, suggesting that the residual follows the normal distribution and contains acceptable normal error components. Figure 4b depicts a residual versus projected value plot of the model response, with a straight line at '0', implying that the expected variance for such a model is constant. Simultaneously, the recommended quadratic model for the current density model appears to be appropriate, and all plots fall between the upper and lower red lines with no discernible pattern.

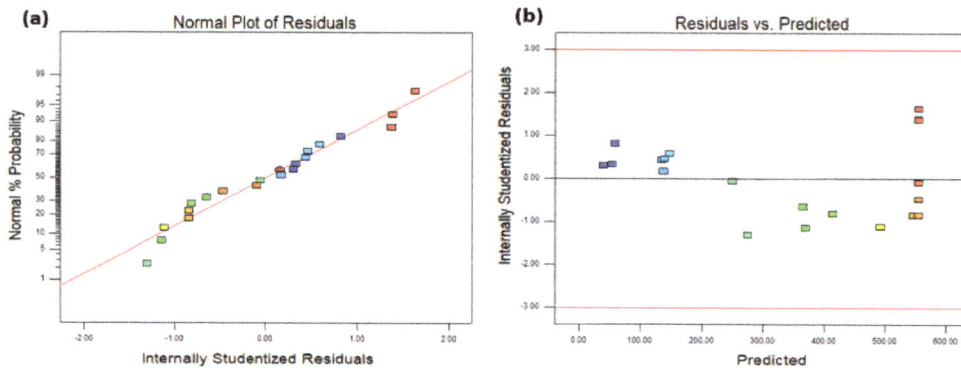

Figure 4. A residual plot for the current density model; (a) normal plot of residual; (b) residual vs. predicted plot.

The predicted vs. actual plot is used to identify values that are challenging for the model to predict [30], which is shown in Figure 5a. The plotted data is just in the centre of the graph and forms a 45° perpendicular line. This requires the model's ability to correctly predict the response. Figure 5b depicts a perturbation plot, which demonstrates how the factors can influence the response. As previously stated, factors A, B, and C are the temperature of the composite, the Pt/Ru ratio, and the catalyst composition, respectively. The actual values are set to the 'coded 0' midpoint for all of the factors: A: 202 °C, B: 1.1, and C: 22 wt%. The perturbation graph is created by changing one factor at a time over the response value. Overall, three factors have a steep slope in the plot, showing that they are all influenced by or sensitive to the experimental response and are important to the system design. Nevertheless, factor A's graph has a little steeper gradient than factors B and C, showing that factor A has a larger impact on the response value.

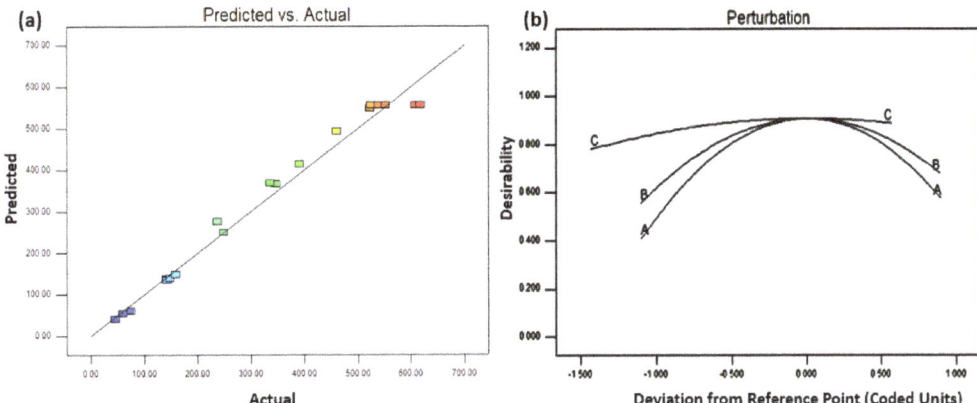

Figure 5. (a) Predicted vs. actual plot, and (b) perturbation plot for the current density model.

Predicting or evaluating the mean reaction is what response surface analysis includes at a specific point in the process [44]. Figure 6 shows the response surface for a contour plot in two dimensions (2D) and a surface plot in three dimensions (3D). The response surface is made up of an analysis of two factors, namely AB, AC, and BC, as well as the current density's response. Nevertheless, Figure 6 depicts an examination of current density factors A and B. The relationship between components A and B as well as the response, as depicted in the 2D contour plot, has a specific result in Figure 6a. The plot shows that increasing both factors increases the response. After a certain point, the response trends begin to decline even as the factor value increases. This is considered the optimum point, when the most response to the model may be correlated by the optimum factors. The other two components, AC and BC, have essentially comparable reaction patterns. In the red area of the contour plot, also referred to as the high response value area, the ideal location for factors can be found. The contour plot displayed the same patterns as the 3D surface plot in Figure 6b, and the distinct peak for all parameters represents an optimum point that reached the maximum response. The AB, AC, and BC factors have a similar pattern. The 3D graph predictive analysis of a second-order model, according to the relevant literature, reveals that the quadratic model is similar to the existing density model.

The following section deals with RSM optimisation evaluation. Alienated numerical optimisation, graphical optimisation, point prediction, and confirmation are the four main categories. The numerical optimisation classifications are required to determine the aspirations and estimate the factors of the optimal condition to yield the maximum response as determined by the model's goals. Following graphical optimisation, Figure 7a,b show the 2D contour plot for desirability with a response prediction value (for instance, in terms of the AB factor). This model's prediction values for desirability and response are 0.908 and 564.866 mA/mg$_{PtRu}$, respectively, as shown in the plot in the high response area. Figure 7 displays the point predictions for each of the model's optimum factors. The graph shows that when two graphs with significant desirability overlap, those three factors achieve an ideal position.

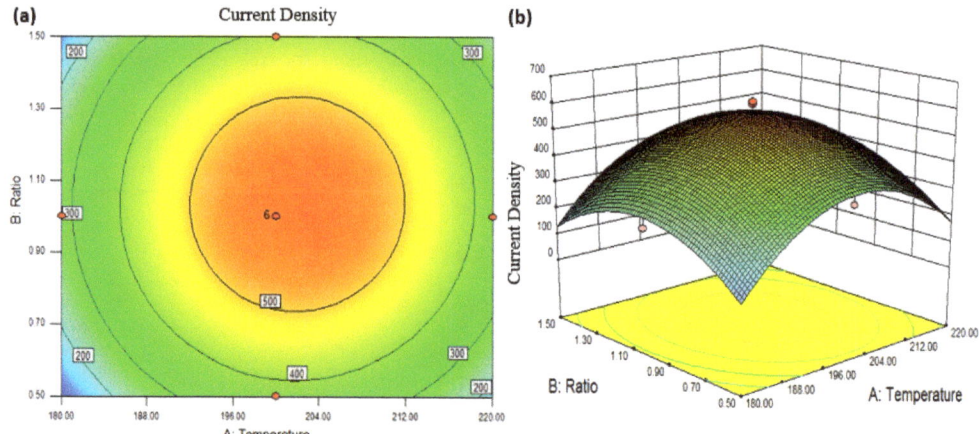

**Figure 6.** Response surface between factors; Pt:Ru ratio and a composite temperature, with the response; current density, (**a**) 2D contour, and (**b**) 3D surface plot.

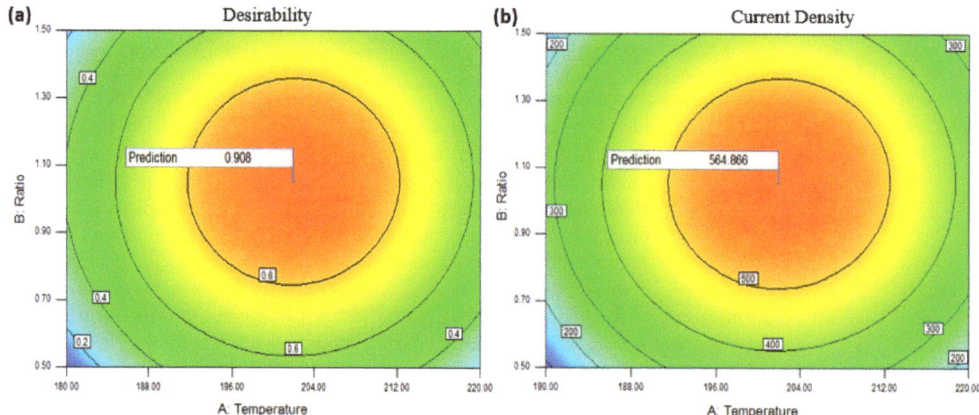

**Figure 7.** 2D contour plots for, (**a**) Desirability and (**b**) Current density in terms of AB factors.

The RSM also examined the optimum factors: A (temperature composite): 202 °C, B (ratio Pt/Ru): 1.1 and C (catalyst composition): 22 wt%. Validation is then used to compare the model's predicted results to the experimental results. To get the average, the validation test with the value of the optimal factor is carried out in triplicate, and the outcome is shown in Table 3.

**Table 3.** Validation of current density model.

| Factor A (°C) | Factor B | Factor C (wt%) | Mass Activity (mA/mg$_{PtRu}$) | | | | | Error (%) |
|---|---|---|---|---|---|---|---|---|
| | | | Prediction | 1 | 2 | 3 | Average | |
| 202 | 1.1 | 22 | 564.87 | 529.40 | 608.17 | 566.89 | 568.15 | 0.6 |

Figure 8 depicts the current density graph from the validation test. The CV test typically provides electrochemical measurements and is used to calculate the response value of this model. Once compared to the RSM anticipated value, the average validation test result was 568.15 mA/mg PtRu, which correlated to a peak potential of 0.65 V vs. Ag/AgCl with just a 0.6% error. The small inaccuracy shows that with the optimal temperature composite, Pt/Ru ratio, and catalyst composition, the best current density response may be achieved. This situation also guarantees that the RSM analysis model is effective and applicable.

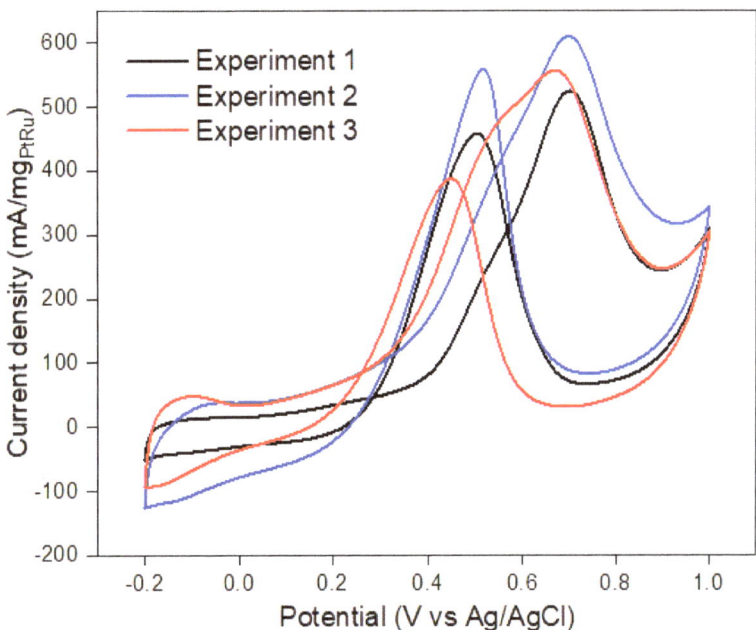

**Figure 8.** Experiment to validate the present current density model.

### 2.4. Interactions of Synthesis PtRu/TiO$_2$-GA Optimisation

#### 2.4.1. Physical Characterisation

Pattern and crystal structure were used to investigate the physical characteristics of the PtRu/TiO$_2$-GA electrocatalyst before and after optimisation compared to PtRu/C. This investigation was carried out utilising an x-ray diffractometer (XRD) in the 5–80° range with 2θ resolution. The PtRu/TiO$_2$-GA electrocatalyst XRD pattern revealed a diffraction peak for all of the included Pt, Ru, TiO$_2$, and C. As a control experiment, PtRu/C was used. The diffraction peak (2θ = 40.3°) does not appear clearly on other electrocatalyst samples after the reduction procedure. Furthermore, no discernible difference is detected in the XRD patterns of PtRu/TiO$_2$-GA and PtRu/TiO$_2$-GA$_{Opt}$. The other peaks are in accordance with the anatase phase of the TiO$_2$ standard spectrum (JCPDS No. 21-1272). TiO$_2$ anatase phase has greater diffraction peaks at 25.4° (1 0 1), 37.7° (2 0 0), 53.8° (2 1 1), 55° (2 0 4), and 63° (1 1 6) than TiO$_2$ rutile, which has 27° (1 1 0), 35.6° (1 0 1), and 39.5° (1 1 1).

Diffraction peaks at planes (1 1 1), (2 0 0), and (2 2 0) at about 2θ are 40.7°, 47.3°, and 67.0°, respectively, which detected the Pt nanoparticle, which is compatible with the face-centred cubic (fcc) structure of Pt [55]. while the diffraction peaks for Ru are in the same plane, namely (1 1 1), (2 0 0), (2 0 0), and (2 2 0). Due to the strong crystallinity of Pt-Ru nanoparticles in the electrocatalyst system, low diffraction peaks of TiO$_2$-GA are difficult to identify following the dispersion of Pt-Ru particles on the TiO$_2$-GA support. The creation of a bimetal or Pt alloy in each synthesised electrocatalyst sample is also indicated by the yield of a single diffraction peak shared by PtRu in each of these electrocatalyst diffractograms [56,57].

The Bragg angle is evident in the range of 25–60° for all electrocatalyst samples, as shown in Figure 9. This suggests that the catalyst has a bimetallic or alloy interaction [58]. Meanwhile, when comparing PtRu/C to PtRu/TiO$_2$-GA before and after optimization, weak and broad peak intensities were detected. Regarding the crystal size values reflecting the high dispersion in the produced sample, no substantial change in peak intensity can be noticed in this situation. Table 3 lists the crystal size values available using Eva software for analysing XRD results.

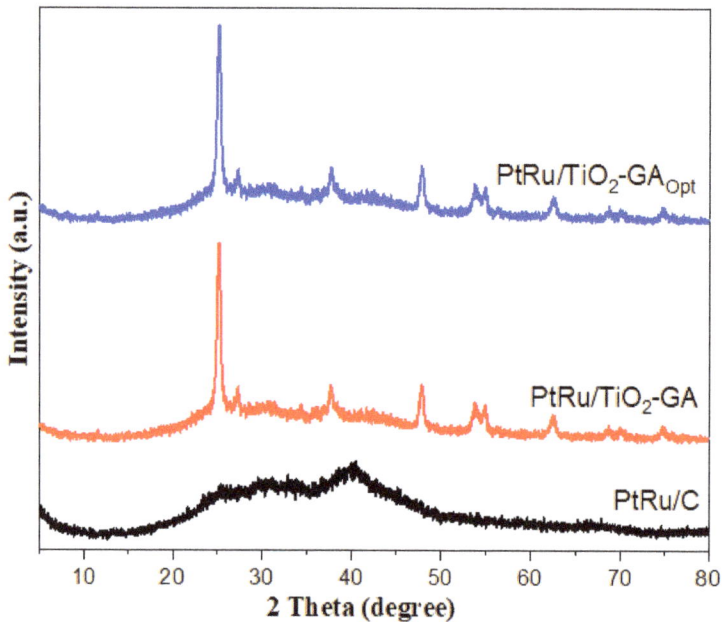

**Figure 9.** XRD patterns for prepared samples.

Additionally, N$_2$ adsorption-desorption isotherms and pore size distribution analyses were performed at a temperature of 77 K to investigate the porous structure and specific surface area of TiO$_2$-GA composites.

The N$_2$ adsorption/desorption plots (Figure 10) show that PtRu/TiO$_2$-GA$_{Opt}$ has a significantly larger surface area than PtRu/TiO$_2$-GA and PtRu/C, as shown in Table 4. The PtRu/TiO$_2$-GA$_{Opt}$ electrocatalyst had the lowest BET surface area of 50.59 m$^2$/g, followed by the PtRu/TiO$_2$-GA and PtRu/C electrocatalysts in that order. The findings of this study are comparable to those of those who found that the PtRu/C electrocatalyst had a significantly greater surface area than the metal oxide composite electrocatalyst. According to the BET adsorption/desorption isothermal curves in Figure 10, all produced electrocatalyst materials had type IV isothermal curves (based on IUPAC classification) with a little indentation of the observed width indicating H3 hysteria [59,60].

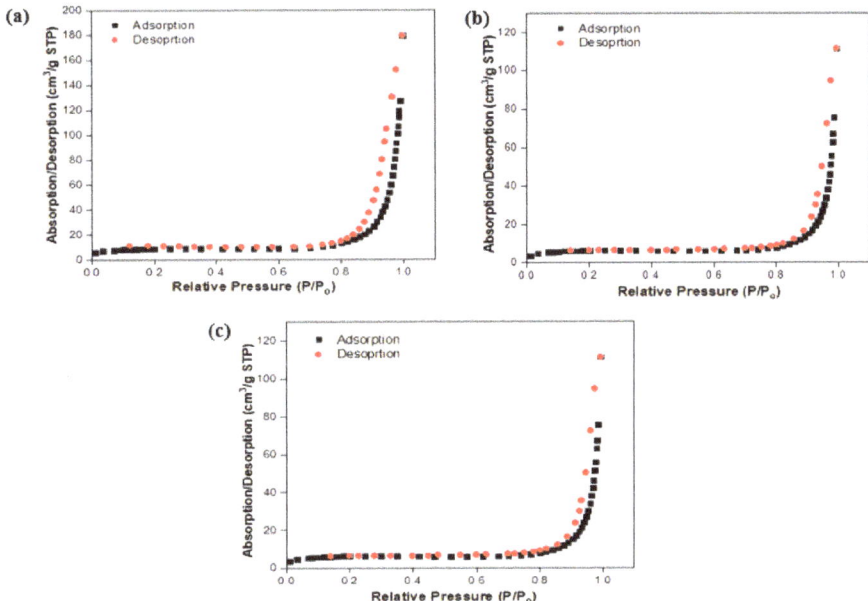

**Figure 10.** Adsorption/desorption linear isotherm: (**a**) PtRu/C; (**b**) PtRu/TiO$_2$-GA; and (**c**) PtRu/TiO$_2$-GA$_{Opt}$.

**Table 4.** BET analysis results for size and porosity properties of PtRu/TiO$_2$-GA$_{Opt}$, PtRu/TiO$_2$-GA, and PtRu/C electrocatalysts.

| Electrocatalyst | S$_{BET}$ (m$^2$ g$^{-1}$) | V$_{Total\ Pore}$ | DPore (nm) | Types of Pore |
|---|---|---|---|---|
| PtRu/TiO$_2$-GA$_{Opt}$ | 19.30 | 0.13 | 24.22 | Meso |
| PtRu/TiO$_2$-GA | 22.36 | 0.45 | 31.05 | Meso |
| PtRu/C | 31.15 | 0.61 | 32.63 | Meso |

Furthermore, at the relative pressure position P/Po > 0.6, the desorption branch was greater than the adsorption branch. While the porosity structure of the electrocatalysts presented is a mesopore type with an average diameter range of 2–50 nm, this can be attributed in large part to the huge gaps seen in the electrocatalyst lattice [61,62]. Additionally, the observations on the nitrogen isotherm adsorption/dehydration graph appear flat at low pressures, i.e., P/Po ≤ 0.6, which could be related to microbe adsorption in the sample. The adsorption of monolayer and/or multilayer nitrogen molecules on the mesostructure increases the sample's adsorption capacity at relatively high-pressure areas (0.6 ≤ P/Po ≤ 1.0) while catalysts spread on the support, which can boost the catalytic activity's stability and performance. The BET pore diameter findings are identical to the XRD particle size analysis results. The smaller the particle, the higher the surface-to-volume ratio, which might result in high surface reactivity and solubility. This notion is supported by the current study, which found that PtRu/TiO$_2$-GA$_{Opt}$ electrocatalysts have smaller pore widths and exhibit stronger MOR activity than PtRu/TiO$_2$-GA and PtRu/C electrocatalysts.

The degree of homogeneity of the distribution of each sample can be identified using FESEM mapping pictures. The interconnected 3D porous networks with open pores ranging in size from a few hundred nanometers to tens of micrometres are visible for both PtRu/TiO$_2$-GA and PtRu/TiO$_2$-GA$_{Opt}$ composites. In the ultra-thin layer of the aerogel matrix, the PtRu/TiO$_2$-GA$_{Opt}$ electrocatalyst has a more prominent porosity structure.

Figure S2a (supplementary document) shows a nanocomposite composed of randomly aggregated, thin, crumpled graphene sheets that are intimately linked. Furthermore, it demonstrates that a nanocomposite composed of randomly aggregated, wrinkling graphene sheets is tightly related to one another. Figure S2b (supplementary document) depicts the distribution of $TiO_2$ and Pt nanoparticles over curly graphene sheets. Furthermore, in FESEM mapping, the uniformity of PtRu/$TiO_2$-$GA_{Opt}$ is higher than that of PtRu/$TiO_2$-GA, which can be related to the uniformity of the PtRu catalyst resulting from its reporter when the reduction process happens. EDX analysis was also performed to further analyse the qualitative elemental analysis of the inserted composite; see Figure S2c,d (supplementary document). The sample contains Ti, Pt, Ru, O, and C, but no other elemental contaminants have been discovered. Based on FESEM and TEM images, i.e., Figure S2 (supplementary document) and Figure S3, no pores are visible on the catalyst nanoparticles, which is consistent with the low BET surface area study because this substance is a metallic element with high porosity. However, the surface area of the BET does not provide an overall view of the surface of the electrocatalyst's active site. As a result, electrochemical analysis can provide precise information about the active surface area of the catalyst.

The TEM images of PtRu/$TiO_2$-GA before and after optimisation to determine the morphology of all samples created using the microwave-assisted alcohol reduction process are shown in the figure. The image clearly shows the spherical shape of the carbon support material. This spherical carbon could be associated with graphene sheets to some extent, helping to increase the pores of the aerogel by cooperative dispersion, resulting in increased porosity [63]. Both samples have an estimated diameter size of 2–5 nm, allowing the samples to be classified in nano size parallel to the XRD crystal particle size as given in Table 5. When comparing the PtRu/$TiO_2$-GA sample area to the PtRu/$TiO_2$-$GA_{Opt}$ sample area, TEM examination revealed only tiny clumps. This situation indirectly affects the electrocatalyst's maximum function in the methanol oxidation reaction (MOR). Van der Waals bonding in the particles may cause agglomeration. In conclusion, the reduction approach for synthesising PtRu/$TiO_2$-GA using ethylene glycol solvent and microwave technology was successful in a short time and was effective in the manufacture of dual-function nanoparticles on $TiO_2$-GA support without isolating the support.

Table 5. The porosity properties of samples.

| Properties Sample Name | $S_{BET}$ ($m^2\ g^{-1}$) | $V_{Total\ pore}$ | XRD (Crystallite Size (nm)) | Types of Pores |
|---|---|---|---|---|
| PtRu/C | 26.11 | 0.17 | 3.5 | Meso |
| PtRu/$TiO_2$-GA | 19.30 | 0.11 | 3.1 | Meso |
| PtRu/$TiO_2$-$GA_{Opt}$ | 17.15 | 0.10 | 2.5 | Meso |

2.4.2. Electrochemical Evolution

The most important aspects of CV and CA are needed in electrochemical testing. CV is necessary to measure electrocatalytic performance in this part, whereas CA is required to test electrocatalytic durability and stability. Figure 11a depicts the CV profiles of all catalysts in a 0.5 M $H_2SO_4$ solution at potentials ranging from −0.2 to 1.0 V. ECSA could be used to calculate the surface area of PtRu nanoparticles in electrocatalysts [64]. This happens when there is limited adsorption at the activation site, requiring electrode current cycles within a voltage range. The total charge required for monolayer adsorption and desorption identifies the reactive surface site for ECSA in the range of −0.2 to 0.1 V [65]. Table 6 contains the ECSA results, and CV measurements were used to evaluate ECSA using the equation below:

$$ECSA\ \left(m^2 g_{Pt}^{-1}\right) = \frac{Q}{\Gamma \cdot W_{Pt}}$$

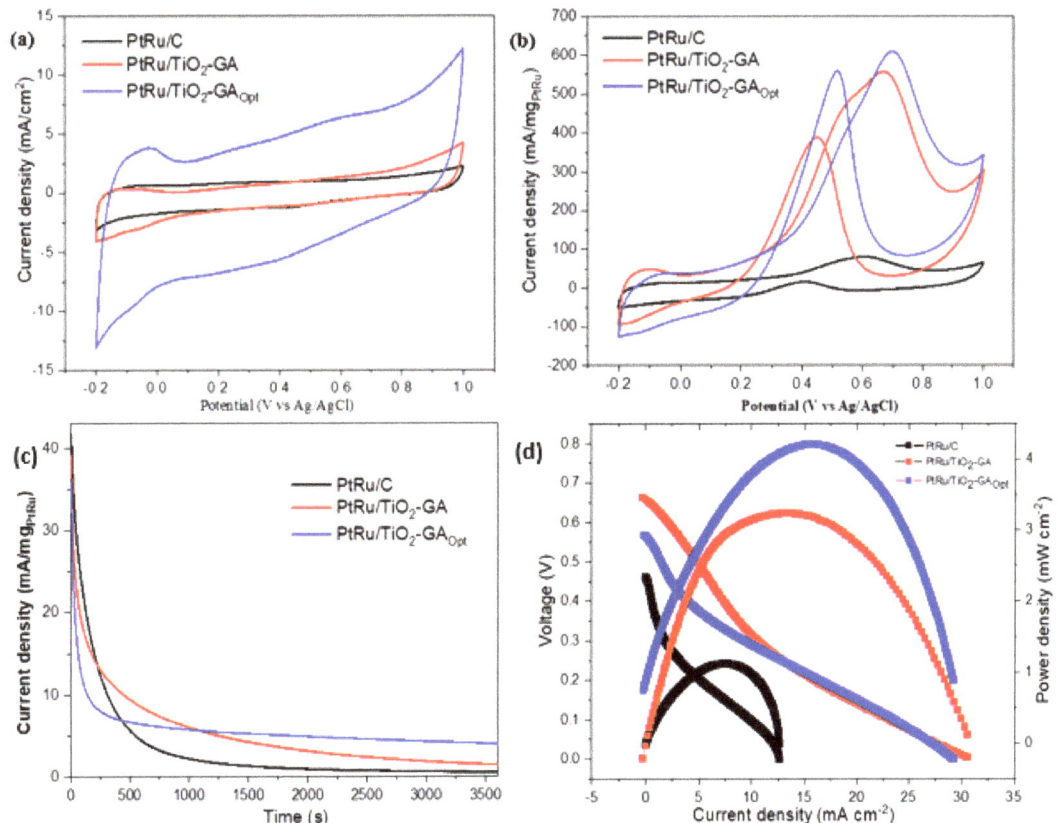

**Figure 11.** Represents the (**a**) CV scan of PtRu catalyst in 0.5 M $H_2SO_4$ electrolyte at 50 mV/s; (**b**) CV scan in 0.5 M $H_2SO_4$ and 2.0 M $CH_3OH$ aqueous at 50 mV/s; (**c**) chronoamperometry curve carried out in 0.5 M $H_2SO_4$ + 2.0 M $CH_3OH$ for commercial PtRu/C and different electrocatalysts at 0.4 V vs. Ag/AgCl; (**d**) current–voltage curve for electrocatalyst in 2 M methanol at room temperature.

**Table 6.** Evaluation of the current density results with the different catalyst supports.

| Electrocatalysts | Peak Potential (V vs. Ag/AgCl) | Onset Potential (V vs. Ag/AgCl) | Peak Current Density (mA mg$^{-1}$) | ECSA (m$^2$ g$^{-1}$) | CO Tolerance, $I_f/I_b$ |
|---|---|---|---|---|---|
| PtRu/C | 0.613 | 0.419 | 79.93 | 3.25 | 5.31 |
| PtRu/TiO$_2$-GA | 0.676 | 0.336 | 554.67 | 23.43 | 1.44 |
| PtRu/TiO$_2$-GA$_{Opt}$ | 0.693 | 0.346 | 568.15 | 30.63 | 1.09 |

Based on the equation, Q is the charge density or area under the graph ((C) of the experimental CV) $\Gamma$ (2.1 $Cm_{Pt}{}^{-2}$) is the constant for the charge required to reduce the proton monolayer on the Pt, and $W_{Pt}$ is the Pt loading ($g_{Pt}$) on the electrode. According to the ECSA calculations, the synthesised electrocatalyst, PtRu/TiO2-GA$_{Opt}$, has the highest value of 30.63 m$^2$g$^{-1}$, compared to PtRu/C (3.25 m$^2$g$^{-1}$). This occurred due to the PtRu crystallite size, as shown in Table 5 from XRD analysis; the PtRu crystallite size for PtRu/TiO$_2$-GA$_{Opt}$ is the smallest and has a high ECSA value. The catalyst and reaction surface area can be enhanced by using the smallest crystallite size. The crystallite size trend is paralleled by the ECSA value trend for PtRu/TiO$_2$-GA$_{Opt}$ and PtRu/C. The various CV curves in

Figure 11b depict the inverted scan, with the minor oxidation peak appearing between 0.3 and 0.6 V vs. Ag/AgCl. During the initial oxidation peak, modest oxidation on the reversed scan, also known as the reversed oxidation peak, occurred, resulting in the formation of incompletely oxidised carbonaceous species. According to the CV data in Table 6, if the value of ($I_f/I_b$) can be formulated, the optimised electrocatalyst combination is lower than the commercial. This is due to particle aggregation. However, the catalytic activity of PtRu/TiO$_2$-GA$_{Opt}$ was 7.6 times that of PtRu/C. This discovery suggests that using an aerogel structure and a metal oxide combination in the electrocatalyst has a high potential for usage in DMFC applications.

As shown in Figure 11b, the electrocatalytic performance of the synthesised electrocatalyst was evaluated using CV. At room temperature, the CV curves for the electrocatalysts PtRu/TiO$_2$-GA, PtRu/TiO2-GA$_{Opt}$, and PtRu/C are measured in 2 M methanol with 0.5 M H$_2$SO$_4$ and saturated N$_2$ gas. The various curves are measured between −0.2 and 1.0 V vs. Ag/AgCl. The peak current density was PtRu/TiO$_2$-GA$_{Opt}$ > PtRu/TiO$_2$-GA > PtRu/C in decreasing order. In comparison to Ag/AgCl, the peak current density of PtRu/TiO$_2$-GAOpt for the MOR appeared to be 0.639 V. Table 6 shows the peak current density and other CV values for all of the samples. PtRu/TiO$_2$-GA$_{Opt}$ has a current density of 608.17 mA mg$^{-1}$, which is 1.1 and 7.6 times greater than that of PtRu/TiO$_2$-GA and commercial electrocatalyst PtRu/C, respectively. This is due to TiO$_2$-GA's unusual 3D structures, which allow for a quick ion/charge transfer channel [66]. This one-of-a-kind characteristic benefits surface chemical processes and increases the activity of the electrocatalyst. Furthermore, the TiO$_2$-GA structure enables Pt and Ru nanoparticles to attach more firmly to the TiO$_2$-GA surface, as illustrated in the surface morphology section in Supplementary Figures S2 and S3.

The superior MOR performance of PtRu/TiO$_2$-GA$_{Opt}$ can be attributed to the XRD discovery that the lowest particle size in the electrocatalyst yields high ECSA values, as proven by CV analysis. Furthermore, the FESEM analysis revealed that the distribution of TiO$_2$-GA$_{Opt}$ support particles was more uniform than the others. Furthermore, the BET study demonstrated that the inclusion of TiO$_2$-GA$_{Opt}$ support, which has good electrical conductivity, a unique network structure, and a high surface area, increases electrocatalyst area and activity. Therefore, this electrocatalyst combination is promising for future usage as a catalyst for methanol electrooxidation in DMFC applications.

The CA results would be used to conduct studies on electrocatalyst durability and stability. Figure 11c compares the anodic peaks in terms of the mass activity of PtRu/TiO$_2$-GA and PtRu/C commercial electrocatalysts towards MOR in 0.5 M sulfuric acid and 2.0 M methanol at a scan rate of 50 mV s$^{-1}$. The electrocatalysts all displayed a quick and dramatic reduction in current density before slowly becoming horizontal during the stability test using CA measurement. The retention value (percent) (Table 7) in the pattern rose after 3600 min, such as PtRu/TiO$_2$-GA$_{Opt}$ < PtRu/TiO$_2$-GA < PtRu/C. The lowering current density ratio of the PtRu/TiO$_2$-GA$_{Opt}$ electrocatalyst was somewhat greater than that of the PtRu/C, however, and this electrocatalyst obtained the maximum current density of all the electrocatalysts in Table 6. This was attributable to the catalyst support's superior dispersion as well as the enhanced utilisation of catalysis [51,67,68].

Table 7. Retention rates of electrocatalysts.

| Electrocatalyst | $j_i$ (mA cm$^{-2}$) | $J_f$ (mA cm$^{-2}$) | Retention Rates (%) |
|---|---|---|---|
| PtRu/C | 41.7 | 0.55 | 98.7 |
| PtRu/TiO$_2$-GA | 38.8 | 1.53 | 96.1 |
| PtRu/TiO$_2$-GA$_{Opt}$ | 35.9 | 4.05 | 88.7 |

*2.5. Passive Single Cell Performance*

Based on the single-cell findings presented in the figure, PtRu/TiO$_2$-GA after optimisation had a positive effect on the single-cell performance test of direct methanol fuel cells. As a result, considerable alterations in methanol oxidation occur. PtRu/TiO$_2$-GA evaluated the single-cell performance of passive mode DMFC and was published for the first time. The current-voltage (I-V) polarisation curves for 22% PtRu/TiO$_2$-GA and 22% PtRu/C electrocatalysts tested using a single DMFC cell at room temperature, as well as 20% PtRu/TiO$_2$-GA electrocatalysts before optimisation are shown in Figure 11d, along with the 3.2 mW cm$^{-2}$ achievements. These molecules could obstruct the surface location of the catalyst where methanol would react, resulting in decreased performance. Since the methanol flow rate cannot be controlled in passive mode, the tank is easily contaminated with methanol intermediate products, which can block GDL and impair DMFC performance. Table 8 compares the performance of the PtRu/TiO$_2$-GA electrocatalyst current density with prior studies in passive mode DMFC. In contrast to before optimisation, the PtRu/TiO$_2$-GA catalyst's optimum conditions demonstrated a considerable improvement in DMFC performance. Once all of the results were compared, the PtRu/TiO$_2$-GA electrocatalyst was found to be comparable to electrocatalysts previously reported by other researchers. The PtRu/TiO$_2$-GA electrocatalyst has significant potential and could be used as an anode catalyst for passive-mode DMFC in the future.

Table 8. Comparison of the single cell performance results.

| Study | Electrocatalyst | Power Density (mW/cm$^2$) |
|---|---|---|
| This study | PtRu/TiO$_2$-GAOpt | 4.2 |
| This study | PtRu/TiO$_2$-GA | 3.2 |
| This study | PtRu/C | 1.1 |
| Abdullah et al. [69] | PtRu/TiO2-CNF | 3.8 |
| Ramli et al. [64] | PtRu/CNC | 3.35 |
| Shimizu et al. [70] | PtRu/C | 3.0 |
| Hashim et al. [71] | PtRu/C | 3.3 |

## 3. Experimental Section

### 3.1. Materials and Chemicals

The Pt precursor, H$_2$PtCl$_6$, with 20% content, was kindly supplied by Merck, Darmstadt, Germany, while the Ru precursor, RuCl$_3$ (45–55% content), supplied by Sigma-Aldrich Co., St. Louis, MO, USA, was used in the synthesis catalyst. Ethylene glycol (EG), Nafion solution, isopropyl alcohol, ethanol, and methanol were obtained from Sigma-Aldrich. Titanium isopropoxide (TiPP, 97%) was obtained from Sigma-Aldrich Co. Graphene oxide was obtained from GO Advanced Solution Sdn. Bhd (Malaysia).

### 3.2. Preparation of the PtRu/TiO$_2$-GA Electrocatalyst

Fabrication of a TiO$_2$-GA composite by a combination of hydrothermal and freeze-dried methods that utilised moderate ultrasonication for 2 h, 50 mg of TiO$_2$ was homogeneously dispersed into 100 mL of GO solution (2 mg/mL). The dilution was then placed in a Teflon-lined 100 mL autoclave and kept at 200 °C for 12 h. To make a TiO$_2$/GA composite, the synthesised hydrogel was carefully washed with deionised water and freeze-dried for 24 h. The electrocatalyst fabrication was carried out using the standard atomic ratio of 1:1 of PtRu catalyst with 20 wt% loaded to TiO$_2$-GA support, as shown in Figure 12. The formation started with the weighing of chloroplatinic acid (Pt source) and ruthenium chloride (Ru source) precursors, followed by blending with sonicated ethylene glycol solutions for 15 min. TiO$_2$-GA was added to the precursor solution after it had been thoroughly mixed, and it was stirred for about 30 min while the pH solution (1 M NaOH) was used to modify the solution to 10. To ensure that the reduction process for this formation was completed, the microwave reduction approach was designed for one minute and then off for one minute, repeated two times. Eventually, the composite was washed and filtered several times with DI water and ethanol and then dried in an oven for 3 h at 120 °C.

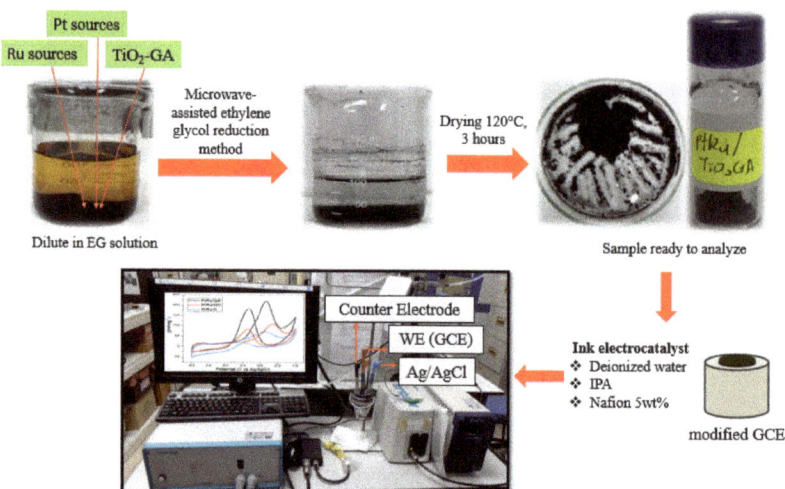

**Figure 12.** Preparation of PtRu/TiO$_2$-GA electrocatalyst.

*3.3. Electrocatalyst Characterisation and Electrochemical Testing*

The crystallinity and presence of the metal alloy in the electrocatalyst samples were characterised using X-ray diffraction (XRD) (D8 Advance, Bruker, Bremen, Germany)). XRD was used to provide the particle size distribution. The pore characteristics were measured using a Brunauer-Emmett-Teller (BET) analysis on a Micromeritics ASAP 2020 instrument (Beijing, China) at 77 K under nitrogen gas flow. A FESEM analysis with a SUPRA 55 VP (Zeiss, Birmingham, UK) was used to examine the surface morphology of Pt-based bimetallic electrocatalyst-supported TiO$_2$-GAs. In addition, energy-dispersive X-ray (EDX) and mapping were performed on all prepared electrocatalysts to estimate the distribution of the electrocatalyst element in samples. Transmission electron microscopy (Tecnai G2 F20 X-Twin, FEI, Hillsboro, USA) was used to measure the catalyst particle size.

For this research, CV and CA electrochemical tests were conducted on the samples and carried out in an electrolyte containing 0.5 M H$_2$SO$_4$ and 2.0 M CH$_3$OH as the fuel solution. At the moment, a three-electrode system was being used at room temperature, with a working electrode consisting of a glassy carbon electrode (3 mm Ø), a counter electrode containing platinum, and Ag/AgCl electrodes serving as reference electrodes. To make preparations for the electrocatalyst ink for the working electrode, 50 µL (5 wt%) Nafion solution, 150 µL DI water, and 150 µL IPA are sorted with 2.5 mg electrocatalyst. Then, 2.5 µL electrocatalyst ink was carefully dropped onto a glassy carbon electrode and allowed to dry. The Autolab electrochemical workstation was used to measure CV with a scan rate of 50 mVs$^{-1}$ and potentials ranging from −0.2 to 1.0 V vs. Ag/AgCl.

*3.4. Experimental Design*

3.4.1. One-Factor-At-A-Time (OFAT)

OFAT is among the approaches that change each parameter at a time while keeping the others constant [72]. Before final optimisation using the RSM method, the OFAT concept is used to reach a target of required factor values. This study focuses on three parameters, the most important of which is the electrocatalyst variable of the temperature composite, TiO$_2$-GA, manufactured with the catalyst. A second parameter investigated is the ratio of Pt to Ru, and the last parameter is catalyst composition for optimised half-cell performance. The variables were selected for this research because they have a robust impact on the results of electrocatalyst elements of DMFC, which is consistent with the findings of other researchers [32,33,51–54]. Depending on metal oxide and/or carbon nanostructure electrocatalysts supported by PtRu, the values for the variables were chosen as guidance for screening levels via other researchers [22,32,51–54]. The temperature composite, Pt/Ru ratio, and catalyst composition are ranged in this analysis in the ranges of 180–220 °C, 0.5–2.0, and 10–30 wt%.

### 3.4.2. Response Surface Methodology

The central composite design (CCD) was preferred to evaluate the influence of various factors on a specific response. The three factors investigated in this study were the temperature of the composition $TiO_2$-GA, the ratio of Pt to Ru, and the catalyst composition, whereas half-cell DMFC performance was measured using a current density. Design Expert 8.0.6 was used to carry out this method (Stat-Ease Inc., Minneapolis, MN, USA). The factors were graded on a five-point scale ($-\alpha$, $-1$, $0$, $+1$, $+\alpha$). Three factors generate 20 experimental studies with CCD, of which five are performed at a face centre. The second-order polynomial model agrees with this assessment [37]:

$$Y = \beta 0 + \beta 1A + \beta 2B + \beta 12AB + \beta 11A2 + \beta 22B2 \tag{6}$$

where Y is the dependent variable (current density), A and B are the limit values of the independent variable, $\beta o$, $\beta_1$, $\beta_2$, and $\beta_{11}$ are constant coefficients, o are quadratic constants, $\beta_1$ and $\beta_2$ are coefficients for linear terms, $\beta_{11}$ and $\beta_{12}$ are the coefficients for quadratic terms, and $\beta_{22}$ is the coefficient for the second-order interaction terms. The coefficient of regression, analysis of variance (ANOVA), F-value, and p-value were used to analyse the constructed regression model. The determination coefficient, $R^2$, measures the fit effectiveness of the polynomial equation model. The optimal values for these three variables were then evaluated by running validation experiments and comparing the values projected by the model. Thus, every test was performed in triplicate to produce an average value.

### 3.5. Preparation of MEA

The anode, membrane, and cathode are three important parts in the production of the membrane electrode assembly (MEA). The membrane Nafion 117 is chosen, and it is treated with hydrogen peroxide ($H_2O_2$) and DI water as shown in a study by Hasran et al. [73] to remove impurities. Besides, the electrocatalyst anode layer was PtRu/$TiO_2$-GA, and the cathode layer was Pt/C based on water-impermeable porous carbon paper. Thin plastic blades helped load the carbon slurry (HiSPEC 1000 carbon powder and 9% PTFE solution) onto the carbon paper surface. After that, the coated carbon layer was allowed to cool before being sintered in a furnace at 350 °C for 60 min to allow a diffusion layer to form at a load of 2 mg cm$^{-2}$. The catalyst is then properly dispersed on the coated carbon support layer using PtRu/$TiO_2$-GA catalyst ink on the anode side and Pt/C on the cathode side, respectively. In this case, carbon cloth is used to prepare the catalyst layer, which is rubbed with carbon slurry and dried in an oven at 100 °C for 1 h. Hence, DI water (64 mg), IPA (96 mg), and Nafion dispersion (170 mg) with an 8 mg cm$^{-2}$ loading were added. In the homogenizer, the solution is dispersed and cast onto the carbon cloth. The anode and cathode are dried at 100 °C for 1 h in the oven. According to the suitability of the research guided by Zainoodin et al. [74], the method of preparing MEA was modified. Finally, clamping the anode, PtRu, and cathode, Pt/C, in the middle is a commercial MEA (Nafion 117 from DuPont) prepared using a heater at 135 °C and a pressure of 15 kg cm$^{-2}$ for 3 min.

### 3.6. DMFC Performance Test

In this study, the DMFC performance testing of PtRu/$TiO_2$-GA before and after optimism was studied and compared with commercial PtRu/C via a polarisation graph using a potentiostat or galvanostat while an MEA with a 4 cm$^{-2}$ active area was installed. The anode was then filled with 10.0 mL of 2.0 M methanol as fuel and tested. PtRu/$TiO_2$-

GA single-cell experiments were carried out at room temperature in passive conditions with PtRu/C.

## 4. Conclusions

In conclusion, the improved PtRu/TiO$_2$-GA electrocatalyst outperformed PtRu/TiO$_2$-GA and PtRu/C in terms of fuel cell performance, which was the major goal of this study. This research is significant since no previous study on adding modified PtRu/TiO$_2$-GA electrocatalysts for DMFC applications has been published. The input factors were chosen as the temperature of the composite (°C), the ratio of Pt to Ru, and the catalyst composition employed, and the output variables were chosen as the current density. Second-order quadratic models were obtained by CCD optimisation in RSM. These models showed a strong connection between projected and experimental outcomes. The best-optimised PtRu/TiO$_2$-GA electrocatalyst was manufactured at a TiO$_2$-GA temperature of 202 °C, a Pt/Ru ratio of 1.1:1, and a catalyst composition of 22 wt%. The PtRu/TiO$_2$-GA electrocatalyst had the best physicochemical performance, with temperatures of 200 °C, 1:1, and 22 wt% for input variables, respectively. The validation test with optimum factors yields a current density of 568.15 mA/mg$_{PtRu}$, with just a 0.6% deviation from the predicted value (564.87 mA/mg$_{PtRu}$). PtRu/TiO$_2$-GA had a maximum power density of 4.2 mW cm$^{-2}$ in the passive single DMFC test, which was 3.5 times greater than PtRu/C. The comparison of electrocatalyst performance and passive DMFC single-cell performance of the current work and other recent research is summarised in Tables 4 and 8. As a result, it was demonstrated that by optimising the PtRu/TiO$_2$-GA electrocatalyst, good performance may be attained to replace PtRu/C in DMFC applications. As a result, the unique combination of PtRu and TiO$_2$-GA 3D materials demonstrated notable activity and stability. Furthermore, this research approach is regarded as a practical, quick, and low-cost way of producing graphene aerogel-based hybrid nano-electrocatalysts, and this form of the catalyst is promising for low-temperature fuel cell applications.

**Supplementary Materials:** The following supporting information can be downloaded at: https://www.mdpi.com/article/10.3390/catal13061001/s1, Figure S1: Activity of the catalyst during the CV electrochemical test; Figure S2. Surface morphology for (a) FESEM of TiO2-GA, (b) FESEM of PtRu/TiO2-GA, (c) EDX analysis of PtRu/TiO2-GA; Figure S3. TEM images of the prepared (a) PtRu/TiO2-GA and (b) PtRu/TiO2-GA$_{Opt}$.

**Author Contributions:** Formal analysis, S.H.O.; Investigation, S.H.O.; Writing—original draft, S.H.O.; Writing—review & editing, S.K.K.; Supervision, S.K.K., S.B. and N.A.K.; Project administration, S.K.K.; Funding acquisition, S.K.K. All authors have read and agreed to the published version of the manuscript.

**Funding:** The study received financial support from the Universiti Kebangsaan Malaysia: DIP-021-028.

**Data Availability Statement:** The data will not be shared due to privacy and confidentiality for the purpose of patent filling.

**Acknowledgments:** The authors gratefully acknowledge the financial support given for this work by the Universiti Kebangsaan Malaysia under: DIP-021-028.

**Conflicts of Interest:** We confirm that the work described has not been published before, it is not under consideration for publication anywhere else, and publication has been approved by all co-authors and the responsible authorities at the institute(s) where the work was carried out. The authors declare that they have no competing interests.

## References

1. Mallakpour, S.; Zhiani, M.; Barati, A.; Rostami, H. Improving the direct methanol fuel cell performance with poly(vinyl alcohol)/titanium dioxide nanocomposites as a novel electrolyte additive. *Int. J. Hydrogen Energy* **2013**, *38*, 12418–12426. [CrossRef]
2. Basri, S.; Kamarudin, S.K.; Daud, W.R.W.; Ahmad, M.M. Non-linear optimization of passive direct methanol fuel cell (DMFC). *Int. J. Hydrogen Energy* **2010**, *35*, 1759–1768. [CrossRef]

3. Sharaf, O.Z.; Orhan, M.F. An overview of fuel cell technology: Fundamentals and applications. *Renew. Sustain. Energy Rev.* **2014**, *32*, 810–853. [CrossRef]
4. Andújar, J.M.M.; Segura, F. Fuel cells: History and updating. A walk along two centuries. *Renew. Sustain. Energy Rev.* **2009**, *13*, 2309–2322. [CrossRef]
5. Yuan, W.; Zhou, B.; Deng, J.; Tang, Y.; Zhang, Z.; Li, Z. Overview on the developments of vapor-feed direct methanol fuel cells. *Int. J. Hydrogen Energy* **2014**, *39*, 6689–6704. [CrossRef]
6. Niu, H.; Xia, C.; Huang, L.; Zaman, S.; Maiyalagan, T.; Guo, W.; You, B.; Xia, B. Rational design and synthesis of one-dimensional platinum-based nanostructures for oxygen-reduction electrocatalysis. *Chin. J. Catal.* **2022**, *43*, 1459–1472. [CrossRef]
7. Ye, F.; Liu, H.; Feng, Y.; Li, J.; Wang, X.; Yang, J. A solvent approach to the size-controllable synthesis of ultrafine Pt catalysts for methanol oxidation in direct methanol fuel cells. *Electrochim. Acta* **2014**, *117*, 480–485. [CrossRef]
8. Mehmood, A.; Ha, H.Y.H. Performance restoration of direct methanol fuel cells in long-term operation using a hydrogen evolution method. *Appl. Energy* **2014**, *114*, 164–171. [CrossRef]
9. Mekhilef, S.; Saidur, R.; Safari, A. Comparative study of different fuel cell technologies. *Renew. Sustain. Energy Rev.* **2012**, *16*, 981–989. [CrossRef]
10. Ball, M.; Wietschel, M. The future of hydrogen—Opportunities and challenges. *Int. J. Hydrogen Energy* **2009**, *34*, 615–627. [CrossRef]
11. Abdullah, N.; Kamarudin, S.K. Titanium dioxide in fuel cell technology: An overview. *J. Power Sources* **2015**, *278*, 109–118. [CrossRef]
12. Thiam, H.S.; Daud, W.R.W.; Kamarudin, S.K.; Mohamad, A.B.; Kadhum, A.A.H.; Loh, K.S.; Majlan, E.H. Nafion/Pd-SiO$_2$ nanofiber composite membranes for direct methanol fuel cell applications. *Int. J. Hydrogen Energy* **2013**, *38*, 9474–9483. [CrossRef]
13. Gustavsson, M.; Ekström, H.; Hanarp, P.; Eurenius, L.; Lindbergh, G.; Olsson, E.; Kasemo, B. Thin film Pt/TiO$_2$ catalysts for the polymer electrolyte fuel cell. *J. Power Sources* **2007**, *163*, 671–678. [CrossRef]
14. Muthuraman, N.; Guruvaiah, P.; Gnanabaskara Agneeswara, P. High performance carbon supported Pt-WO3 nanocomposite electrocatalysts for polymer electrolyte membrane fuel cell. *Mater. Chem. Phys.* **2012**, *133*, 924–931. [CrossRef]
15. Dou, M.; Hou, M.; Liang, D.; Lu, W.; Shao, Z.; Yi, B. SnO$_2$ nanocluster supported Pt catalyst with high stability for proton exchange membrane fuel cells. *Electrochim. Acta* **2013**, *92*, 468–473. [CrossRef]
16. Puigdollers, A.R.; Schlexer, P.; Tosoni, S.; Pacchioni, G. Increasing oxide reducibility: The role of metal/oxide interfaces in the formation of oxygen vacancies. *ACS Catal.* **2017**, *7*, 6493–6513. [CrossRef]
17. Yu, T.; Liu, L.; Li, L.; Yang, F. A self-biased fuel cell with TiO$_2$/g-C$_3$N$_4$ anode catalyzed alkaline pollutant degradation with light and without light—What is the degradation mechanism? *Electrochim. Acta* **2016**, *210*, 122–129. [CrossRef]
18. Shahid, M.; Zhan, Y.; Sagadeven, S.; Akermi, M.; Ahmad, W.; Hatamvand, M.; Oh, W.C. Platinum doped titanium dioxide nanocomposite an efficient platform as anode material for methanol oxidation. *J. Mater. Res. Technol.* **2021**, *15*, 6551–6561. [CrossRef]
19. Zhao, L.; Wang, Z.B.; Liu, J.; Zhang, J.J.; Sui, X.L.; Zhang, L.M.; Gu, D.M. Facile one-pot synthesis of Pt/graphene-TiO$_2$ hybrid catalyst with enhanced methanol electrooxidation performance. *J. Power Sources* **2015**, *279*, 210–217. [CrossRef]
20. Xia, B.Y.; Wu, H.B.; Chen, J.S.; Wang, Z.; Wang, X.; Lou, X.W. Formation of Pt–TiO$_2$–rGO 3-phase junctions with significantly enhanced electro-activity for methanol oxidation. *Phys. Chem. Chem. Phys.* **2011**, *14*, 473–476. [CrossRef]
21. Yan, D.J.; Zhu, X.D.; Mao, Y.C.; Qiu, S.Y.; Gu, L.L.; Feng, Y.J.; Sun, K.N. Hierarchically organized CNT@TiO$_2$@Mn3O4 nanostructures for enhanced lithium storage performance. *J. Mater. Chem. A* **2017**, *5*, 17048–17055. [CrossRef]
22. Xia, T.; Zhang, W.; Wang, Z.; Zhang, Y.; Song, X.; Murowchick, J.; Battaglia, V.; Liu, G.; Chen, X. Amorphous carbon-coated TiO$_2$ nanocrystals for improved lithium-ion battery and photocatalytic performance. *Nano Energy* **2014**, *6*, 109–118. [CrossRef]
23. Chen, Y.; Li, Z.; Shi, S.; Song, C.; Jiang, Z.; Cui, X. Scalable synthesis of TiO$_2$ crystallites embedded in bread-derived carbon matrix with enhanced lithium storage performance. *J. Mater. Sci. Mater. Electron.* **2017**, *28*, 9206–9220. [CrossRef]
24. Mo, R.; Lei, Z.; Sun, K.; Rooney, D. Facile Synthesis of Anatase TiO$_2$ Quantum-Dot/Graphene-Nanosheet Composites with Enhanced Electrochemical Performance for Lithium-Ion Batteries. *Adv. Mater.* **2014**, *26*, 2084–2088. [CrossRef]
25. Wang, Z.; Sha, J.; Liu, E.; He, C.; Shi, C.; Li, J.; Zhao, N. A large ultrathin anatase TiO$_2$ nanosheet/reduced graphene oxide composite with enhanced lithium storage capability. *J. Mater. Chem. A* **2014**, *2*, 8893–8901. [CrossRef]
26. Song, T.; Han, H.; Choi, H.; Lee, J.W.; Park, H.; Lee, S.; Park, W.; Kim, S.; Liu, L.; Paik, U. TiO$_2$ nanotube branched tree on a carbon nanofiber nanostructure as an anode for high energy and power lithium ion batteries. *Nano Res.* **2015**, *7*, 491–501. [CrossRef]
27. Li, X.; Chen, Y.; Zhou, L.; Mai, Y.W.; Huang, H. Exceptional electrochemical performance of porous TiO$_2$–carbon nanofibers for lithium ion battery anodes. *J. Mater. Chem. A* **2014**, *2*, 3875–3880. [CrossRef]
28. Wang, B.; Xin, H.; Li, X.; Cheng, J.; Yang, G.; Nie, F. Mesoporous CNT@TiO$_2$-C Nanocable with Extremely Durable High Rate Capability for Lithium-Ion Battery Anodes. *Sci. Rep.* **2014**, *4*, 3729. [CrossRef]
29. Ding, H.; Zhang, Q.; Liu, Z.; Wang, J.; Ma, R.; Fan, L.; Wang, T.; Zhao, J.; Lu, X.; Yu, X.; et al. TiO$_2$ quantum dots decorated multi-walled carbon nanotubes as the multifunctional separator for highly stable lithium sulfur batteries. *Electrochim. Acta* **2018**, *284*, 314–320. [CrossRef]
30. Basri, S.; Kamarudin, S.K.; Daud, W.R.W.W.; Yaakob, Z.; Kadhum, A.A.H.H. Novel anode catalyst for direct methanol fuel cells. *Sci. World J.* **2014**, *2014*, 547604. [CrossRef]

31. Kim, H.J.; Kim, D.Y.; Han, H.; Shul, Y.G. PtRu/C-Au/TiO₂ electrocatalyst for a direct methanol fuel cell. *J. Power Sources* **2006**, *159*, 484–490. [CrossRef]
32. Abdullah, N.; Kamarudin, S.K.; Shyuan, L.K.; Karim, N.A. Synthesis and optimization of PtRu/TiO₂-CNF anodic catalyst for direct methanol fuel cell. *Int. J. Hydrogen Energy* **2019**, *44*, 30543–30552. [CrossRef]
33. Abdullah, N.; Saidur, R.; Zainoodin, A.M.; Aslfattahi, N. Optimization of electrocatalyst performance of platinum–ruthenium induced with MXene by response surface methodology for clean energy application. *J. Clean. Prod.* **2020**, *277*, 123395. [CrossRef]
34. Ahmad, M.M.M.; Kamarudin, S.K.K.; Daud, W.R.W.R.W.; Yaakub, Z. High power passive μDMFC with low catalyst loading for small power generation. *Energy Convers. Manag.* **2010**, *51*, 821–825. [CrossRef]
35. Shaari, N.; Kamarudin, S.K. ScienceDirect Performance of crosslinked sodium alginate / sulfonated graphene oxide as polymer electrolyte membrane in DMFC application : RSM optimization approach. *Int. J. Hydrogen Energy* **2018**, *43*, 22986–23003. [CrossRef]
36. You, P.Y.; Kamarudin, S.K.; Masdar, M.S.; Zainoodin, A.M. Modification and optimization of rice husk ash bio-filler in sulfonated polyimide membrane for direct methanol fuel cell. *Sains Malays.* **2020**, *49*, 3125–3143. [CrossRef]
37. Bezerra, M.A.; Santelli, R.E.; Oliveira, E.P.; Villar, L.S.; Escaleira, L.A. Response surface methodology (RSM) as a tool for optimization in analytical chemistry. *Talanta* **2008**, *76*, 965–977. [CrossRef] [PubMed]
38. Fletcher, R. *Practical Methods of Optimization*; John Wiley & Sons: Hoboken, NJ, USA, 2000; pp. 1–436. [CrossRef]
39. Asfaram, A.; Ghaedi, M.; Agarwal, S.; Tyagi, I.; Gupta, V.K. Removal of basic dye Auramine-O by ZnS:Cu nanoparticles loaded on activated carbon: Optimization of parameters using response surface methodology with central composite design. *RSC Adv.* **2015**, *5*, 18438–18450. [CrossRef]
40. Dharma, S.; Masjuki, H.H.; Ong, H.C.; Sebayang, A.H.; Silitonga, A.S.; Kusumo, F.; Mahlia, T.M.I. Optimization of biodiesel production process for mixed Jatropha curcas–Ceiba pentandra biodiesel using response surface methodology. *Energy Convers. Manag.* **2016**, *115*, 178–190. [CrossRef]
41. Danmaliki, G.I.; Saleh, T.A.; Shamsuddeen, A.A. Response surface methodology optimization of adsorptive desulfurization on nickel/activated carbon. *Chem. Eng. J.* **2017**, *313*, 993–1003. [CrossRef]
42. Sulaiman, N.S.; Hashim, R.; Mohamad Amini, M.H.; Danish, M.; Sulaiman, O. Optimization of activated carbon preparation from cassava stem using response surface methodology on surface area and yield. *J. Clean. Prod.* **2018**, *198*, 1422–1430. [CrossRef]
43. Caponi, N.; Collazzo, G.C.; Da Silveira Salla, J.; Jahn, S.L.; Dotto, G.L.; Foletto, E.L. Optimisation of crystal violet removal onto raw kaolin using response surface methodology. *Int. J. Environ. Technol. Manag.* **2019**, *22*, 85–100. [CrossRef]
44. Myers, R.H.; Montgomery, D.C.; Anderson-Cook, C.M. *Response Surface Methodology: Process and Product Optimization Using Designed Experiments*, 4th ed.; John Wiley & Sons: Hoboken, NJ, USA, 2016.
45. Khatti, T.; Naderi-Manesh, H.; Kalantar, S.M. Application of ANN and RSM techniques for modeling electrospinning process of polycaprolactone. *Neural Comput. Appl.* **2019**, *31*, 239–248. [CrossRef]
46. Deivaraj, T.C.; Lee, J.Y. Preparation of carbon-supported PtRu nanoparticles for direct methanol fuel cell applications—A comparative study. *J. Power Sources* **2005**, *142*, 43–49. [CrossRef]
47. Neațu, Ș.; Neațu, F.; Chirica, I.M.; Borbáth, I.; Tálas, E.; Tompos, A.; Somacescu, S.; Osiceanu, P.; Folgado, M.A.; Chaparro, A.M.; et al. Recent progress in electrocatalysts and electrodes for portable fuel cells. *J. Mater. Chem. A* **2021**, *9*, 17065–17128. [CrossRef]
48. Wakisaka, M.; Mitsui, S.; Hirose, Y.; Kawashima, K.; Uchida, H.; Watanabe, M. Electronic structures of Pt-Co and Pt-Ru alloys for CO-tolerant anode catalysts in polymer electrolyte fuel cells studied by EC-XPS. *J. Phys. Chem. B* **2006**, *110*, 23489–23496. [CrossRef]
49. Li, Y.; Zhang, Y.; Qian, K.; Huang, W. Metal-Support Interactions in Metal/Oxide Catalysts and Oxide-Metal Interactions in Oxide/Metal Inverse Catalysts. *ACS Catal.* **2022**, *12*, 1268–1287. [CrossRef]
50. Yi, Q.; Chen, A.; Huang, W.; Zhang, J.; Liu, X.; Xu, G.; Zhou, Z. Titanium-supported nanoporous bimetallic Pt-Ir electrocatalysts for formic acid oxidation. *Electrochem. Commun.* **2007**, *9*, 1513–1518. [CrossRef]
51. Kolla, P.; Smirnova, A. Methanol oxidation on hybrid catalysts: PtRu/C nanostructures promoted with cerium and titanium oxides. *Int. J. Hydrogen Energy* **2013**, *38*, 15152–15159. [CrossRef]
52. Ito, Y.; Takeuchi, T.; Tsujiguchi, T.; Abdelkareem, M.A.; Nakagawa, N. Ultrahigh methanol electro-oxidation activity of PtRu nanoparticles prepared on TiO₂-embedded carbon nanofiber support. *J. Power Sources* **2013**, *242*, 280–288. [CrossRef]
53. Chen, W.; Wei, X.; Zhang, Y. A comparative study of tungsten-modified PtRu electrocatalysts for methanol oxidation. *Int. J. Hydrogen Energy* **2014**, *39*, 6995–7003. [CrossRef]
54. Tsukagoshi, Y.; Ishitobi, H.; Nakagawa, N. Improved performance of direct methanol fuel cells with the porous catalyst layer using highly-active nanofiber catalyst. *Carbon Resour. Convers.* **2018**, *1*, 61–72. [CrossRef]
55. Yang, Z.; Choi, D.; Kerisit, S.; Rosso, K.M.; Wang, D.; Zhang, J.; Graff, G.; Liu, J. Nanostructures and lithium electrochemical reactivity of lithium titanites and titanium oxides: A review. *J. Power Sources* **2009**, *192*, 588–598. [CrossRef]
56. Antolini, E.; Cardellini, F. Formation of carbon supported PtRu alloys: An XRD analysis. *J. Alloys Compd.* **2001**, *315*, 118–122. [CrossRef]
57. Nassr, A.B.A.A.; Bron, M. Microwave-Assisted Ethanol Reduction as a New Method for the Preparation of Highly Active and Stable CNT-Supported PtRu Electrocatalysts for Methanol Oxidation. *ChemCatChem* **2013**, *5*, 1472–1480. [CrossRef]

58. Cordero-Borboa, A.E.; Sterling-Black, E.; Gómez-Cortés, A.; Vázquez-Zavala, A. X-ray diffraction evidence of the single solid solution character of bi-metallic Pt-Pd catalyst particles on an amorphous $SiO_2$ substrate. *Appl. Surf. Sci.* **2003**, *220*, 169–174. [CrossRef]
59. Argyle, M.D.; Bartholomew, C.H. Heterogeneous Catalyst Deactivation and Regeneration: A Review. *Catalysts* **2015**, *5*, 145–269. [CrossRef]
60. Alothman, Z.A. A Review: Fundamental Aspects of Silicate Mesoporous Materials. *Materials* **2012**, *5*, 2874–2902. [CrossRef]
61. Pal, N.; Bhaumik, A. Mesoporous materials: Versatile supports in heterogeneous catalysis for liquid phase catalytic transformations. *RSC Adv.* **2015**, *5*, 24363–24391. [CrossRef]
62. Bagheri, S.; Muhd Julkapli, N.; Bee Abd Hamid, S. Titanium dioxide as a catalyst support in heterogeneous catalysis. *Sci. World J.* **2014**, *2014*, 727496. [CrossRef]
63. Wang, J.; Fu, C.; Wang, X.; Yao, Y.; Sun, M.; Wang, L.; Liu, T. Three-dimensional hierarchical porous $TiO_2$/graphene aerogels as promising anchoring materials for lithium–sulfur batteries. *Electrochim. Acta* **2018**, *292*, 568–574. [CrossRef]
64. Ashutosh Tiwari, M.S. *Graphene Materials: Fundamentals and Emerging Applications*; Wiley: Hoboken, NJ, USA, 2015; pp. 1–424. Available online: https://books.google.com.my/books?hl=en&lr=&id=0up8CAAAQBAJ&oi=fnd&pg=PP1&dq=Tiwari+A.;+Syväjärvi+M.;+(2015)+Graphene+materials:+fundamentals+and+emerging+applications.+Hoboken,+Wiley.&ots=GixKmanSs8&sig=G9K3kfMbBnXN-zApXdOeQQ4JwXw#v=onepage&q&f=false (accessed on 17 May 2022).
65. Cooper, K. *Situ PEM FC Electrochemical Surface Area And Catalyst Utilization Measurement*; Scribner Associates Inc.: Southern Pines, NC, USA, 2009.
66. Reghunath, S.; Pinheiro, D.; KR, S.D. A review of hierarchical nanostructures of $TiO_2$: Advances and applications. *Appl. Surf. Sci. Adv.* **2021**, *3*, 100063. [CrossRef]
67. Ali, H.; Zaman, S.; Majeed, I.; Kanodarwala, F.K.; Nadeem, M.A.; Stride, J.A.; Nadeem, M.A. Porous Carbon/rGO Composite: An Ideal Support Material of Highly Efficient Palladium Electrocatalysts for the Formic Acid Oxidation Reaction. *ChemElectroChem* **2017**, *4*, 3126–3133. [CrossRef]
68. Zaman, S.; Wang, M.; Liu, H.; Sun, F.; Yu, Y.; Shui, J.; Chen, M.; Wang, H. Carbon-based catalyst supports for oxygen reduction in proton-exchange membrane fuel cells. *Trends Chem.* **2022**, *4*, 886–906. [CrossRef]
69. Abdullah, N.; Kamarudin, S.K.; Shyuan, L.K.; Karim, N.A. Fabrication and Characterization of New Composite Tio2 Carbon Nanofiber Anodic Catalyst Support for Direct Methanol Fuel Cell via Electrospinning Method. *Nanoscale Res. Lett.* **2017**, *12*, 613. [CrossRef] [PubMed]
70. Shimizu, T.; Momma, T.; Mohamedi, M.; Osaka, T.; Sarangapani, S. Design and fabrication of pumpless small direct methanol fuel cells for portable applications. *J. Power Sources* **2004**, *137*, 277–283. [CrossRef]
71. Hashim, N.; Kamarudin, S.K.; Daud, W.R.W. Design, fabrication and testing of a PMMA-based passive single-cell and a multi-cell stack micro-DMFC. *Int. J. Hydrog. Energy* **2009**, *34*, 8263–8269. [CrossRef]
72. Jamal, P.; Wan Nawawi, W.M.F.; Alam, M.Z. Optimum medium components for biosurfactant production by Klebsiella pneumoniae WMF02 utilizing sludge palm oil as a substrate. *Aust. J. Basic Appl. Sci.* **2012**, *6*, 100–108.
73. Hasran, U.A.; Kamarudin, S.K.; Daud, W.R.W.; Majlis, B.Y.; Mohamad, A.B.; Kadhum, A.A.H.; Ahmad, M.M. Optimization of hot pressing parameters in membrane electrode assembly fabrication by response surface method. *Int. J. Hydrogen Energy* **2013**, *38*, 9484–9493. [CrossRef]
74. Zainoodin, A.M.; Kamarudin, S.K.; Masdar, M.S.; Daud, W.R.W.; Mohamad, A.B.; Sahari, J. High power direct methanol fuel cell with a porous carbon nanofiber anode layer. *Appl. Energy* **2014**, *113*, 946–954. [CrossRef]

**Disclaimer/Publisher's Note:** The statements, opinions and data contained in all publications are solely those of the individual author(s) and contributor(s) and not of MDPI and/or the editor(s). MDPI and/or the editor(s) disclaim responsibility for any injury to people or property resulting from any ideas, methods, instructions or products referred to in the content.

*Article*

# Three-Dimensional Graphene Aerogel Supported on Efficient Anode Electrocatalyst for Methanol Electrooxidation in Acid Media

Siti Hasanah Osman [1], Siti Kartom Kamarudin [1,2,*], Sahriah Basri [1] and Nabilah A. Karim [1]

[1] Fuel Cell Institute, Universiti Kebangsaan Malaysia, Bangi 43600 UKM, Selangor, Malaysia; hasanahosman17@gmail.com (S.H.O.); sahriah@ukm.edu.my (S.B.); nabila.akarim@ukm.edu.my (N.A.K.)
[2] Department of Chemical and Process Engineering, Faculty of Engineering and Built Environment, Universiti Kebangsaan Malaysia, Bangi 43600 UKM, Selangor, Malaysia
* Correspondence: ctie@ukm.edu.my

**Abstract:** This work attempted to improve the catalytic performance of an anodic catalyst for use in direct methanol fuel cells by coating graphene aerogel (GA) with platinum nanoparticles. A hydrothermal, freeze-drying, and microwave reduction method were used to load Pt–Ru bimetallic nanoparticles onto a graphene aerogel. The mesoporous structure of a graphene aerogel is expected to enhance the mass transfer in an electrode. XRD, Raman spectroscopy, SEM, and TEM described the as-synthesized PtRu/GA. Compared to commercial PtRu/C with the same loading (20%), the electrocatalytic performance of PtRu/GA presents superior stability in the methanol oxidation reaction. Furthermore, PtRu/GA offers an electrochemical surface area of 38.49 $m^2g^{-1}$, with a maximal mass activity/specific activity towards methanol oxidation of 219.78 $mAmg^{-1}$/0.287 $mAcm^{-2}$, which is higher than that of commercial PtRu/C, 73.11 $mAmg^{-1}$/0.187 $mAcm^{-2}$. Thus, the enhanced electrocatalytic performance of PtRu/GA for methanol oxidation proved that GA has excellent potential to improve the performance of Pt catalysts and tolerance towards CO poisoning.

**Keywords:** graphene aerogel; hydrothermal; freeze-drying; catalyst; anode; direct methanol fuel cell

## 1. Introduction

Fuel cells have been predicted to play a vital role in the hydrogen economy, as they offer significantly better energy efficiency with zero or minimal greenhouse gas emissions [1,2]. Direct methanol fuel cells (DMFCs) are a potential type of fuel cell for mobile applications due to their simple system fabrication and easy liquid fuel storage [3,4]. However, cost and slow methanol oxidation reaction limit the commercialization of DMFCs. Fuel cells and internal combustion engines can run on the liquid alcohol known as methanol. Methanol offers various benefits beyond other fuels, such as a high energy density and the capacity to be made from renewable sources. One problem is methanol crossover, which is the diffusion of methanol across the membrane in a direct methanol fuel cell (DMFC) [5,6]. Therefore, intensive research has been carried out focusing on Pt-based catalysts to improve the catalyst performance using minimal Pt metal [7,8]. Currently, the best catalyst for DMFCs is bimetallic platinum ruthenium (PtRu). Although bimetallic PtRu has better tolerance for CO than pure Pt, the electrochemical activity of bimetallic PtRu remains low [7].

Ceramic and carbon materials are commonly used to support catalysts in chemical processes, but each has unique properties and performance. Ceramic materials like alumina, silica, and zirconia are sturdy, heat-resistant, and chemically inert. These characteristics make them perfect for high-temperature applications like catalytic converters in cars and offer plenty of space for catalyst deposition. Mahmood and his team [9] examined how methanol oxidation behaved on thin films made of nickel oxide and composites of nickel oxide with zirconium and yttrium oxides. They used a simple dip-coating technique to make NiO-ZrO$_2$/FTO and NiO-Y$_2$O$_3$/FTO thin films. The results showed that these thin films were more efficient than commercial electrodes, with only a small loss in efficiency. This suggests that adding ZrO$_2$ and Y$_2$O$_3$ to NiO reduces the risk of CO poisoning, which is often seen in commercial electrodes. Liaqat et al. [10] fabricated thin films of Nickel Oxide incorporated with the metal Copper (Cu) and Chromium (Cr) for the electrochemical oxidation of methanol. The films, NiO−CuO and Ni$_{0.95}$Cr$_{0.05}$O$_2$+δ, were created using a simple dip-coating method, and their efficiency in methanol oxidation was compared to pure NiO and CuO films. Results showed that the NiO−CuO and Ni$_{0.95}$Cr$_{0.05}$O$_2$ + δ thin films exhibited significantly higher efficiency in methanol oxidation than the pure NiO and CuO films. However, the addition of Cr$_2$O$_3$ was found to be ineffective in methanol electro-oxidation. However, ceramics are brittle and may break when subjected to stress.

Recently, researchers have focused on fabricating new carbon materials such as activated carbon, graphene, and carbon nanotubes are conductive, heat-resistant, and have a vast surface area, making them suitable for electrochemical reactions and high-temperature reactions an effective approach to improving catalytic activity while reducing the need for Pt-based catalysts [11,12]. Carbon supports are lightweight and flexible, making them less prone to mechanical failure than ceramics. However, carbon materials can interact with the catalyst or reaction products, affecting the catalyst's overall performance. Nguyen et al. [13] discovered a highly conductive surface and achieved excellent performance using carbon black as a catalytic support for Pt. Moreover, based on their superb electrical conductivity, porous structure, and high surface area, various carbon materials, such as carbon nanotubes (CNTs), carbon black (CB), and mesoporous carbon, have been broadly investigated as support materials for electrocatalysts to improve the catalytic activity, stability, and performance of DMFCs. Subsequently, the discovery of graphene as single layers of carbon atoms by Geim et al. in 2004 sparked the curiosity of scientists worldwide [14,15]. Graphene nanosheets have consistent physical and chemical properties, virtuous electrical conductivity, and a high surface area [16].

In addition, the catalytic activity can be increased by reducing the surface area of densely adherent graphene by including distinct graphene sheets [17]. Other carbon compounds can be used as spacers to bond with the sheets and reduce accumulation to give graphene as separate sheets [18,19]. The 3D porous architectures and excellent inherent characteristics of graphene and three-dimensional (3D) graphene aerogels (GAs) have recently gained much attention. 3D GAs have special features, including a large surface area, multidimensional electron transport, and porous structure, making them even more desirable as catalytic supports when compared to graphene [20,21]. GA preserves the inherent features of 2D graphene sheets while exhibiting certain desirable functionalities with increased performance. Due to the potential to eliminate CO intermediate species that develop throughout methanol electrooxidation, PtRu alloy catalysts are frequently employed as DMFC anodes within Pt-based bimetallic electrocatalysts [22]. Dong et al. [20] presented PtRu/graphene for fuel cell applications with improved performance, but more improvement is needed to achieve uniform PtRu particle sizes.

In many situations, GA support is a desirable material in DMFCs [23], supercapacitors [24], Li-ion batteries [25], and other applications [26,27] that utilize this type of material. However, no work has been presented for bimetallic platinum-ruthenium (PtRu) with a graphene aerogel support as an electrocatalyst in methanol oxidation. Therefore, this study aims to investigate the effectiveness of PtRu nanoparticles on GA. Moreover, aerogels are one of the ten most advanced structures used today and have a specific purpose in industry or research, contributing to this study's novelty. Morphological and structural properties of the materials were obtained from various analytical techniques, including X-ray diffraction (XRD), Raman spectroscopy, field emission scanning electron microscopy (FESEM), and transmission electron microscopy were used to study the physical properties of the catalysts (TEM). In addition, cyclic voltammetry (CV) and chronoamperometric (CA) tests were used to study the activity and stability of the catalysts for the electrooxidation of methanol. Finally, this work portrays a significant improvement of anodic catalysts for application in DMFCs.

## 2. Results and Discussion

*2.1. Characterization of Materials*

The XRD analysis of PtRu/GA, GA and GO is shown in Figure 1a. The dominant peak for GA and the XRD pattern with a gap between layers of 3.42 Å was found at 26°. This value is greater than that for the graphite plane, i.e., 3.34 Å, but much smaller than the value of 8.7 Å obtained from the GO peak of 10.6° in 2θ. By removing several oxygen-containing functional groups from GO, a decrease in the distance between layers from GO to GA can be observed. In addition, GA has a broader gap between layers than graphite, indicating residual functional groups on its surface.

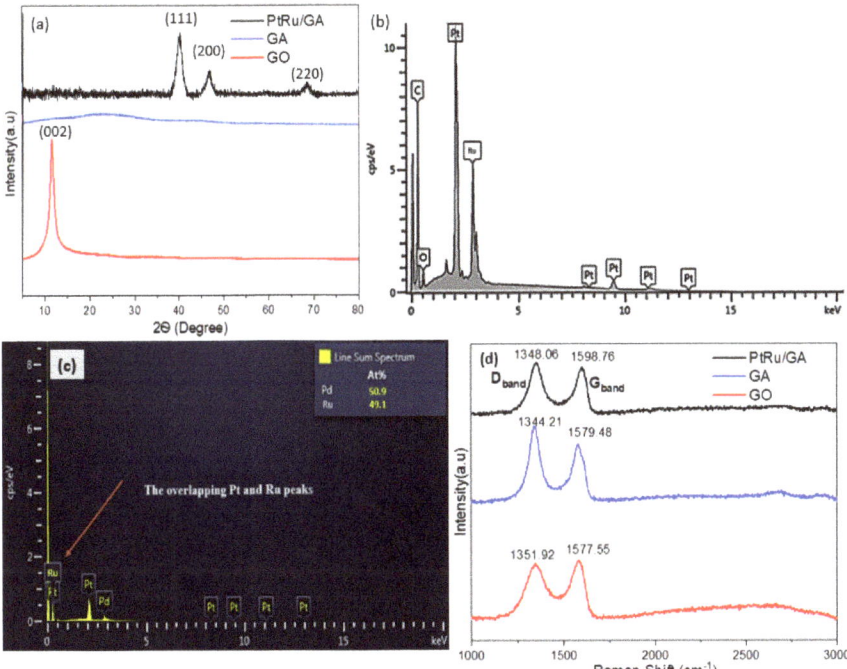

**Figure 1.** (a) XRD patterns and (b) typical EDX spectrum of PtRu/GA nanocomposite catalyst (c) the EDX spectrum of individual Pt-Ru alloy nanoparticle (d) Raman spectra of GO, GA, and PtRu/GA.

GA has a broad peak, and the baseline is not obstructed. This suggests that GA contains a significant amount of amorphous structure [27]. The PtRu/GA XRD patterns exhibit Pt diffraction peaks at 39.7° (111), 46.2° (200), and 67.5° (220). The spectrum for the catalyst shows Ru diffraction peaks at 40.7° (111), 47° (200), and 69° (220). In this case, we assume that the PtRu in the crystal structure has a face-centred cubic (fcc) structure. GA does not appear to result in any noticeable changes in the crystallization of PtRu nanoparticles [28]. Furthermore, this difference suggests that graphene nanosheets are closely packed, whereas PtRu/GA without stacked graphene nanosheets preserves its three-dimensional structure.

Based on the XRD pattern, it can be concluded that the PtRu/GA nanocomposite catalyst prepared in this study has an FCC structure, which is consistent with the SAED pattern shown in Figure 2f. This indicates that the Pt and Ru precursors were effectively loaded onto the GA without impurities. The diffraction peaks from the PtRu/GA composite are located between the positions of pure FCC-structured monometallic Pt (JCPDS-04-0802) and Ru (JCPDS-46-1043), suggesting the formation of Pt-Ru alloy nanoparticles [29]. The (111) plane dominates the orientation, as evidenced by the most intense peak. Moreover, the width of the diffraction peaks indicates the average size of the nanoparticles, with broader peaks indicating smaller particles. The intensity of the (111) orientation suggests that it was the dominant orientation during the growth process, likely due to a high concentration of atoms and the strong affinity of Pt atoms for that plane. This orientation is highly reactive and has the same d-spacing. The strong and narrow peak indicates that the particles have a high level of crystallinity.

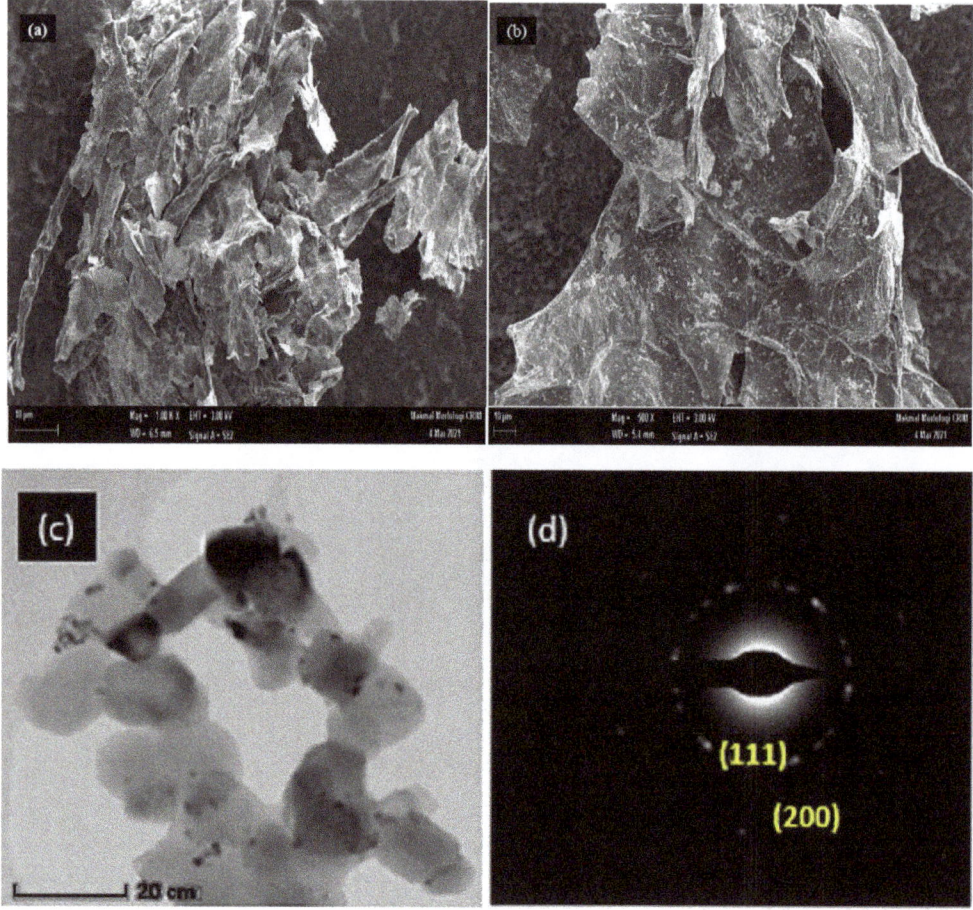

**Figure 2.** *Cont.*

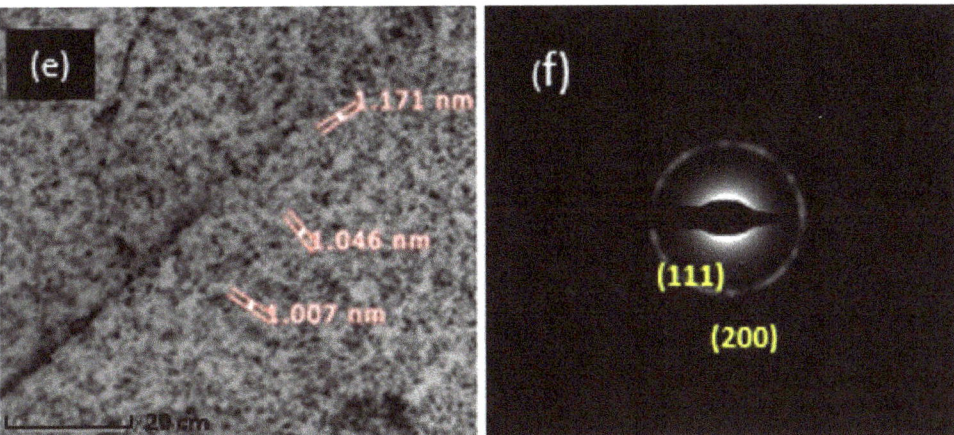

**Figure 2.** FESEM images (**a**) PtRu/GO, (**b**) PtRu/GA, TEM image (**c**) PtRu/GO, (**e**) PtRu/GA SAED pattern (**d**) PtRu/GO, (**f**) PtRu/GA.

Additionally, no impurity peaks were detected, confirming the formation of pure crystalline platinum. Figure 1b displays the EDX spectrum of the PtRu/GA nanocomposite catalyst, indicating the presence of Pt, Ru, C, and O elements. The appearance of characteristic peaks for these elements indicates the successful formation and deposition of Pt-Ru alloy nanoparticles onto GA. The C peak corresponds to GA, while the O peak results from the remaining oxygen-containing functional groups in GA due to incomplete chemical reduction from GO to GA. The EDX spectrum of the Pt-Ru alloy Figure 1c reveals overlapping peaks for Pt and Ru, providing strong evidence of the formation of a Pt-Ru alloy structure. The respective atomic ratios of the elements are known or taken into consideration. Therefore, EDX can be used to determine the composition of a Pt-Ru alloy. This can be accomplished by contrasting the alloy's X-ray emissions with those of a fictitious alloy with a known or assumed composition. However, these estimates might not be extremely precise or accurate because of EDX's limitations. Therefore, the use of a more quantitative approach, such as ICP-AES, is advised for atomic ratios that are more exact. Additionally, the atomic ratio of Pt to Ru, assumptions by their EDX line scanning profiles, is approximately 51:49, which closely matches the theoretical composition of a 1:1 ratio. This further confirms the successful formation of the Pt-Ru alloy.

Interestingly, analysis of the Raman spectrum enables the identification of the carbon species in the sample, which can be used to investigate further the electrocatalysts produced. The structural change in GO, which occurs during the hydrothermal process, was studied by Raman spectroscopy. Figure 1d illustrates the bands observed at approximately 1344–1351 cm$^{-1}$ and 1577–1598 cm$^{-1}$, corresponding to the D and G bands for GO, GA, and PtRu/GA to the existence of poly-crystalline graphite [30]. The G band generally corresponds to the vibrational modes of the sp2 hybridized carbon atoms in the graphite layer. At the same time, the large D-band peaks indicate the low crystallinity of the GA support. Analysis of the Raman spectrum also shows that the higher the value of $I_D/I_G$, the more defects are present in the graphite. The $I_D/I_G$ intensity ratio for the bands was calculated from the Raman spectra: PtRu/GA (1.02), GA (1.10), and GO (0.94). The degree of graphitization can be calculated from the relative intensity ratio of the D and G bands ($I_D/I_G$), which varies depending on the type of graphite material. The higher the ($I_D/I_G$) value, the more flaws there are in graphite. However, the $I_D/I_G$ value for the PtRu/GA electrocatalyst was only slightly increased in this instance. Graphite in the sample has very little flaw or disorder, which is meant [31]. The PtRu metal addition, which may have slightly altered the GA structure, can also be linked to this issue. The intensity ratio for the $I_D/I_G$

bands for GO (0.94) is increased compared to GA (1.10), which is similar to a previous description [32]. This increase reveals that the average sp2 domain size is decreased during the synthesis process, which is most likely due to the production of a significant number of new smaller graphitic domains [33]. Furthermore, even after the catalyst is disseminated on the GA support, the resultant ($I_D/I_G$) ratio is similar to that obtained for the electrocatalyst samples in this work. Consequently, the carbon layers in these electrocatalyst samples in the disordered and structured portions of the GA structure are not significantly different.

Figure 2a,b, FESEM shows that the electrocatalyst has a well-developed 3D link porosity network, consistent with hydrothermally produced 3D graphene [34]. In addition, the pore structure of the generated samples will facilitate the mass transit of reactants and products [35]. The graphene sheets are stacked in Figure 2a, and this discovery aligns with the findings shown in Figure 1a. As shown in Figure 2b, a porous 3D structure is maintained on the PtRu/GA electrocatalyst, which allows reactant transport. At the same time, a negative influence on the surface of the platinum particles is limited by the structure of the graphene sheets. Platinum particles with a size of 1–3 nm are uniformly distributed on a graphene nanosheet, as shown in Figure 2c,e. Based on Figure 2f, the particles have the same size as the PtRu/GA particles, which indicates that GA and GO have similarities in terms of the absorption and scattering ability of the particles. The selected area electron diffraction (SAED) pattern in Figure 2d,f clearly shows the face-centred cubic (FCC) or crystalline structure of the as-prepared Pt-Ru alloy nanoparticles. The (111), (200) and (220) planes of the crystalline FCC structure are attributed to the three concentric rings and dots formed in the SAED evaluation from the inside to the outside. The fabrication result for PtRu/GA provides a concentric ring light spot, indicating that the resulting sample has a highly crystalline structure. These results are consistent with the XRD data in Figure 1a, which show that PtRu crystals have crystal defects and multiple domains.

## 2.2. Electrochemical Evaluation

The potential and effectiveness of all catalysts were electrochemically characterized and determined to be anodic catalysts in DMFCs. This segment includes two main measurements: cyclic voltammetry (CV) and chronoamperometry (CA), which analyse electrocatalytic performance. This was carried out to determine how stable and durable the samples are in the long term. Figure 3a depicts the CV profiles measured for all catalysts in a 0.5 M $H_2SO_4$ solution with potentials ranging from −0.2 V to 1.0 V. The hydrogen adsorption/desorption area is also reported on the scale of −0.2 V to 0.1 V, which was used to determine the electrochemical active surface area (ECSA). The ECSA is a method for calculating the surface area of PtRu nanoparticles in an electrocatalyst [36]. Charge transfer reactions at the activation sites are adsorption limited; therefore, this approach requires a cycle of electrode current in the voltage range. ECSA employs the full charge necessary for monolayer adsorption and desorption as reactive surface sites [37].

In general, the larger ECSA of a catalyst usually possesses higher electro-catalytic activity. The methanol oxidation reaction (MOR) performance of anode catalyst is a significant factor influencing DMFC's efficiency. The common methanol electrooxidation reaction steps with Pt-based alloy catalyst are given in Equations (1)–(3) as shown below [38]:

$$PtRu + CH_3OH \rightarrow PtRu\text{-}CO_{ad} + 4H^+ + 4e^- \quad (1)$$

$$PtRu + H_2O \rightarrow PtRu\text{-}OH_{ad} + H^+ + e^- \quad (2)$$

$$PtRu\text{-}CO_{ad} + Pt\text{-}Ru\text{-}OH_{ad} \rightarrow CO_2 + 2PtRu + H^+ + e^- \quad (3)$$

From Equation (1), the intermediate species of carbon monoxide (CO) is produced during MOR and then further adsorbed onto the surface of the Pt-Ru catalyst (PtRu-$CO_{ad}$). The produced $CO_{ad}$ block the surface of the Pt-Ru catalyst, thereby suppressing the continuous MOR. The PtRu-$CO_{ad}$ can be oxidized by the hydroxyl group (OH) to form carbon dioxide ($CO_2$). GA acts as Pt-PRu catalyst support which can prevent the agglomeration of the nanoparticles during MOR and enhance the durability of the catalyst. The calculated ECSA for the CV observation is shown in Table 1. The ECSAs for PtRu/GA, PtRu/GO, and PtRu/C were computed to be approximately 38.49 $m^2g^{-1}$, 20.44 $m^2g^{-1}$, and 19.65 $m^2g^{-1}$, respectively. PtRu/GA has an ECSA of approximately two times that of PtRu/C. This finding demonstrates that preserving a 3D structure while raising the ECSA for PtRu nanoparticles and improving catalyst utilization is possible with GA.

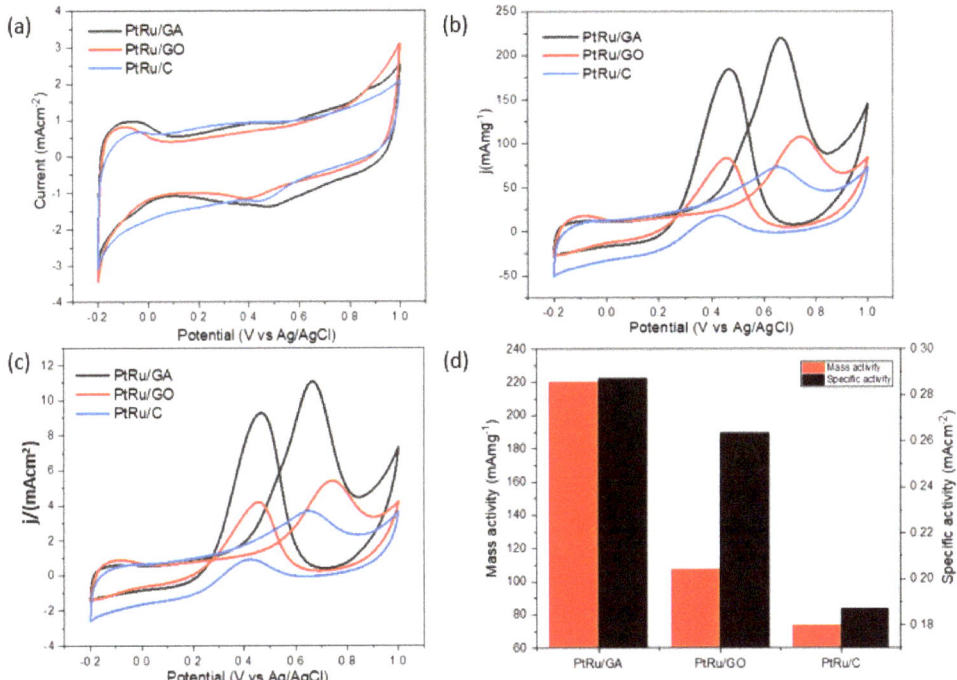

**Figure 3.** Represent (**a**) the $H_2$ absorption and desorption of PtRu catalyst in 0.5 M $H_2SO_4$, (**b**,**c**) CV curves in 0.5 M $H_2SO_4$ and 1 M $CH_3OH$ aqueous at 50 mVs$^{-1}$ and (**d**) histogram of mass activity and specific activity for electrocatalyst.

**Table 1.** Comparison of the current density results with the different electrocatalysts.

| Eletrocatalyst | ECSA ($m^2$/$g_{PtRu}$) | Peak Potential (V vs. Ag/AgCl) | Onset Potential (V vs. Ag/AgCl) | Mass Activity (mA/mg$_{PtRu}$) | Specific Activity (mA/$cm^2_{PtRu}$) | CO Tolerance $I_f/I_b$ Ratio |
|---|---|---|---|---|---|---|
| PtRu/GA | 38.49 | 0.67 | 0.205 | 219.78 | 0.287 | 1.19 |
| PtRu/GO | 20.44 | 0.739 | 0.211 | 107.06 | 0.263 | 1.29 |
| PtRu/C | 19.65 | 0.649 | 0.387 | 73.11 | 0.187 | 3.94 |

CV was used to examine the catalytic activity of the synthesized electrocatalyst and other electrocatalysts, as shown in Figure 3b. The CV curves obtained for the PtRu/GA, PtRu/GO, and PtRu/C electrocatalysts were evaluated in 0.5 M $H_2SO_4$ with 1 M $CH_3OH$ with saturated $N_2$ gas at room temperature. The potential range for the several curves is −0.2 to 1.0 V versus Ag/AgCl. The data show well-known methanol oxidation characteristics, with identifiable anode peaks at 0.60 V for all catalysts. For the oxidation of methanol at PtRu, a kinetically controlled reaction occurs because of this process, a huge anodic peak amplitude is presented, and a larger anodic current density will result from a large electrocatalytic activity. PtRu/GA has a peak current density of 219.78 mAmg$^{-1}$ PtRu, which is 2.1 times that of PtRu/GO (107.06 mAmg$^{-1}$ PtRu) and three times that of commercial PtRu/C (73.11 mAmg$^{-1}$ PtRu). This indicates that PtRu/GA has significantly higher electrochemical activity than PtRu/GO and Pt/C, owing to the increased ECSA of PtRu nanoparticles on PtRu/GA.

Table 1 illustrates the CV curves for CO tolerance, onset oxidation potential, and forward oxidation potential for all electrocatalysts generated in the MOR. Throughout the forward scan, the forward oxidation peak ($I_f$) reflects the oxidation of specific adsorbed species of methanol, while the elimination of carbonaceous species owing to incomplete oxidation is shown by the reverse oxidation peak ($I_b$) [39]. The $I_f$ and $I_b$ ratios also indicated in Table 1, can be utilized to determine the ability of electrocatalysts to tolerate CO intermediate molecules during methanol oxidation. Based on the minimal number of carbonaceous species remaining, a high $I_f/I_b$ ratio in the forward scan suggests an effective electrocatalyst in methanol oxidation. According to the CV data given in Table 1, the $I_f/I_b$ value for the PtRu/C electrocatalyst is the largest (3.94) when compared to the PtRu/GO (1.29) and PtRu/GA (1.19) electrocatalysts, indicating that the PtRu/C electrocatalyst has greater CO tolerance. This finding demonstrates that using a nanostructure structure and a metal oxide combination in an electrocatalyst can help to solve the critical challenge for DMFC technology and has strong potential to replace the commercial supports that are now used in this technology.

Electrocatalysts on manufactured PtRu/GA were compared to various PtRu supports for electrocatalysts in DMFCs, such as nanostructured catalysts and metal oxide combinations, as shown in Table 2. The results demonstrate that PtRu/GA has the highest peak current density among all electrocatalysts.

**Table 2.** Comparison of the performance results with the previous study.

| Authors | Electrocatalyst | Peak Potential (V vs. RHE) | Peak Current Density (mA/mg$_{PtRu}$) |
|---|---|---|---|
| This study | PtRu/GA | 0.864 | 0.287 mA/cm$^2$ 219.78 mA/mg |
| Nishanth et al. [40] | PtRu/TiO$_2$-C | 0.761 | 151.47 |
| Lin et al. [41] | PtRu/CNT | 0.857 | 66.69 |
| Chen et al. [42] | PtRuWO$_x$/C | 0.913 | 56.02 |
| Basri et al. [43] | PtRuNiFe/MWCNT | 0.941 | 31 |
| Guo et al. [44] | PtRu$_{0.7}$(CeO$_2$)$_{0.3}$/C | 0.191 | 21.43 |

To highlight the distinction, this study used mass activities (mAmg$^{-1}$) and specific activity (mAcm$^2$), as illustrated in Figure 3 to give a standard evaluation by normalizing the current for the PtRu mass loading and the ECSA, respectively as to make a comparison for all. Notably, the specific activities of the catalysts were evaluated for relevance to the ECSA in the current context. The PtRu mass utilization efficiency can be determined using a specific activity (mA/cm$^2$$_{PtRu}$) for which PtRu/GA has an abnormally high mass activity of 219.78 mA/mg tiny amount of specific activity (mA/cm$^2$$_{PtRu}$). For the histogram shown in Figure 3d, PtRu has a utilization of approximately 0.287 mA/cm$^2$ compared to other Pt electrocatalysts with high utilization of Pt mass, in line with the histogram for a particular activity. This indicates that PtRu/GA is successful in achieving excellent catalytic activity. PtRu/GA has a higher electrochemical activity because of the well-distributed and uniform size of PtRu particles and a higher ECSA value, which results in many electrochemically active sites.

To provide a deeper understanding of how methanol oxidation works at a molecular level, it is important to consider the variations in structure and mechanical stability between different samples. Figure 4a offers a clear visualization of how the electrical properties of these samples vary, as it depicts the relationship between resistivity and material density. Notably, there are significant disparities between the performance of PtRu/GA and commercial samples. PtRu/GA demonstrates low resistivity and a strong potential for use in anodic catalyst reactions due to its tendency towards the high current. PtRu/GA, PtRu/GO, and PtRu/C catalysts were chronoamperometrically tested for 3600 s in 0.5 M $H_2SO_4$ containing 1.0 M $CH_3OH$ at a fixed potential of 0.60 V to determine their long-term electrocatalytic activity and stability against methanol oxidation. Two electrocatalysts were compared to commercial PtRu/C in this study. These tests are crucial for assessing activity and ensuring electrocatalytic activity against methanol oxidation reactions over a time-consuming need. Instead, Figure 4b shows their corresponding curves, while Table 3 shows the long-term stability analysis results. Initially, all the catalysts were found to have a high current value, which may be attributable to numerous active sites on their surfaces.

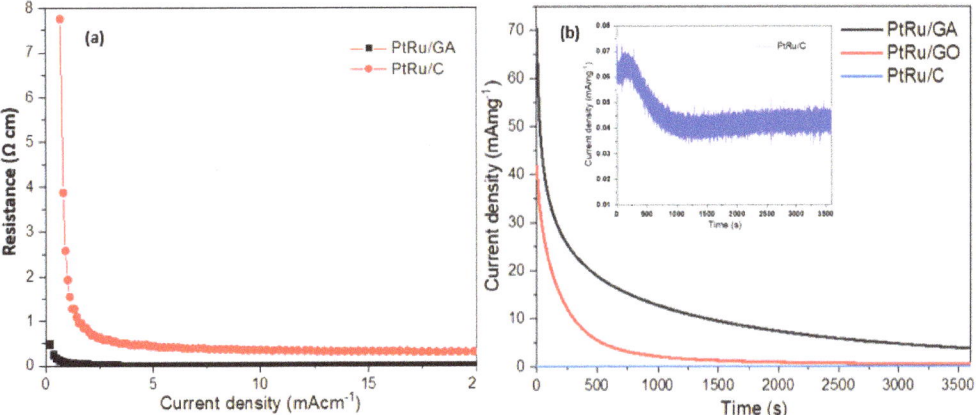

**Figure 4.** (a) Resistance curves of commercial PtRu/GA and PtRu/C (b) Chronoamperometric curves in 0.5 M $H_2SO_4$ + 1 M $CH_3OH$ solution.

**Table 3.** Summary results of the long-term stability test for nanocomposite catalysts.

| Electrocatalyst | Initial Current Density (mAcm$^{-2}$) | Final Current Density at 3600 s (mAcm$^{-2}$) | Current Density Decline (%) |
|---|---|---|---|
| PtRu/GA | 70.26 | 3.79 | 94.6 |
| PtRu/GO | 42.21 | 0.57 | 98.6 |
| PtRu/C | 0.07 | 0.04 | 42.9 |

Continuous oxidation of methanol fuel on the catalyst surface at a fixed potential (0.60 V) occurs during the MOR. As a result, various intermediate-adsorbed CO species begin to form on the catalyst surface. As demonstrated, the initial current for the PtRu/GA nanocomposite catalyst is substantially higher than that for the PtRu/GO and PtRu/C catalysts, implying higher double-layer charging [45]. Within the first 1000 s of the experiment, the PtRu/GA nanocomposite catalyst demonstrates a faster current decay than the PtRu/C catalyst, with an approximately 80% loss compared to 40% [46]. Even though a decline in current was observed, the PtRu/GA nanocomposite catalyst showed a much higher current than the PtRu/GO and PtRu/C nanocomposite catalysts for the entire period. PtRu/GA nanocomposite catalysts have a final current density of 3.79 mA/cm$^2$, which is higher than that of the PtRu/GO (0.57 mA/cm$^2$) and PtRu/C (0.04 mA/cm$^2$) catalysts (see Figure 4 for a better version).

Furthermore, compared to PtRu/GO (98.6%) and PtRu/GA (94.6%), the PtRu/C nanocomposite catalyst shows the lowest calculated decrease in current density (42.9%), demonstrating that the Pt-Ru alloy nanocomposite catalyst is more tolerant of toxins during MOR. The current decays slowly or gradually over time, indicating that the catalyst has good anti-poisoning properties [8]. The PtRu/GA nanocomposite catalyst displays a slower current decline than a conventional catalyst. As a result, the PtRu/GA nanocomposite catalyst outperforms the PtRu/GO and PtRu/C catalysts regarding the electrocatalytic activity and stability of the methanol oxidation process. The outcomes show that oxidation processes are encouraged by acidic conditions because they promote an environment that increases the likelihood that they will take place in the methanol oxidation reaction in an acidic medium. In addition, acidic conditions, which aid in neutralizing these charges and improve their stability, can stabilize the positively charged intermediate species created during oxidation processes. As synthesized, the PtRu/GA nanocomposite catalyst seems promising for DMFC anode catalytic applications.

## 3. Experimental Section

### 3.1. Materials

Merck provided a platinum precursor ($H_2PtCl_6$) with a Pt concentration of 40%. Graphene oxide was obtained from GO Advanced Solution Sdn. Bhd. Ruthenium precursor ($RuCl_4$), isopropyl alcohol, ethylene glycol (EG), Nafion solution, methanol, and ethanol were obtained from Sigma–Aldrich (St. Louis, MO, USA).

### 3.2. Formulation of the GA Support

An 80 mg GO was put into 40 mL of distilled water at the start of production to create a composite produced in the lab. This solution was prepared using ultrasonic, sonic mixing to achieve a homogeneous solution before being transported to a Teflon-coated autoclave with a stainless-steel enclosure. The graphene hydrogel was then subjected to hydrothermal treatment for 12 h in an oven at 200 °C. The autoclave was then allowed to cool to room temperature before performing a freeze-drying process for 24 h to generate a graphene aerogel sample. Figure 5 depicts the creation of graphene aerogels using a combination of hydrothermal and freeze-drying techniques.

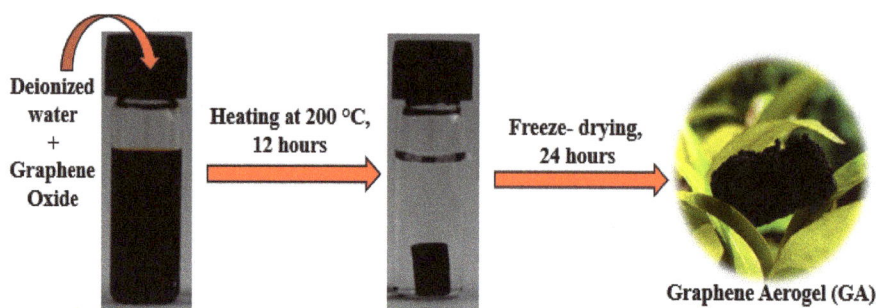

**Figure 5.** Graphene aerogel (GA) synthesis.

### 3.3. Synthesis of the PtRu/GA Electrocatalyst

The study will select the optimal GA support for doping with a Pt-Ru catalyst. The GA support will be loaded with 20 wt% of Pt-Ru in a 1:1 atomic ratio. Initially, the Pt source, chloroplatinic acid, and the Ru source, ruthenium chloride, will be mixed with Ethylene Glycol (EG) solutions in a 70:30 ratio of EG to DI ($v/v$) and sonicated for 15 min. Subsequently, the precursor solutions will be added to the synthesized GA powder and stirred for 30 min until well mixed. Next, the pH of the solution was changed to a value of 10 using a 1 M NaOH solution. In addition, the mixture was heated in a microwave for 1 min and then turned off twice for 1 min to complete the reduction process. Finally, the sample was dried in an oven at 120 °C for 3 h, filtered, and washed numerous times with DI water and ethanol.

### 3.4. Preparation of the Working Electrode

An Autolab electrochemical workstation was used to examine the electrochemical measurements, as shown in Figure 6. The catalyst slurry was produced by ultrasonically dispersing 2.5 mg catalyst in a solution of 150 µL deionized water, 150 µL isopropyl alcohol, and 50 µL Nafion (5 wt%) solution for 20 min. In this experiment, a working electrode known as a glassy carbon electrode (GCE) with a diameter of 3 mm was used (this electrode was glossy with an alumina suspension), where 2.5 µL of catalyst ink was drop cast onto the top GCE surface and dried at room temperature overnight before testing the performance the next day.

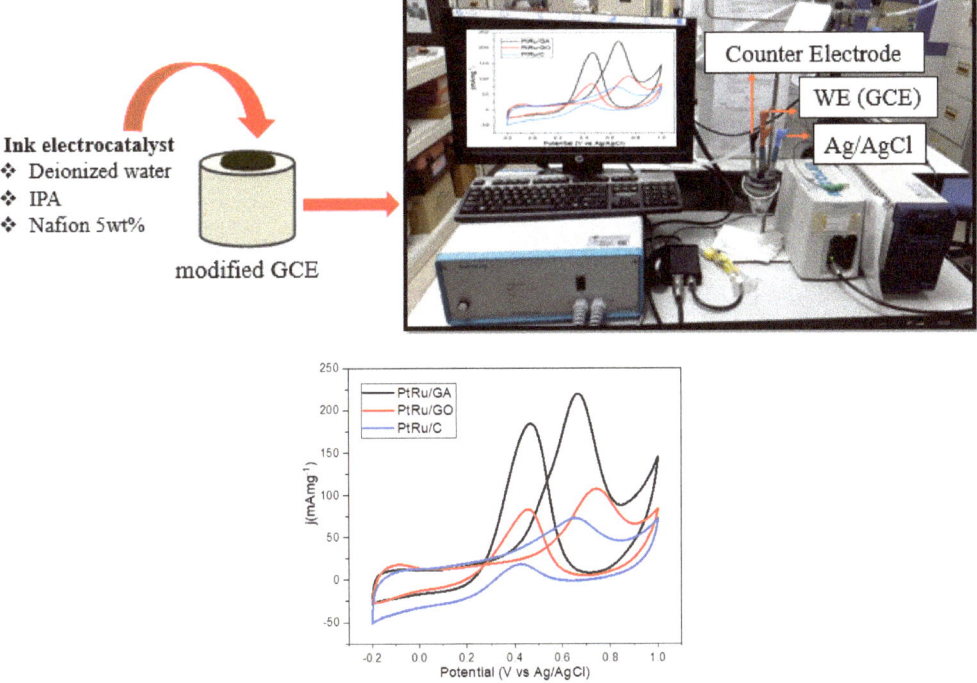

**Figure 6.** Electrochemical test with Autolab electrochemical workstation.

### 3.5. Structural Characterization

Sample-ray diffraction (XRD) is a powerful physical characterization technique that may be utilized to demonstrate manufactured materials' crystallinity and determine their physical characteristics. The size of the catalyst particles and the phase of the prepared catalysts were determined through X-ray diffraction (XRD) analysis using a D-8 Advance diffractometer (Bruker) with Cu Kα (λ = 0.154056 nm) radiation source. This analysis was carried out at room temperature, and the voltage and current were set at 40 kV and 30 mA, respectively. The 2q range values between 5 and 100 were used to obtain the X-ray diffractogram. Scherer's equation was employed to calculate the crystallite size of the catalyst from the XRD data.

$$d = \frac{K\lambda}{\beta \cos \theta} \quad (4)$$

The average diameter of the catalyst can be calculated in nanometers using Scherer's constant (K = 0.89), the wavelength of the radiation (l = 0.154056 nm), the full width at half maximum of the diffraction peak (111) in radians (b), and the Bragg diffraction angle (q).

Concurrently, the degree of graphitization can be determined via Raman spectrum analysis. The surface morphology and forms of the samples were studied using a field emission scanning electron microscope (FESEM). In contrast, transmission electron microscopy (TEM) was performed using a JEM-1010 JEOL apparatus with a voltage of 100 kV.

### 3.6. Electrochemical Characterization

For this study, the catalyst's electrochemical surface area (ECSA) was evaluated in a 0.5 M $H_2SO_4$ solution, and the electrocatalytic activity for methanol oxidation was investigated in a 0.5 M $H_2SO_4$ + 1 M $CH_3OH$ solution. Next, an electrode electrochemical analyzer was utilized to record all data using cyclic voltammetry (CV) at a scan rate of 50 mVs$^{-1}$. Table 1 contains the ECSA results, and CV measurements were used to evaluate ECSA using the equation below.

$$\text{ECSA } (m^2 g_{Pt}^{-1}) = \frac{Q}{\Gamma \cdot W_{Pt}} \quad (5)$$

Based on the equation, Q is the charge density or area under the graph ((C) of experimental CV), $\Gamma$ (2.1 $Cm_{Pt}^{-2}$) is the constant for the charge required to reduce the proton monolayer on the Pt, and $W_{Pt}$ is the Pt loading ($g_{Pt}$) on the electrode. The electrocatalytic activity of the PtRu/GA electrocatalysts MOR using 0.5 M $H_2SO_4$ + 1 M $CH_3OH$ solution was tested using a cyclic voltammogram (Figure 3). The anodic peak current density for the sweep is associated with the oxidation of freshy chemisorbed species originating from methanol adsorption [47]. Chronoamperometry measurements were performed for 3600 s a with a step potential of 0.6 V [48]. Each experiment was purged for 15 min with high-quality nitrogen gas to achieve oxygen-free content. Gas chromatography-mass spectrometry (GCMS) is a powerful analytical technique used to identify and quantify the chemical compounds present in a gas sample. In addition, each electrochemical experiment was conducted using a fresh electrolyte solution to assure repeatability. In addition to testing the strength of the electrocatalyst, the retention value of the electrocatalyst is checked by CA, which focuses on the level of resistance or force of the electrocatalyst object during operation for a certain period according to the following equation [40];

$$\text{Retention value} = \frac{\text{Initial current density} - \text{Final current density}}{\text{Initial current density}} \times 100\% \quad (6)$$

To further clarify the catalytic performance of the electrocatalysts toward MOR, the poisoning tolerance was evaluated. The forward anodic peak current ($I_F$) to the backward anodic peak current ($I_B$) ratio, $I_F/I_B$, can be employed to evaluate the tolerance to accumulated CO and other adsorbed species [40]. A higher $I_F/I_B$ ratio generally implies a more effective removal of CO and other adsorbed intermediates on the electrocatalyst surface during the anodic scan [42].

### 4. Conclusions

This work describes a novel graphene aerogel support for bimetallic PtRu that was successfully fabricated using an environmentally friendly hydrothermal reaction with a working temperature range of 200 °C and a freeze-drying approach. A PtRu/GA electrocatalyst was fabricated from the best GA utilizing a simple microwave-aided alcohol-reduction approach, and the results obtained from prior investigations of bimetallic catalysts were compared. This work describes the synthesis, characterization, and chemical activities of bimetallic PtRu with a graphene aerogel support and its catalytic abilities in a direct methanol fuel cell. XRD, Raman spectroscopy, FESEM, and TEM studies were used to determine the structure, composition, and morphology of the as-synthesized PtRu/GA.

PtRu/GA also has an excellent 3D porous structure that prevents the active surface area loss found for stacked graphene sheets, according to FESEM images. In addition, all electrocatalysts were compared to PtRu/C. The as-prepared PtRu/GA nanocomposite catalyst showed a significantly greater ECSA, superior electrocatalytic activity, lower onset potential, more significant peak current, and improved stability against the MOR. The high surface area and synergistic impact of the generated PtRu alloy nanoparticles with high dispersion on the surface area of the GA support can contribute to these findings. As a result, this study offers a new avenue for catalyst development, raising the prospect of utilizing the as-developed PtRu/GA nanocomposites as a viable DMFC catalyst. Alternatively, PtRu/GA synthesis can be easily expanded to prepare various GA-integrated nanomaterials with diverse uses in photocatalysis, sensors, batteries, and super-capacitor applications.

**Author Contributions:** Conceptualization, S.H.O., S.K.K. and S.B.; software, N.A.K.; validation, N.A.K. and S.B.; formal analysis, S.B.; investigation, S.K.K.; writing—original draft preparation, Siti Hasanah Osman.; writing—review and editing, S.H.O.; visualization, S.H.O.; supervision, S.K.K.; project administration, S.K.K.; funding acquisition, S.K.K. All authors have read and agreed to the published version of the manuscript.

**Funding:** The study received financial support from the Universiti Kebangsaan Malaysia: DIP-2021-028.

**Data Availability Statement:** Not applicable.

**Acknowledgments:** The authors gratefully acknowledge the financial support given for this work by Universiti Kebangsaan Malaysia under DIP-2021-028.

**Conflicts of Interest:** We confirm that the work described has not been published before, it is not under consideration for publication anywhere else, and publication has been approved by all co-authors and the responsible authorities at the institute(s) where the work was carried out. The authors declare no conflict of interest.

# References

1. Osman, S.H.; Kamarudin, S.K.; Karim, N.A.; Basri, S. Application of graphene in low-temperature fuel cell technology: An overview. *Int. J. Energy Res.* **2021**, *45*, 1–19. [CrossRef]
2. Ramli, Z.A.C.; Kamarudin, S.K. Platinum-Based Catalysts on Various Carbon Supports and Conducting Polymers for Direct Methanol Fuel Cell Applications: A Review. *Nanoscale Res. Lett.* **2018**, *13*, 410. [CrossRef]
3. Hong, W.; Wang, J.; Wang, E. Facile synthesis of PtCu nanowires with enhanced electrocatalytic activity. *Nano Res.* **2015**, *8*, 2308–2316. [CrossRef]
4. Bandapati, M.; Goel, S.; Krishnamurthy, B. Platinum utilization in proton exchange membrane fuel cell and direct methanol fuel cell. *J. Electrochem. Sci. Eng.* **2019**, *9*, 281–310. [CrossRef]
5. Rambabu, G.; Bhat, S.D. Simultaneous tuning of methanol crossover and ionic conductivity of sPEEK membrane electrolyte by incorporation of PSSA functionalized MWCNTs: A comparative study in DMFCs. *Chem. Eng. J.* **2014**, *243*, 517–525. [CrossRef]
6. Rambabu, G.; Sasikala, S.; Bhat, S.D. Nanocomposite membranes of sulfonated poly(phthalalizinone ether ketone)-sulfonated graphite nanofibers as electrolytes for direct methanol fuel cells. *RSC Adv.* **2016**, *6*, 107507–107518. [CrossRef]
7. Lori, O.; Elbaz, L. Recent Advances in Synthesis and Utilization of Ultra-low Loading of Precious Metal-based Catalysts for Fuel Cells. *ChemCatChem* **2020**, *12*, 3434–3446. [CrossRef]
8. Hanifah, M.F.R.; Jaafar, J.; Othman, M.; Ismail, A.; Rahman, M.; Yusof, N.; Aziz, F.; Rahman, N.A. One-pot synthesis of efficient reduced graphene oxide supported binary Pt-Pd alloy nanoparticles as superior electro-catalyst and its electro-catalytic performance toward methanol electro-oxidation reaction in direct methanol fuel cell. *J. Alloy. Compd.* **2019**, *793*, 232–246. [CrossRef]
9. Mahmood, K.; Mansoor, M.A.; Iqbal, M.; Kalam, A.; Iqbal, J.; Jilani, A.; Wageh, S. An Electrochemical Investigation of Methanol Oxidation on Thin Films of Nickel Oxide and Its Composites with Zirconium and Yttrium Oxides. *Crystals* **2022**, *12*, 534. [CrossRef]
10. Liaqat, R.; Mansoor, M.A.; Iqbal, J.; Jilani, A.; Shakir, S.; Kalam, A.; Wageh, S. Fabrication of metal (Cu and Cr) incorporated nickel oxide films for electrochemical oxidation of methanol. *Crystals* **2021**, *11*, 1398. [CrossRef]
11. Tang, S.; Sun, G.; Qi, J.; Sun, S.; Guo, J.; Xin, Q.; Haarberg, G.M. Review of New Carbon Materials as Catalyst Supports in Direct Alcohol Fuel Cells. *Chin. J. Catal.* **2010**, *31*, 12–17. [CrossRef]
12. Hassani, S.S.; Samiee, L. Carbon Nanostructured Catalysts as High Efficient Materials for Low Temperature Fuel Cells. *Handb. Ecomater.* **2018**, *713*, 1–28. [CrossRef]

13. Phong, N.T.P.; Nguyen, C.M.T.; Minh, N.H.; Ngo, T.L. Synthesis of Platin/Carbon XC72R Nanocomposite Using as Electrocatalyst for Direct Methanol Fuel Cells. *Mater. Sci.* **2012**, *6*, 925–929.
14. Geim, A.K.; Novoselov, K.S. The rise of graphene. *Nat. Mater.* **2007**, *6*, 183–191. [CrossRef]
15. Novoselov, K.S.; Geim, A.K.; Morozov, S.V.; Jiang, D.; Katsnelson, M.I.; Grigorieva, I.V.; Dubonos, S.V.; Firsov, A.A. Two-dimensional gas of massless Dirac fermions in graphene. *Nature* **2005**, *438*, 197–200. [CrossRef]
16. Smith, A.T.; LaChance, A.M.; Zeng, S.; Liu, B.; Sun, L. Synthesis, properties, and applications of graphene oxide/reduced graphene oxide and their nanocomposites. *Nano Mater. Sci.* **2019**, *1*, 31–47. [CrossRef]
17. Papageorgiou, D.G.; Kinloch, I.A.; Young, R.J. Mechanical properties of graphene and graphene-based nanocomposites. *Prog. Mater. Sci.* **2017**, *90*, 75–127. [CrossRef]
18. Wang, Y.; Wu, Y.; Huang, Y.; Zhang, F.; Yang, X.; Ma, Y.; Chen, Y. Preventing graphene sheets from restacking for high-capacitance performance. *J. Phys. Chem. C* **2011**, *115*, 23192–23197. [CrossRef]
19. Tiwari, S.K.; Sahoo, S.; Wang, N.; Huczko, A. Graphene research and their outputs: Status and prospect. *J. Sci. Adv. Mater. Devices* **2020**, *5*, 10–29. [CrossRef]
20. Ma, Y.; Chen, Y. Three-dimensional graphene networks: Synthesis, properties and applications. *Natl. Sci. Rev.* **2015**, *2*, 40–53. [CrossRef]
21. Yan, Y.; Nashath, F.Z.; Chen, S.; Manickam, S.; Lim, S.S.; Zhao, H.; Lester, E.; Wu, T.; Pang, C.H. Synthesis of graphene: Potential carbon precursors and approaches. *Nanotechnol. Rev.* **2020**, *9*, 1284–1314. [CrossRef]
22. Wang, Y.-S.; Yang, S.-Y.; Li, S.-M.; Tien, H.-W.; Hsiao, S.-T.; Liao, W.-H.; Liu, C.-H.; Chang, K.-H.; Ma, C.-C.M.; Hu, C.-C. Three-dimensionally porous graphene–carbon nanotube composite-supported PtRu catalysts with an ultrahigh electrocatalytic activity for methanol oxidation. *Electrochim. Acta* **2013**, *87*, 261–269. [CrossRef]
23. Yaqoob, L.; Noor, T.; Iqbal, N. Recent progress in development of efficient electrocatalyst for methanol oxidation reaction in direct methanol fuel cell. *Int. J. Energy Res.* **2020**, *45*, 6550–6583. [CrossRef]
24. Xiong, C.; Li, B.; Lin, X.; Liu, H.; Xu, Y.; Mao, J.; Duan, C.; Li, T.; Ni, Y. The recent progress on three-dimensional porous graphene-based hybrid structure for supercapacitor. *Compos. Part B Eng.* **2019**, *165*, 10–46. [CrossRef]
25. Mo, R.; Li, F.; Tan, X.; Xu, P.; Tao, R.; Shen, G.; Lu, X.; Liu, F.; Shen, L.; Xu, B.; et al. High-quality mesoporous graphene particles as high-energy and fast-charging anodes for lithium-ion batteries. *Nat. Commun.* **2019**, *10*, 1474. [CrossRef]
26. Aldroubi, S.; Brun, N.; Bou Malham, I.; Mehdi, A. When graphene meets ionic liquids: A good match for the design of functional materials. *Nanoscale* **2021**, *13*, 2750–2779. [CrossRef] [PubMed]
27. Thiruppathi, A.R.; Sidhureddy, B.; Boateng, E.; Soldatov, D.V.; Chen, A. Synthesis and electrochemical study of three-dimensional graphene-based nanomaterials for energy applications. *Nanomaterials* **2020**, *10*, 1295. [CrossRef]
28. Chen, Z.; He, Y.-C.; Chen, J.-H.; Fu, X.-Z.; Sun, R.; Chen, Y.-X.; Wong, C.-P. PdCu Alloy Flower-like Nanocages with High Electrocatalytic Performance for Methanol Oxidation. *J. Phys. Chem. C* **2018**, *122*, 8976–8983. [CrossRef]
29. Baronia, R.; Goel, J.; Kaswan, J.; Shukla, A.; Singhal, S.K.; Singh, S.P. PtCo/rGO nano-anode catalyst: Enhanced power density with reduced methanol crossover in direct methanol fuel cell. *Mater. Renew. Sustain. Energy* **2018**, *7*, 27. [CrossRef]
30. Abdullah, M.; Kamarudin, S.K.; Shyuan, L.K. TiO2 Nanotube-Carbon (TNT-C) as Support for Pt-based Catalyst for High Methanol Oxidation Reaction in Direct Methanol Fuel Cell. *Nanoscale Res. Lett.* **2016**, *11*, 553. [CrossRef]
31. Chao, L.; Qin, Y.; He, J.; Ding, D.; Chu, F. Robust three dimensional N-doped graphene supported Pd nanocomposite as efficient electrocatalyst for methanol oxidation in alkaline medium. *Int. J. Hydrogen Energy* **2017**, *42*, 15107–15114. [CrossRef]
32. Kung, C.-C.; Lin, P.-Y.; Xue, Y.; Akolkar, R.; Dai, L.; Yu, X.; Liu, C.-C. Three dimensional graphene foam supported platinum-ruthenium bimetallic nanocatalysts for direct methanol and direct ethanol fuel cell applications. *J. Power Sources* **2014**, *256*, 329–335. [CrossRef]
33. Franceschini, E.A.; Bruno, M.M.; Williams, F.J.; Viva, F.A.; Corti, H.R. High-activity mesoporous Pt/Ru catalysts for methanol oxidation. *ACS Appl. Mater. Interfaces* **2013**, *5*, 10437–10444. [CrossRef] [PubMed]
34. Motshekga, S.C.; Pillai, S.K.; Sinha Ray, S.; Jalama, K.; Krause, R.W.M. Recent trends in the microwave-assisted synthesis of metal oxide nanoparticles supported on carbon nanotubes and their applications. *J. Nanomater.* **2012**, *2012*, 528–544. [CrossRef]
35. Li, Z.; Jaroniec, M.; Papakonstantinou, P.; Tobin, J.M.; Vohrer, U.; Kumar, S.; Attard, G.; Holmes, J.D. Supercritical fluid growth of porous carbon nanocages. *Chem. Mater.* **2007**, *19*, 3349–3354. [CrossRef]
36. Stankovich, S.; Dikin, D.A.; Piner, R.D.; Kohlhaas, K.A.; Kleinhammes, A.; Jia, Y.; Wu, Y.; Nguyen, S.T.; Ruoff, R.S. Synthesis of graphene-based nanosheets via chemical reduction of exfoliated graphite oxide. *Carbon* **2007**, *45*, 1558–1565. [CrossRef]
37. Chadha, N.; Sharma, R.; Saini, P. A new insight into the structural modulation of graphene oxide upon chemical reduction probed by Raman spectroscopy and X-ray diffraction. *Carbon Lett.* **2021**, *31*, 1125–1131. [CrossRef]
38. Zhang, Y.; Wan, Q.; Yang, N. Recent Advances of Porous Graphene: Synthesis, Functionalization, and Electrochemical Applications. *Small* **2019**, *15*, 1903780. [CrossRef]
39. Lin, Y.; Liao, Y.; Chen, Z.; Connell, J.W. Holey graphene: A unique structural derivative of graphene. *Mater. Res. Lett.* **2017**, *5*, 209–234. [CrossRef]
40. Nishanth, K.G.; Sridhar, P.; Pitchumani, S.; Shukla, A.K. Enhanced Methanol Electro-Oxidation on Pt-Ru Decorated Self-Assembled TiO2-Carbon Hybrid Nanostructure. *ECS Trans.* **2011**, *41*, 1139–1149. [CrossRef]
41. Lin, Y.; Cui, X.; Yen, C.H.; Wai, C.M. PtRu/carbon nanotube nanocomposite synthesized in supercritical fluid: A novel electrocatalyst for direct methanol fuel cells. *Langmuir* **2005**, *21*, 11474–11479. [CrossRef] [PubMed]

42. Chen, Y.-W.; Chen, H.-G.; Lo, M.-Y.; Chen, Y.-C.; Lo, M.-Y.; Chen, Y. Modification of Carbon Black with Hydrogen Peroxide for High Performance Anode Catalyst of Direct Methanol Fuel Cells. *Materials* **2021**, *14*, 3902. [CrossRef] [PubMed]
43. Basri, S.; Kamarudin, S.K.; Daud, W.R.W.W.; Yaakob, Z.; Kadhum, A.A.H.H. Novel anode catalyst for direct methanol fuel cells. *Sci. World J.* **2014**, *2014*, 547604. [CrossRef]
44. Guo, J.W.; Zhao, T.S.; Prabhuram, J.; Chen, R.; Wong, C.W. Preparation and characterization of a PtRu/C nanocatalyst for direct methanol fuel cells. *Electrochim. Acta* **2005**, *51*, 754–763. [CrossRef]
45. Kung, C.-C.; Lin, P.-Y.; Buse, F.J.; Xue, Y.; Yu, X.; Dai, L.; Liu, C.-C. Preparation and characterization of three dimensional graphene foam supported platinum-ruthenium bimetallic nanocatalysts for hydrogen peroxide based electrochemical biosensors. *Biosens. Bioelectron.* **2013**, *52*, 1–7. [CrossRef] [PubMed]
46. Arukula, R.; Vinothkannan, M.; Kim, A.R.; Yoo, D.J. Cumulative effect of bimetallic alloy, conductive polymer and graphene toward electrooxidation of methanol: An efficient anode catalyst for direct methanol fuel cells. *J. Alloy. Compd.* **2019**, *771*, 477–488. [CrossRef]
47. Yamada, H.; Yoshii, K.; Asahi, M.; Chiku, M.; Kitazumi, Y. Cyclic Voltammetry Part 2: Surface Adsorption, Electric Double Layer, and Diffusion Layer. *Electrochemistry* **2022**, *90*, 102006. [CrossRef]
48. Ng, J.C.; Tan, C.Y.; Ong, B.H.; Matsuda, A. Effect of Synthesis Methods on Methanol Oxidation Reaction on Reduced Graphene Oxide Supported Palladium Electrocatalysts. *Procedia. Eng.* **2017**, *184*, 587–594. [CrossRef]

**Disclaimer/Publisher's Note:** The statements, opinions and data contained in all publications are solely those of the individual author(s) and contributor(s) and not of MDPI and/or the editor(s). MDPI and/or the editor(s) disclaim responsibility for any injury to people or property resulting from any ideas, methods, instructions or products referred to in the content.

*Article*

# Unveiling the Photocatalytic Activity of Carbon Dots/g-C$_3$N$_4$ Nanocomposite for the O-Arylation of 2-Chloroquinoline-3-carbaldehydes

Ravichandran Manjupriya and Selvaraj Mohana Roopan *

Chemistry of Heterocycles & Natural Product Research Laboratory, Department of Chemistry, School of Advanced Sciences, Vellore Institute of Technology, Vellore 632 014, India
* Correspondence: mohanaroopan.s@vit.ac.in; Tel.: +0416-220-2313

**Abstract:** Visible-light-active, organic, heterogeneous photocatalysts offer an ecologically friendly and sustainable alternative to traditional metal-based catalysts. In this work, we report the microwave synthesis of nanocarbon dots (CDs), loaded with graphitic carbon nitride (g-C$_3$N$_4$). The fabricated nanocomposite was shown to exhibit various properties, such as the Schottky heterojunction. The optical properties, functional group analysis, surface morphology, crystallinity, chemical stability, electronic properties, and pore size distribution of the synthesized nanocomposite were analyzed by Ultraviolet-Diffuse Reflectance Spectroscopy (UV-DRS), Photoluminescence (PL), Fourier Transform Infrared Spectroscopy (FTIR), Transmission Electron Microscopy (TEM), X-Ray Diffraction (XRD), Zeta potential, X-Ray Photoelectron Spectroscopy (XPS), and Brunauer–Emmett–Teller (BET). Until now, to the best of our knowledge, there have been no reports published on the light-assisted synthesis of O-arylation of 2-chloroquinoline-3-carbaldehyde. Therefore, we explored the photocatalytic activity of the fabricated nanocomposite in the production of the O-arylated 2-chloroquinoline-3-carbaldehyde. This facile technique uses a blue LED light source as a non-conventional source and operates under moderate conditions, resulting in useful O-arylated products. The experimental data shows the good recyclability of the catalyst for up to five cycles without a loss in catalytic activity, a simple operational protocol, easy recoverability of the catalyst, and good product yields (65–90%) within 12–24 h. Additionally, the preliminary mechanistic investigations are discussed. The results show that the phenoxy and quinoline-3-carbaldehyde radicals generated upon blue LED irradiation during the course of the reaction are responsible for C-O bond formation, which results in O-arylation. The present study clearly indicates that 0D/2D nanocomposites have a bright future as metal-free, heterogeneous photocatalysts suitable for organic reactions.

**Keywords:** blue LED; O-arylation; photocatalyst; carbon nanodots; CDs/g-C$_3$N$_4$ heterojunction

**Citation:** Manjupriya, R.; Roopan, S.M. Unveiling the Photocatalytic Activity of Carbon Dots/g-C$_3$N$_4$ Nanocomposite for the O-Arylation of 2-Chloroquinoline-3-carbaldehydes. *Catalysts* **2023**, *13*, 308. https://doi.org/10.3390/catal13020308

Academic Editor: Bo Hou

Received: 29 December 2022
Revised: 13 January 2023
Accepted: 28 January 2023
Published: 30 January 2023

**Copyright:** © 2023 by the authors. Licensee MDPI, Basel, Switzerland. This article is an open access article distributed under the terms and conditions of the Creative Commons Attribution (CC BY) license (https:// creativecommons.org/licenses/by/ 4.0/).

## 1. Introduction

Aromatic ethers constitute one of the most basic organic molecular structures. As a result, various O-arylation processes for aromatic ether synthesis have been developed [1,2]. Ullmann coupling and Chan–Lam coupling are the most extensively used and most trustworthy technique of transition-metal-catalyzed O-arylation [3,4]. When transition metal poisoning from products is a concern, O-arylation by a metal-free catalyst is an appealing solution [5]. Because of its synthetic and response flexibility, as well as its wide range of chemical activity, 2-chloroquinoline-3-carbaldehyde chemistry has received a great deal of attention in recent years. These aldehydes are an intriguing class of organic compounds that could be used as synthetic intermediates and building blocks in the development of a wide range of heterocyclic systems, as well as presenting effective antibiotics for microbial and cancer therapy [6]. Furthermore, the quinoline molecule serves as the basic skeleton for many alkaloids found in nature, as well as anticancer treatments [7].

Catalysis is essential for chemical transformations and is fundamental to a vast array of chemical processes [8,9], ranging from academic laboratory research to industrial chemical applications. By employing catalytic reagents, it is possible to control the temperature during synthesis, reduce reagent waste, and enhance reaction selectivity, thereby potentially preventing unwanted side reactions and contributing to the development of green technology [10]. Nano-catalysis is widely used to speed up modern synthesis [11,12]. Typically, these catalysts are composed of nano-dimensional active constituents, distributed on a solid support. Achieving sustainability is the fundamental principle underlying the use of catalysts in organic synthesis [13,14]. Organic synthesis research in photocatalysis, particularly in relation to visible light, is also noteworthy [15,16].

Photochemistry is a branch of chemistry that studies chemical reactions caused by light absorption [17]. Photons are used in these chemical reactions to provide enough energy to convert the starting materials into the final products [18]. The most common visible light photocatalysts are expensive and hazardous polypyridyl complexes of ruthenium and iridium [19]. Furthermore, ligands are required for these species' catalytic activity, and tuning their chemical structures is critical to imparting the desired characteristics to metal-based photocatalysts [20]. Because of these considerations, there is currently a great deal of interest in developing novel, efficient, metal-free organic photocatalytic systems [21]. To meet these needs, these organic photocatalysts must be effective, safe, cheap, and easy to obtain using simple synthesis methods.

Photocatalysts based on graphitic carbon nitride (g-$C_3N_4$) are gaining popularity due to their unique electronic band structure and physicochemical properties [22,23]. With a bandgap of 2.7 eV, g-$C_3N_4$, shortly referred to as CN, is an environmentally friendly metal-free semiconductor [24,25]. Bonding with any metal, nonmetal, or doping agent with CN is necessary to improve its photocatalytic activities [26].

Carbon dots (CDs) have evolved as flexible metal-free substances with a wide range of features in terms of catalytic applications when it comes to metal-free catalysts [27,28]. Because of their inertness, high surface functionality, and chemical stability, CDs are an intriguing choice of support [29]. CDs are a fluorescent, potential group of zero-dimensional carbon nanomaterials composed of carbon cores ($sp^2$/$sp^3$ hybridized carbon), with sizes less than 10 nm, which are amorphous and poorly crystalline [30]. CDs have unique properties, such as outstanding water solubility at the nanoscale [31], strong photostability [32], and biocompatibility [33], in addition to captivating size- and excitation wavelength-dependent photoluminescence activity. All of these characteristics combine to make CDs environmentally friendly, safe, and cost-effective nanomaterials with a wide range of applications [34,35]. Similarly, the ease of their mass production from bioresources increases their versatility of applications [36]. In comparison with other carbon-based nanomaterials, such as carbon nanotubes [37] and graphene [38], the capability of CDs for catalytic processes in synthetic organic chemistry has yet to be fully revealed [39].

The optimum form of light source for photocatalytic applications is a light-emitting diode (LED). In terms of efficiency, energy, flexibility, longevity, and environmental friendliness, blue LEDs exceed other light sources. LEDs are an upgrade over traditional illumination sources due to their distinct properties; they provide designers with sn additional conceptual scope while creating various photochemical reactions [40,41].

In 2017, Samanta and his colleagues [42] made a CDs/$Bi_2MoO_6$ nanocomposite as a photocatalyst and tested its efficiency in the synthesis of benzimidazole. In 2017, Sarma et al. [43] used carbon nanodots with carboxyl groups on them to make several quinazolinones and aza-Michael adducts. In 2018, Majumdar et al. [44] came up with a new way to make quinazolinone derivatives from alcohol and 2-amino benzamide substrates in a single pot, using a CDs-$Fe_3O_4$ composite as a catalyst in an aqueous medium. Our research team has also recently published a review on carbon-dot-based organic transformations [45].

In terms of previous work, we report in this study for the first time on the photoirradiation of glucose-derived CDs/CN for promoting the O-arylation of 2-chloroquinoline-3-carbaldehydes. The Schottky-like heterojunction formed by the CDs/CN provides electron mediation through the CDs, enhancing their photocatalytic activity. When compared to pristine CDs and CN, the prepared CDs/CN nanocomposite demonstrated higher photocatalytic activity and stability. UV-DRS, PL, FTIR, TEM, XRD, zeta potential, XPS, and BET were used to investigate the optical properties, functional group analysis, surface morphology, crystallinity, chemical stability, electronic properties, and pore size distribution of the samples. This study will pave the way for future advances in the field of photocatalysts in organic synthesis.

## 2. Results and Discussion
### 2.1. Characterization of the Prepared Nanocomposite
#### 2.1.1. XRD Analysis

The XRD pattern revealed the crystalline nature and structure of the prepared materials. The XRD pattern of CN and various loading percentages of CD-modified CN nanocomposite are shown in Figure 1a. The pristine g-$C_3N_4$ has two strong peaks at about 13 and 27.3 degrees, which correspond to the (100) and (002) planes of graphitic carbon nitride (JCPDS No. 87-1526) [46]. No crystalline CDs were observed in the XRD patterns of the various as-fabricated loading percentages of CDs/g-$C_3N_4$ nanocomposite (Figure 1a), which may be related to the low CD doping level. The primary diffraction peaks of the CN, however, are similar to the diffraction peaks of 3CDCN, 6CDCN, and 9CDCN. This demonstrates that the addition of CD did not alter or disrupt the structure of CN. Meanwhile, the major peaks of 3CDCN, 6CDCN, and 9CDCN shifted slightly, compared to CN, confirming that CDs were attached to the CN matrix.

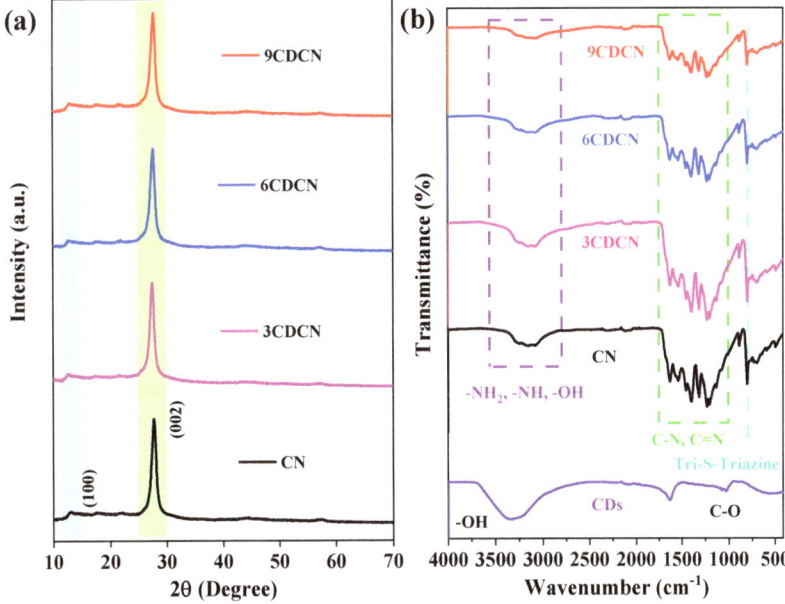

**Figure 1.** (a) The XRD pattern of CN, 3CDCN, 6CDCN, and 9CDCN, and (b) FT−IR spectrum of CDs, CN, 3CDCN, 6CDCN, and 9CDCN.

#### 2.1.2. FT-IR Studies

Figure 1b depicts the functional groups of CN and the various CDs/CN nanocomposite materials. The broad peaks in the 3600–3000 cm$^{-1}$ range are ascribed to the presence of −NH and −OH groups. The N−H stretching vibrational peaks in CN originate from the terminal −$NH_2$ or N−H groups, and O−H stretching or water molecule-absorbing peaks in CDs [47,48]. At 1750 cm$^{-1}$ and 1250 cm$^{-1}$, respectively, vibrational absorption bands corresponding to C=O and C-O-C were recorded on CDs [49]. The characteristic sharp peak at around 800 cm$^{-1}$ corresponds to the triazine ring or vibration of the tri-s-triazine subunit [50]. As the percentage of CDs increases, the peaks shift slightly to the lower wavenumber side. The addition of CDs weakened the major characteristic peaks of CN in the FTIR spectrum of the CDs/CN nanocomposite, which was consistent with the XRD results.

2.1.3. XPS Studies

Surveys and elemental analysis are used in Figure 2a to further identify the amounts of C, N, and O. The results show that 3CDCN has a carbon/nitrogen mass ratio of 0.98, which is greater than that of CN (0.95), indicating that the carbon dots were successfully incorporated. In the XPS analysis of a 3CDCN nanocomposite, a trace amount of O (4.7%) is detected. The survey spectra in Figure 2a show that CN and 3CDCN contained only C, N, and O elements and no other impurities. The HR-XPS of C1s is shown in Figure 2b, where two distinct peaks at 288.1 and 285.0 eV are associated with carbon (sp$^2$-hybridized) in the aromatic structure (N−C=N) and graphitic C−C, respectively, in CN [51,52].

A mixture of three peaks located at 287.1, 285.9, and 284.4 eV, which exhibit C atoms in the (N−C=N), C-O, and C-C, existing in the structure of 3CDCN (Figure 2b), can fit the C 1s XPS signal [53]. The HR-XPS N 1s spectra of CN (401.3, 399.8, 398.5 eV) and 3CDCN (400.3, 398.7, 397.5 eV) show several N species that can be attributed to amino functions (−NH), tertiary N bonded to carbon atoms (N−(C)3), and sp$^2$-hybridized aromatic N (C=N −C) [54,55].

Meanwhile, the C1s and N1s orbital binding energies (CN) in the 3CDCN sample had shifted (Figure 2b,c). The results clearly show that when CN and CDs are combined, they form a heterojunction photocatalytic system with the highest electron density, and localize this charge in the C and N atoms [56].

2.1.4. TEM: Morphological Aspects

Figure 3 shows how the different HR-TEM magnifications and representative HR-TEM images were used to study the 3CDCN photocatalyst's characteristic shape and CD distribution. Figure 3a–c shows TEM images of the CN, CDs, and 3CDCN. The wrinkled 2D lamellar structure of the CN nanosheets with exfoliated layers can be seen in Figure 3a. The TEM images of the synthesized CDs are shown in Figure 3b. As can be seen, partial ethanol combustion produces relatively monodisperse CDs, with average sizes of 10 nm (Figure S1a in the Supplementary Materials). Figure 3c,e shows good CDs dispersion on CN nanosheets. The CDs in CN had an average size of 9.3 nm (Figure S1b in the Supplementary Materials), which was slightly smaller than pristine CDs. The HR-TEM images (Figure 3d) show that CDs have crystalline nature, with a lattice spacing of around 0.31 nm (Figure S1c in the Supplementary Materials). Furthermore, highly dispersed CDs are clearly visible (Figure 3f), implying that CDs were successfully incorporated into CN and formed a heterojunction.

2.1.5. BET Surface Area

Surface area is a key metric that influences photocatalytic activity because the process is surface−dependent. The active sites are bigger and have stronger photocatalytic activity when the surface area is larger [57]. The BET−specific surface area of 3CDCN is observed by the N$_2$ adsorption and desorption isotherms, as shown in Figure 4a. The BET−specific surface area of 3CDCN (108.532 m$^2$/g) is significantly greater than that of CN [58], indicating that the CDs are attached to the surface of CN nanosheets. A larger BET surface area generates more active sites, which is beneficial for substrate adsorption and photocatalytic performance.

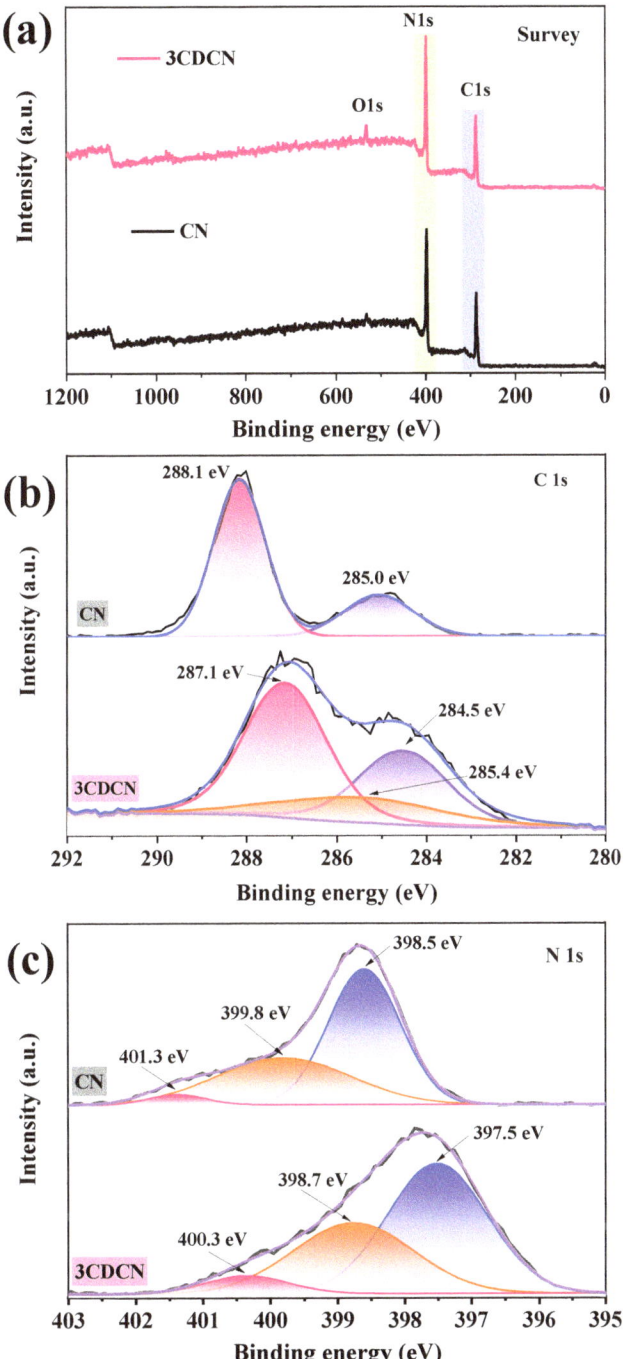

**Figure 2.** XPS spectra of CN and 3CDCN (**a**) survey spectra, (**b**) the HR-XPS of carbon, and (**c**) the HR-XPS of nitrogen.

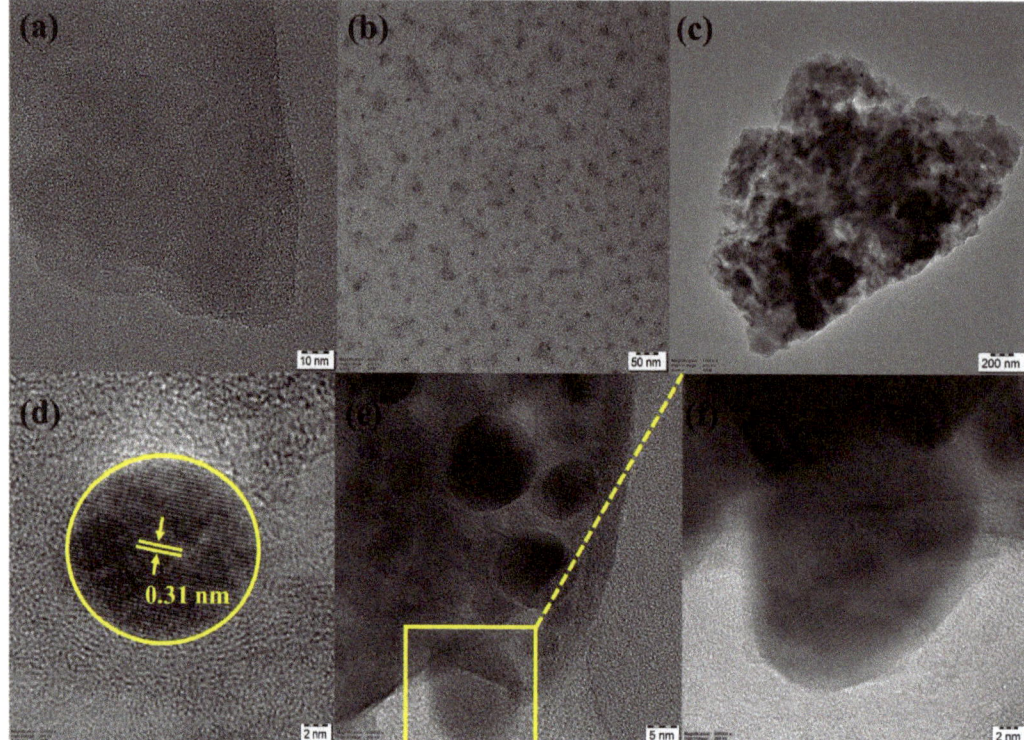

**Figure 3.** TEM images of (**a**) CN, (**b**) CDs, (**c**) 3CDCN, HR-TEM of (**d**) CDs, (**e**) 3CDCN at a 5 nm scale, and (**f**) 3CDCN at a 2 nm scale.

2.1.6. Zeta Surface Charge Potential

The zeta potential value of the pristine CN and CD-loaded nanocomposites was measured, and the results are shown in Figure 4b. CN nanosheets have a zeta potential value of −21.9 mV, indicating that their surface is negatively charged, whereas the zeta potential value of pristine CDs was found to be 20.3 mV. CDs loaded with 3CDCN, 6CDCN, and 9CDCN have zeta potentials of 13.9 mV, 25.6 mV, and 24.1 mV, respectively. This means that the CDs were successfully designed on CN surfaces. Electrostatic attraction aids in the formation of uniform distribution and deposits CDs in the 3CDCN nanocomposite.

2.1.7. Thermal Stability of 3CDCN

The thermal stability of the nanocomposite was characterized by the TGA and DTA curves. The initial decomposition of CN was recorded at 400 °C and the complete weight loss was observed at 720 °C, which is due to the combustion of carbon, while for 3CDCN, it started to decompose earlier at 300 °C than CN, and complete decomposition occurred as with CN. This clearly indicates the incorporation of CDs into CN. The experimental findings revealed that 3CDCN was thermally stable up to 600 °C (Figure 4c).

2.1.8. Optical Properties

Changes in the UV-vis absorption spectra could be used to detect surface functionality variations. As a result, Figure 5a depicts the UV-vis spectra of glucose and as-synthesized CDs, emphasizing the distinct influence of reaction time. Finally, 30 min of microwave irradiation resulted in the formation of a new absorbance peak at 280 nm, with non-bonding orbitals such as C=O bonds attributed to the n → π* transitions. The optical properties were analyzed using UV-vis DRS spectra (Figure 5c), and the relevant band gap was estimated using Tauc's plot (Figure 5c). At about 470 nm, pure g-C$_3$N$_4$ exhibited a fundamental absorption edge. As the number of CDs that were being loaded increased, the absorption edge in red gradually migrated from the beginning at 470 nm. This finding implies that when CN is hybridized with CDCN, it can harvest more visible light, which is crucial for improving photocatalytic performance because better visible light utilization can result in producing more effective photo-generated electron-hole pairs. The band gap of pure CN was calculated to be approximately 2.77 eV, which is very similar to previous g-C$_3$N$_4$ results [59]. The band gaps of 3CDCN, 6CDCN, and 9CDCN were 2.4, 2.2, and 1.9 eV, respectively, indicating that they narrowed progressively from 2.4 to 1.9 eV when increasing the loading of CDs.

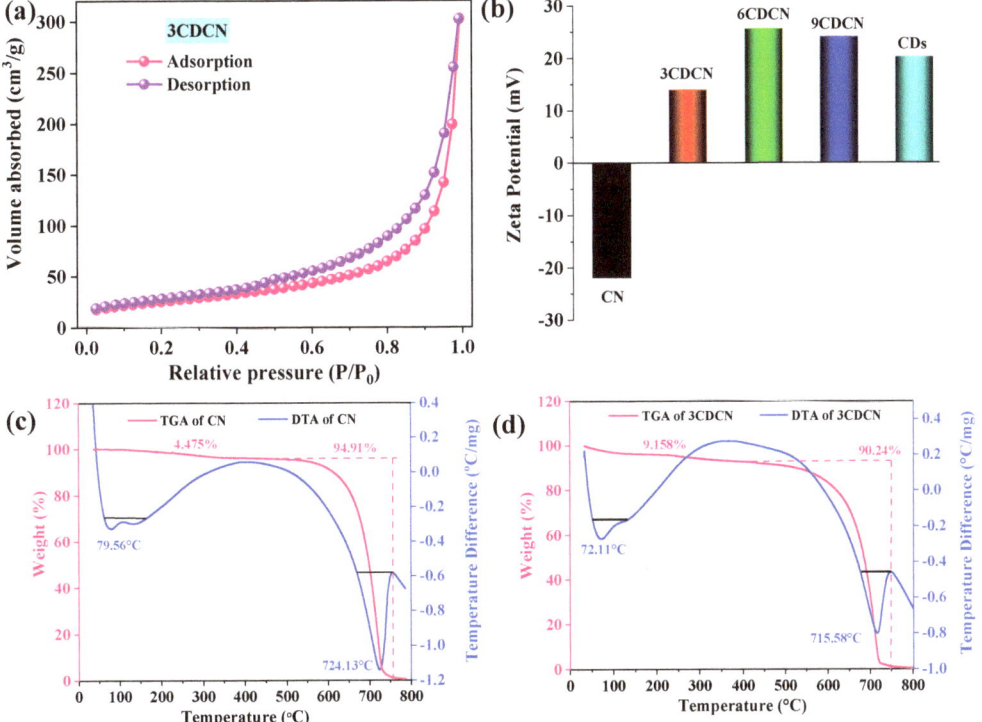

**Figure 4.** (**a**) BET isotherm of the 3CDCN nanocomposite, (**b**) Zeta potential of CN, 3CDCN, 6CDCN and 9CDCN, CDs and TGA, and the DTA graph of (**c**) CN and (**d**) 3CDCN.

The photogenerated charge carrier separation efficiency in semiconductors is frequently investigated, using PL emission spectroscopy [60]. The PL emission signal is produced by the combination of excited electrons and holes. Figure 5d depicts the PL emission spectra of the as-prepared photocatalysts, stimulated at 390 nm. A broad emission peak with a center wavelength of about 458 nm, as seen in all CDCN photocatalysts, could be attributed to the band–band PL phenomenon of the photo-induced charge carriers for CN. The 3CDCN, 6CDCN, and 9CDCN composites exhibited significant PL quenching when compared to the CN, due to the effective charge transfer at the heterostructure interface [61].

## 2.2. Photocatalytic Activity of CDs/g-C$_3$N$_4$ in the O-Arylation of 2-Chloroquinoline-3-carbaldehyde

After all the characterizations of CDCN were performed, the photocatalytic application of CDCN was incorporated into the O-arylation of quinoline-3-carbaldehyde. The reaction of 2-chloroquinoline-3-carbaldehyde (**1**) and 4-bromophenol (**2**) was chosen as a model reaction (Scheme 1). Initially, reaction parameters such as catalyst, base, solvent, and light parameters were optimized. Various carbon materials were used in the catalyst selection process. All the chosen materials demonstrated a significant yield. However, 3CDCN was shown to have a higher yield in terms of efficiency and catalyst separation. When the loading weight percentage of 3CDCN was examined, it was shown that 10 wt % (9.6 mg) catalysts loading produced the highest yield (Table 1). When the loading weight percentage of 3CDCN exceeds 10 wt %, there is no change in the yield of the product and turbidity also occurs in the reaction. Turbidity affects the yield of the reaction by screening the interaction between the light source and the reactants in the reaction mixture. Therefore, we fixed the optimized catalyst loading as 10 wt %. Therefore, further reaction parameters were optimized with the use of a 10 wt % catalyst.

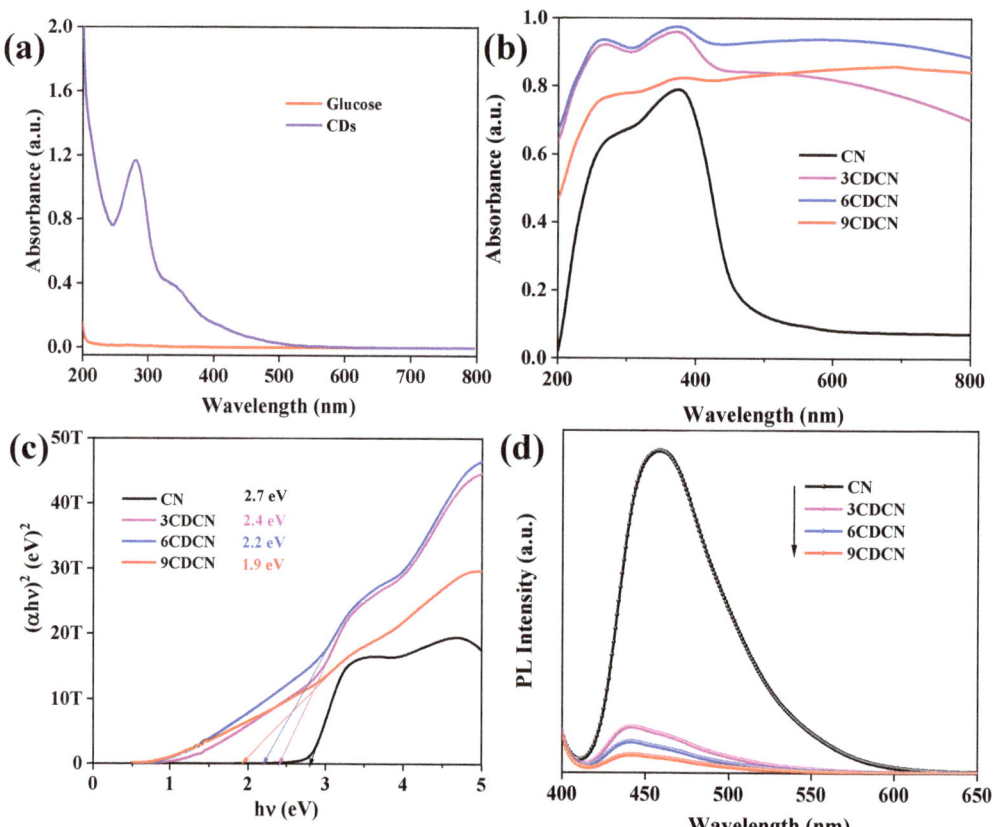

**Figure 5.** (**a**) UV absorption spectra of glucose and CDs. Prepared nanomaterials: CN, 3CDCN, 6CDCN, and 9CDCN. (**b**) UV-vis DRS spectra, (**c**) band gap, and (**d**) PL spectra.

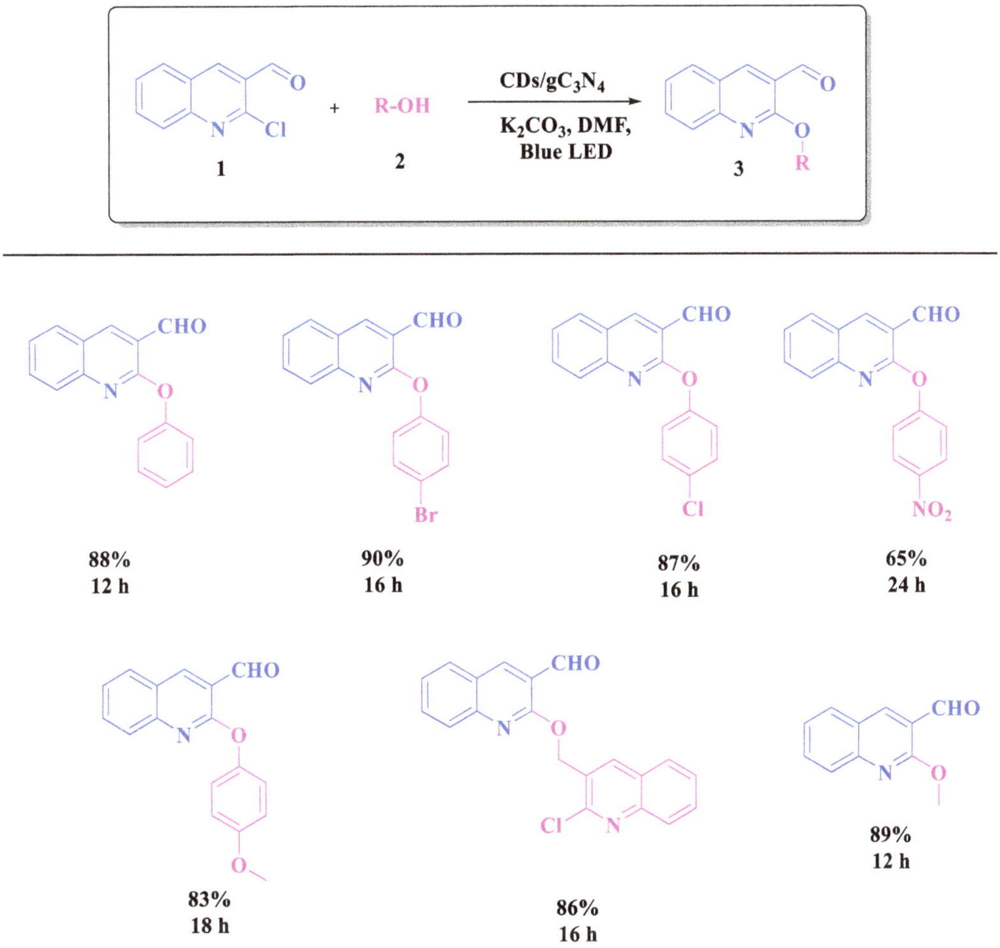

**Scheme 1.** Synthesis of 2-substituted-quinoline-3-carbaldehyde.

**Table 1.** The selection of photocatalysts.

| S. No. | Catalyst | Catalyst Loading | Yield [a] (%) |
|---|---|---|---|
| 1. | Activated Carbon | 10 wt % | 25% |
| 2. | Graphite | 10 wt % | 16% |
| 3. | Graphene oxide | 10 wt % | 19% |
| 4. | CDs | 100 μL | 55% |
| 5. | GCN | 10 wt % | 60% |
| 6. | 3CDCN | 10 wt % | 90% |
| 7. | 6CDCN | 10 wt % | 88% |
| 8. | 9CDCN | 10 wt % | 85% |
| 9. | 3CDCN | 15 wt % | 88% |

Conditions: 2-chloroquinoline-3-carbaldehyde (0.5 mmol, 96 mg, 1 equiv.), 4-bromophenol (0.5 mmol, 86 mg, 1 equiv.), K$_2$CO$_3$ (1 mmol, 138 mg, 2 equiv.), DMF (2 mL), blue LED (12 W, 450–495 nm), time 12 h, ambient temperature, [a] isolated yields.

In terms of base optimization, the reaction was initially carried out in DMF without the use of a base. The reaction condition did not produce the desired product. This is due to the importance of the base in proton abstraction. Furthermore, the reaction conditions were tuned using weak bases, such as $Na_2CO_3$ and $K_2CO_3$, and strong bases, such as NaOH and KOH. All the bases that were chosen demonstrated a prominent yield. Mild bases were used to produce a significant yield, particularly $K_2CO_3$ (Table 2). Therefore, solvent and light optimization was carried out in the presence of $K_2CO_3$ as a base. Non-polar, polar protic, and aprotic solvents were used to investigate the effect of solvent on the O-arylation of quinolines. It has been observed that non-polar solvents, such as hexane, tend to suppress the reaction. The effect of the solvent was further investigated by experimenting with polar protic and aprotic solvents.

**Table 2.** Optimization of the base.

| S. No. | Base | Catalyst | Solvent Medium | Yield [a] (%) |
|---|---|---|---|---|
| 1. | - | 3CDCN | DMF | <5% |
| 2. | $K_2CO_3$ | 3CDCN | DMF | 90% |
| 3. | $Na_2CO_3$ | 3CDCN | DMF | 59% |
| 4. | KOH | 3CDCN | DMF | 55% |
| 5. | t-BuOK | 3CDCN | DMF | 35% |

Conditions: 2-chloroquinoline-3-carbaldehyde (0.5 mmol, 96 mg, 1 equiv.), 4-bromophenol (0.5 mmol, 86 mg, 1 equiv.). Base (1 mmol, 138 mg, 2 equiv.), DMF (2 mL), 3CDCN (9.6 mg, 10 wt %), blue LED (12 W, 450–495 nm), time 12 h, ambient temperature, [a] isolated yields.

The reaction did not proceed well and did not give the desired product with the use of polar protic solvents (EtOH and MeOH). The formation of by-products was observed. The reaction went smoothly and produced the desired product when polar aprotic solvents (DMF, DMSO, and ACN) were used. Reactions with $H_2O$ as a reaction medium failed to initiate the conversion of the product. However, DMF tended to have a better yield (Table 3).

**Table 3.** Optimization of Solvent.

| S. No. | Solvent Medium | Base | Catalyst | Yield [a] (%) |
|---|---|---|---|---|
| 1. | EtOH | $K_2CO_3$ | 3CDCN | NR |
| 2. | MeOH | $K_2CO_3$ | 3CDCN | NR |
| 3. | $H_2O$ | $K_2CO_3$ | 3CDCN | NR |
| 4. | Hexane | $K_2CO_3$ | 3CDCN | NR |
| 5. | THF | $K_2CO_3$ | 3CDCN | Trace |
| 6. | ACN | $K_2CO_3$ | 3CDCN | 41% |
| 7. | DMSO | $K_2CO_3$ | 3CDCN | 82% |
| 8. | $DMF/H_2O$ | $K_2CO_3$ | 3CDCN | 20% |
| 9. | DMF | $K_2CO_3$ | 3CDCN | 90% |

Conditions: 2-chloroquinoline-3-carbaldehyde (0.5 mmol, 96 mg, 1 equiv.), 4-bromophenol (0.5 mmol, 86 mg, 1 equiv.), $K_2CO_3$ (1 mmol, 138 mg, 2 equiv.), solvent (2 mL), 3CDCN (9.6 mg, 10 wt %), blue LED (12 W, 450–495 nm), time 12 h, ambient temperature, [a] isolated yields.

To generate UV or visible light, classical photocatalysis depends mostly on xenon or high-pressure mercury lamps. Although this provides a consistent light source, it consumes a great deal of energy and generates a large amount of heat. As a result, in recent years, LED light sources have attracted attention in terms of photocatalysis, due to their low-energy, practical, and dependable qualities. Table 4 discusses the optimization of visible light of different powers (150, 300, and 500 W) and blue LED (12 W, 450–495 nm). Reactions carried out under visible light had a prominent yield (42–69%), while reactions with blue LED gave a higher yield of 90%. This is because the photocatalyst under study had a band edge in the region of 400–500 nm; this is in accordance with the UV-vis DRS plot (Figure 5b). Thus, the photocatalyst could absorb light effectively between 400 and 500 nm. Out of the diverse lights considered in this study, the blue LED showed better photocatalytic activity in terms of lower energy consumption and allows the photocatalyst to absorb light in this spectral region. With the light-optimized conditions in hand (Table 4), the scope of the diverse substrates was carried out. Unsubstituted phenol gave the corresponding O-arylation product (**3.1**), with an 88% yield. The reactions of phenols containing halogens, such as 4-Br (**3.2**) and 4-Cl (**3.3**), proceeded well, with high to excellent yields (87–90%). However, the reactions of phenol with electron-withdrawing substituents, such as 4-NO$_2$ (**3.4**), gave a moderate yield of 65%. The reactions of phenols with electron-donating groups similar to 4-OMe (**3.5**) proceeded substantially well, giving good yields (83%). Furthermore, an aliphatic alcohol substrate yielded the corresponding ether (**3.7**) at an 89% yield. Furthermore, we strove to examine the catalyst potential of unusual alcohols instead of the substituted phenols (Scheme 1).

**Table 4.** Optimization of the light source.

| S. No. | Catalyst | Base | Solvent Medium | Condition | Yield $^a$ (%) |
|---|---|---|---|---|---|
| 1. | 3CDCN | K$_2$CO$_3$ | DMF | Visible light (150 W) | 42% |
| 2. | 3CDCN | K$_2$CO$_3$ | DMF | Visible light (300 W) | 69% |
| 3. | 3CDCN | K$_2$CO$_3$ | DMF | Visible light (500 W) | 55% |
| 4. | 3CDCN | K$_2$CO$_3$ | DMF | Blue LED (12 W, 450–495 nm) | 90% |

Note: 2-chloroquinoline-3-carbaldehyde (0.5 mmol, 96 mg, 1 equiv.), 4-bromophenol (0.5 mmol, 86 mg, 1 equiv.), K$_2$CO$_3$ (1 mmol, 138 mg, 2 equiv.), DMF (2 mL), 3CDCN (9.6 mg, 10 wt %), light source, time 12 h, ambient temperature, $^a$ isolated yields.

The formation of the product was confirmed by $^1$H NMR, $^{13}$C NMR, GCMS, HRMS, and FTIR analyses (Figures S2–S30 in the Supplementary Materials). The formation of compound **3.6** was identified by $^1$H NMR, $^{13}$C NMR, HR-MS, and FTIR. The signal for the -OH proton in $^1$H NMR vanished around 9–10 ppm, confirming the formation of the C-O bond between quinoline and the alcohol moiety. Further, by comparing the FTIR plots of the reactant (alcohol moiety) and the product, **3.6**, we can see that the disappearance of the -OH stretching (3404 cm$^{-1}$) in the product, **3.6,** strongly indicates the O-arylation of quinoline with the alcohol moiety. Moreover, HR-MS revealed the molecular mass of the predicted compound.

2.2.1. Plausible Mechanism

Relying on all these findings, we propose a plausible mechanism, as illustrated in Figure 6. Under light irradiation, the 3CDCN nanocomposite absorbs light, resulting in photoinduced electron-hole pairs in the conduction and valence bands. The produced charge carriers are then counterbalanced by recombination, when a few of them move to the composite's surface [62]. Thus, the photocatalyst becomes excited. Then, the excited electrons in the conduction band can easily flow through CDs. Thus, a Schottky-like heterojunction is formed [63]. This excited photocatalyst thus induces the homolytic C-Cl bond cleavage of **1**, producing **1a**. This radical generation is induced by the single-electron transfer (SET) of the photoexcited catalyst, where the photoreduction takes place [64]. Accordingly, the intermediate, **2a**, is found to be produced by the proton abstraction of **2** by the base. Then, photo-oxidation takes place, thus generating the phenoxy radical, **2b** [65]. Finally, the radical-generated moieties, **1b** and **2b**, couple together to produce the desired O-arylated product, **3**, by the elimination of KCl.

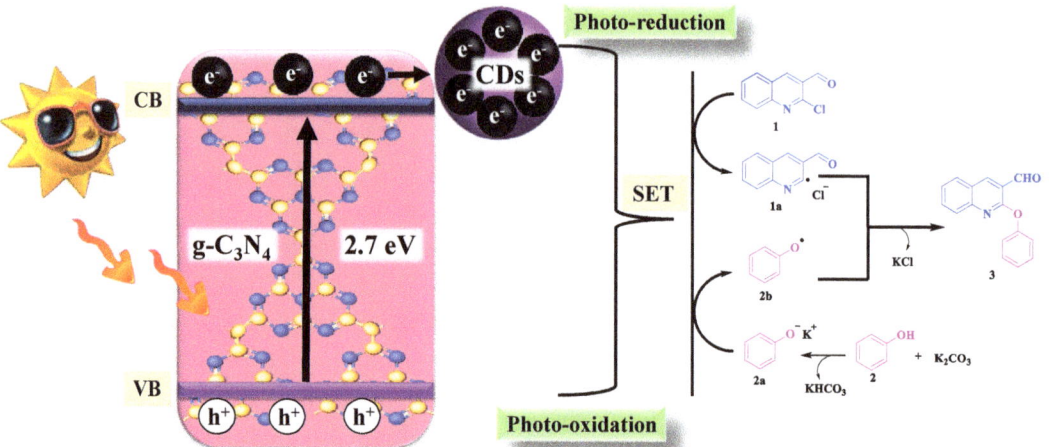

**Figure 6.** The charge transfer mechanism of the CDCN nanocomposite and a plausible mechanism.

2.2.2. Radical Scavenging Study

The identified suppression of the reaction in an oxygen atmosphere was congruent with the proposed radical mechanism. This was further supported by a study performed in the presence of a free radical scavenger, 2,2,6,6-tetramethylpiperidine-1-oxyl (TEMPO, 1 equiv). As expected, the desired product, **3**, has not been detected. This proves the generation of radicals during the course of the reaction.

2.2.3. Reusability and Stability Tests

Figure 7 depicts the 3CDCN photocatalyst recovery and reuse activities. In the current study, 3CDCN was collected by centrifugation, washed with acetone, and reused in the subsequent round of photocatalytic reactions. According to Figure 7a, the photocatalytic performance slightly declined after four trials, to 80%. This small decline in 3CDCN photocatalytic activity could be due to the material losses that might occur during the recovery process (washing and drying), which would result in a reduced dose in the following cycle, hence reducing the specific surface catalytic activity and lowering the performance. In order to determine the reason for this decrease in the photocatalytic performance of 3CDCN during its recycling, in terms of the photocatalytic O-arylation of 2-chloroquinoline-3-carbaldehydes, the physicochemical properties of the fresh and recycled 3CDCN nanocomposite were compared via XRD and XPS analyses. After the process was completed, the 3CDCN photocatalyst's XRD patterns and XPS (C1s and N1s) findings remained consistent, demonstrating that no significant structural or chemical changes occurred in the 3CDCN photocatalyst (Figure 7b,d). Thus, it can be said that the CDs/g-$C_3N_4$ nanocomposite photocatalyst was stable and had the potential to be used in industrial applications.

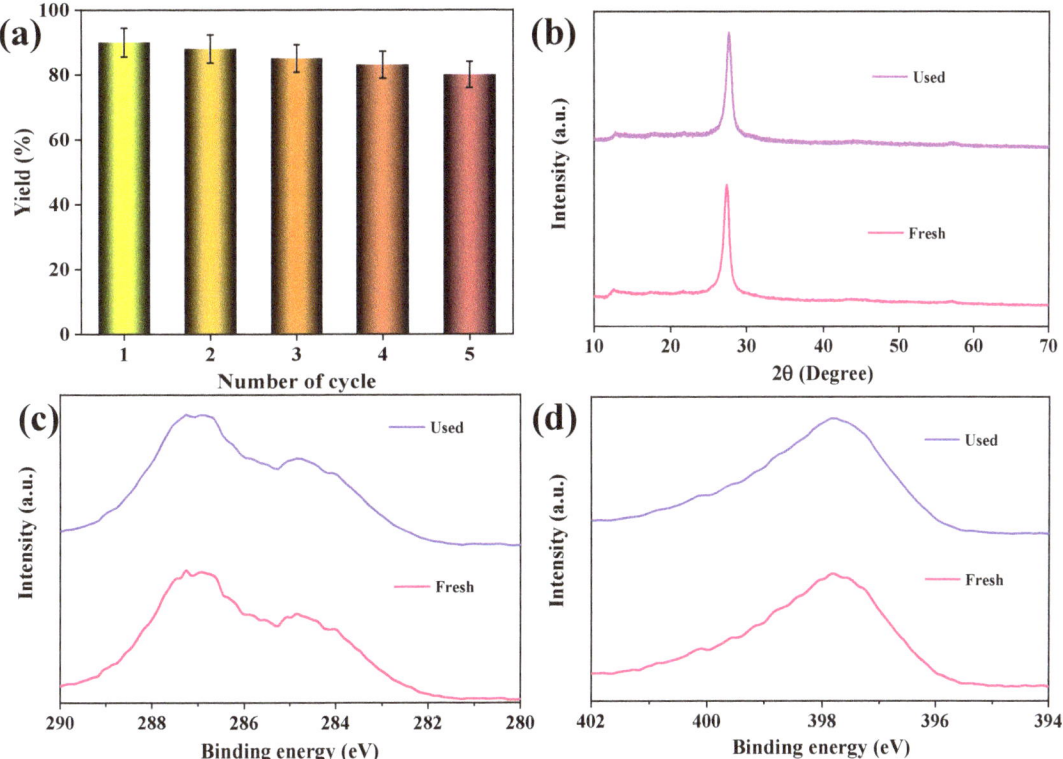

**Figure 7.** (**a**) Reusability of the 3CDCN photocatalyst and the stability of fresh and used 3CDCN, (**b**) XRD pattern, (**c**) XPS of C1s, and (**d**) XPS of N1s.

## 3. Materials and Methods

### 3.1. Details of the Materials Used

The details are available in the supporting document (Text S1).

### 3.2. Microwave-Assisted Synthesis of Carbon Dots

The CDs were made from glucose in a single step using a microwave. In a beaker, 3 g of glucose was dissolved in 10 mL of deionized water. Then, 6 mL of concentrated HCl solution was gradually added to the glucose solution. The mixture was then poured into a three-neck microwave vial for microwave heating (90 °C). The entire mixture was placed in a microwave oven with 500 watts of power for 30 min. As the CDs formed, the color of the mixture changed from colorless to dark brown. Before dialysis, the brown solution was centrifuged at 8000 rpm for 30 min and then filtered through a 0.22 μm membrane after naturally cooling to room temperature. Finally, the obtained CDs were dialyzed for 10 h against ultra-pure water and were stored at 3 °C in the refrigerator before being used in the subsequent experiments.

### 3.3. Preparation of g-$C_3N_4$ Nanosheets

First, 10 g of melamine was calcined at 550 °C in a muffle furnace for 3 h (10 °C min$^{-1}$). The bulk g-$C_3N_4$ sample was then ground into powder and exfoliated using ultrasonication, similar to the procedure described previously in the literature [58].

## 3.4. Fabrication of CDs/g-C₃N₄ Composite

Scheme 2 depicts an overall composite preparation schematic and representation diagram. Under constant stirring, 1 g of g-$C_3N_4$ nanosheets was dispersed in 10 mL of deionized water. The g-$C_3N_4$ suspension was then treated with microwave-assisted synthesized CD solutions of varying concentrations (300, 600, and 900 µL). Finally, the mixture was subjected to a piezoelectric process for 30 min before being heated at 180 °C for 2 h. CD concentrations of 300, 600, and 900 µL were labeled as 3CDCN, 6CDCN, and 9CDCN, respectively.

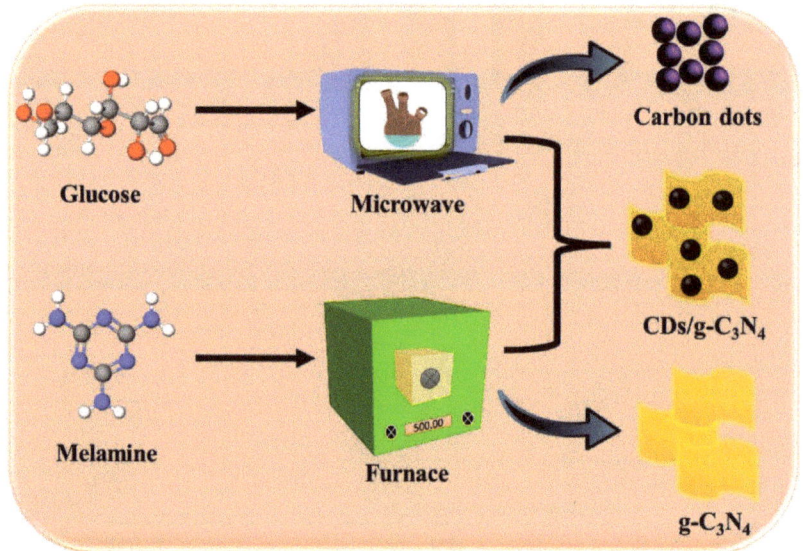

**Scheme 2.** The schematic synthesis procedure of the CDs/g-$C_3N_4$ nanocomposite.

## 3.5. Characterization Details

The details are available in the supporting document (Text S2).

## 3.6. General Procedure for the Synthesis of O-Arylated 2-Chloroquinoline-3-carbaldehydes (3)

The compound **(1)** was prepared and compared with the method in the literature [66]. Yield 79%, m.p. 148 °C. For the O-arylation of 2-chloroquinoline-3-carbaldehydes, the reaction vial was charged with 2-chloroquinoline-3-carbaldehyde (0.5 mmol, 96 mg, 1 equiv.), substituted phenol (0.5 mmol, 1 equiv.), $K_2CO_3$ (2 mmol, 138 mg, 2 equiv.), DMF (2 mL), and photocatalyst (9.6 mg, 10 wt %). The mixture was irradiated under blue LED (12W, 495 nm) for 12–24 h until the product was formed. The reaction was monitored by TLC. After completion of the reaction, the reaction mixture was poured into crushed ice, filtered, dried, and column-purified, using ethyl acetate/petroleum ether (1:9) as an eluent.

## 4. Conclusions

In summary, we prepared a facile CDs/g-$C_3N_4$ heterojunction and explored its optical properties, surface morphology, surface functionality, crystallinity, chemical stability, electronic properties, and pore-size distribution through various characterization techniques. The constructed photocatalyst has been explored for its photocatalytic activity in the successful O-arylation of 2-chloro-3-formyl quinolines. The efficiency of the catalyst could be extended to diverse substrates. Based on the research findings, 10 wt % of CDs/g-$C_3N_4$ nanocomposite exhibited the best photocatalytic activity when compared to pristine CDs and g-$C_3N_4$. The reaction has been carried out with the irradiation of low energy-consumption blue LEDs. The catalyst that was taken into account exhibited excellent recyclability, easy recoverability, and good reusability. We hope that this will spark a wave of investigation and open up a theme that we anticipate will emerge as one of the most prevalent concepts in the coming years.

**Supplementary Materials:** The following supporting information can be downloaded at: https://www.mdpi.com/article/10.3390/catal13020308/s1, Figure S1: Particle size histogram of (a) CDs, (b) 3CDCN, and (c) HRTEM d-spacing value of 3CDCN calculation; Figures S2–S30: FT-IR, $^1$H, $^{13}$C NMR and HR-MS spectra of substituted quinoline-3-carbaldehydes 3.1-3.7. Text S1: Materials details. Text S2: Characterization details.

**Author Contributions:** Investigation, data curation, and original draft preparation, R.M.; conceptualization, formal analysis, writing—review and editing, and supervision, S.M.R. All authors discussed the results and contributed to the final manuscript. All authors have read and agreed to the published version of the manuscript.

**Funding:** This research received no external funding.

**Data Availability Statement:** Not applicable.

**Acknowledgments:** Our profound thanks go to the management of the Vellore Institute of Technology for providing a Sophisticated Instrument Facility (SIF) and for the fellowship offered to Manjupriya to aid in the research endeavor.

**Conflicts of Interest:** The authors declared no potential conflict of interest.

# References

1. Theil, F. Synthesis of Diaryl Ethers: A Long-Standing Problem Has Been Solved. *Angew. Chemie Int. Ed.* **1999**, *38*, 2345–2347. [CrossRef]
2. Zhang, H.; Chen, L.; Oderinde, M.S.; Edwards, J.T.; Kawamata, Y.; Baran, P.S. Chemoselective, Scalable Nickel-Electrocatalytic O-Arylation of Alcohols. *Angew. Chem.* **2021**, *133*, 20868–20873. [CrossRef]
3. Grushin, V.V.; Alper, H. Transformations of Chloroarenes, Catalyzed by Transition-Metal Complexes. *Chem. Rev.* **1994**, *94*, 1047–1062. [CrossRef]
4. Collman, J.P.; Zhong, M. An Efficient Diamine·Copper Complex-Catalyzed Coupling of Arylboronic Acids with Imidazoles. *Org. Lett.* **2000**, *2*, 1233–1236. [CrossRef] [PubMed]
5. Zhao, J.; Zhao, Y.; Fu, H. Transition-Metal-Free Intramolecular Ullmann-Type O-Arylation: Synthesis of Chromone Derivatives. *Angew. Chem. Int. Ed.* **2011**, *50*, 3769–3773. [CrossRef]
6. Sangu, K.; Fuchibe, K.; Akiyama, T. A Novel Approach to 2-Arylated Quinolines: Electrocyclization of Alkynyl Imines via Vinylidene Complexes. *Org. Lett.* **2004**, *6*, 353–355. [CrossRef]
7. Rajakumar, P.; Raja, R.; Selvam, S.; Rengasamy, R.; Nagaraj, S. Synthesis and Antibacterial Activity of Some Novel Imidazole-Based Dicationic Quinolinophanes. *Bioorg. Med. Chem. Lett.* **2009**, *19*, 3466–3470. [CrossRef]
8. Radini, I.; Elsheikh, T.; El-Telbani, E.; Khidre, R. New Potential Antimalarial Agents: Design, Synthesis and Biological Evaluation of Some Novel Quinoline Derivatives as Antimalarial Agents. *Molecules* **2016**, *21*, 909. [CrossRef]
9. Chellapandi, T.; Madhumitha, G. Montmorillonite Clay-Based Heterogenous Catalyst for the Synthesis of Nitrogen Heterocycle Organic Moieties: A Review. *Mol. Divers.* **2021**, *26*, 2311–2339. [CrossRef]
10. Shiri, L.; Ghorbani-Choghamarani, A.; Kazemi, M. S-S Bond Formation: Nanocatalysts in the Oxidative Coupling of Thiols. *Aust. J. Chem.* **2017**, *70*, 9. [CrossRef]
11. Xu, H.-J.; Wan, X.; Geng, Y.; Xu, X.-L. The Catalytic Application of Recoverable Magnetic Nanoparicles-Supported Organic Compounds. *Curr. Org. Chem.* **2013**, *17*, 1034–1050. [CrossRef]
12. Niakan, M.; Masteri-Farahani, M.; Shekaari, H.; Karimi, S. Pd Supported on Clicked Cellulose-Modified Magnetite-Graphene Oxide Nanocomposite for C-C Coupling Reactions in Deep Eutectic Solvent. *Carbohydr. Polym.* **2021**, *251*, 117109. [CrossRef] [PubMed]
13. Dhakshinamoorthy, A.; Garcia, H. Catalysis by Metal Nanoparticles Embedded on Metal-Organic Frameworks. *Chem. Soc. Rev.* **2012**, *41*, 5262. [CrossRef] [PubMed]
14. Nagarjun, N.; Jacob, M.; Varalakshmi, P.; Dhakshinamoorthy, A. UiO-66(Ce) Metal-Organic Framework as a Highly Active and Selective Catalyst for the Aerobic Oxidation of Benzyl Amines. *Mol. Catal.* **2021**, *499*, 111277. [CrossRef]
15. König, B. Photocatalysis in Organic Synthesis—Past, Present and Future. *Eur. J. Org. Chem.* **2017**, *2017*, 1979–1981. [CrossRef]
16. Arunachalapandi, M.; Roopan, S.M. Visible Light-Activated $Cu_3TiO_4$ Photocatalyst for the One-Pot Multicomponent Synthesis of Imidazo-Pyrimido Acridines. *Inorg. Chem. Commun.* **2023**, *148*, 110310. [CrossRef]
17. Gisbertz, S.; Pieber, B. Heterogeneous Photocatalysis in Organic Synthesis. *ChemPhotoChem* **2020**, *4*, 456–475. [CrossRef]
18. Romero, N.A.; Nicewicz, D.A. Organic Photoredox Catalysis. *Chem. Rev.* **2016**, *116*, 10075–10166. [CrossRef]

19. Dutta, S.; Erchinger, J.E.; Schäfers, F.; Das, A.; Daniliuc, C.G.; Glorius, F. Chromium/Photoredox Dual Catalyzed Synthesis of α-Benzylic Alcohols, Isochromanones, 1,2-Oxy Alcohols and 1,2-Thio Alcohols. *Angew. Chem. Int. Ed.* **2022**, *61*, e202212136. [CrossRef]
20. Kundu, S.; Roy, L.; Maji, M.S. Development of Carbazole-Cored Organo-Photocatalyst for Visible Light-Driven Reductive Pinacol/Imino-Pinacol Coupling. *Org. Lett.* **2022**, *24*, 9001–9006. [CrossRef]
21. Bhuyan, S.; Gogoi, A.; Basumatary, J.; Gopal Roy, B. Visible-Light-Promoted Metal-Free Photocatalytic Direct Aromatic C−H Oxygenation. *Eur. J. Org. Chem.* **2022**, *16*, e202200148. [CrossRef]
22. Alaghmandfard, A.; Ghandi, K. A Comprehensive Review of Graphitic Carbon Nitride (g-$C_3N_4$)–Metal Oxide-Based Nanocomposites: Potential for Photocatalysis and Sensing. *Nanomaterials* **2022**, *12*, 294. [CrossRef] [PubMed]
23. Mun, S.J.; Park, S.-J. Graphitic Carbon Nitride Materials for Photocatalytic Hydrogen Production via Water Splitting: A Short Review. *Catalysts* **2019**, *9*, 805. [CrossRef]
24. Arunachalapandi, M.; Roopan, S.M. Environment Friendly G-$C_3N_4$-Based Catalysts and Their Recent Strategy in Organic Transformations. *High Energy Chem.* **2022**, *56*, 73–90. [CrossRef]
25. Wang, J.; Wang, S. A Critical Review on Graphitic Carbon Nitride (g-$C_3N_4$)-Based Materials: Preparation, Modification and Environmental Application. *Coord. Chem. Rev.* **2022**, *453*, 214338. [CrossRef]
26. Chen, J.; Fang, S.; Shen, Q.; Fan, J.; Li, Q.; Lv, K. Recent Advances of Doping and Surface Modifying Carbon Nitride with Characterization Techniques. *Catalysts* **2022**, *12*, 962. [CrossRef]
27. Arcudi, F.; Đorđević, L.; Prato, M. Design, Synthesis, and Functionalization Strategies of Tailored Carbon Nanodots. *Acc. Chem. Res.* **2019**, *52*, 2070–2079. [CrossRef]
28. Lim, S.Y.; Shen, W.; Gao, Z. Carbon Quantum Dots and Their Applications. *Chem. Soc. Rev.* **2015**, *44*, 362–381. [CrossRef]
29. Wang, Y.; Hu, A. Carbon Quantum Dots: Synthesis, Properties and Applications. *J. Mater. Chem. C* **2014**, *2*, 6921. [CrossRef]
30. Du, Y.; Guo, S. Chemically Doped Fluorescent Carbon and Graphene Quantum Dots for Bioimaging, Sensor, Catalytic and Photoelectronic Applications. *Nanoscale* **2016**, *8*, 2532–2543. [CrossRef]
31. Dey, D.; Bhattacharya, T.; Majumdar, B.; Mandani, S.; Sharma, B.; Sarma, T.K. Carbon Dot Reduced Palladium Nanoparticles as Active Catalysts for Carbon-Carbon Bond Formation. *Dalt. Trans.* **2013**, *42*, 13821. [CrossRef]
32. Zhi, B.; Gallagher, M.J.; Frank, B.P.; Lyons, T.Y.; Qiu, T.A.; Da, J.; Mensch, A.C.; Hamers, R.J.; Rosenzweig, Z.; Fairbrother, D.H.; et al. Investigation of Phosphorous Doping Effects on Polymeric Carbon Dots: Fluorescence, Photostability, and Environmental Impact. *Carbon* **2018**, *129*, 438–449. [CrossRef]
33. Boakye-Yiadom, K.O.; Kesse, S.; Opoku-Damoah, Y.; Filli, M.S.; Aquib, M.; Joelle, M.M.B.; Farooq, M.A.; Mavlyanova, R.; Raza, F.; Bavi, R.; et al. Carbon Dots: Applications in Bioimaging and Theranostics. *Int. J. Pharm.* **2019**, *564*, 308–317. [CrossRef]
34. Dhenadhayalan, N.; Lin, K.; Saleh, T.A. Recent Advances in Functionalized Carbon Dots toward the Design of Efficient Materials for Sensing and Catalysis Applications. *Small* **2020**, *16*, 1905767. [CrossRef]
35. Rajendran, S.; UshaVipinachandran, V.; Badagoppam Haroon, K.H.; Ashokan, I.; Bhunia, S.K. A Comprehensive Review on Multi-Colored Emissive Carbon Dots as Fluorescent Probes for the Detection of Pharmaceutical Drugs in Water. *Anal. Methods* **2022**, *14*, 4263–4291. [CrossRef] [PubMed]
36. Meng, W.; Bai, X.; Wang, B.; Liu, Z.; Lu, S.; Yang, B. Biomass-Derived Carbon Dots and Their Applications. *Energy Environ. Mater.* **2019**, *2*, 172–192. [CrossRef]
37. Dai, H. Carbon Nanotubes: Opportunities and Challenges. *Surf. Sci.* **2002**, *500*, 218–241. [CrossRef]
38. Li, X.; Yu, J.; Wageh, S.; Al-Ghamdi, A.A.; Xie, J. Graphene in Photocatalysis: A Review. *Small* **2016**, *12*, 6640–6696. [CrossRef]
39. Gao, J.; Zhu, M.; Huang, H.; Liu, Y.; Kang, Z. Advances, Challenges and Promises of Carbon Dots. *Inorg. Chem. Front.* **2017**, *4*, 1963–1986. [CrossRef]
40. Das, A. LED Light Sources in Organic Synthesis: An Entry to a Novel Approach. *Lett. Org. Chem.* **2022**, *19*, 283–292. [CrossRef]
41. Wang, C.; Sun, Z.; Zheng, Y.; Hu, Y.H. Recent Progress in Visible Light Photocatalytic Conversion of Carbon Dioxide. *J. Mater. Chem. A* **2019**, *7*, 865–887. [CrossRef]
42. Samanta, S.; Khilari, S.; Srivastava, R. Stimulating the Visible-Light Catalytic Activity of $Bi_2MoO_6$ Nanoplates by Embedding Carbon Dots for the Efficient Oxidation, Cascade Reaction and Photoelectrochemical $O_2$ Evolution. *ACS Appl. Nano Mater.* **2018**, *1*, 426–441. [CrossRef]
43. Majumdar, B.; Mandani, S.; Bhattacharya, T.; Sarma, D.; Sarma, T.K. Probing Carbocatalytic Activity of Carbon Nanodots for the Synthesis of Biologically Active Dihydro/Spiro/Glyco Quinazolinones and Aza-Michael Adducts. *J. Org. Chem.* **2017**, *82*, 2097–2106. [CrossRef] [PubMed]
44. Majumdar, B.; Sarma, D.; Jain, S.; Sarma, T.K. One-Pot Magnetic Iron Oxide–Carbon Nanodot Composite-Catalyzed Cyclooxidative Aqueous Tandem Synthesis of Quinazolinones in the Presence of Tert -Butyl Hydroperoxide. *ACS Omega* **2018**, *3*, 13711–13719. [CrossRef]
45. Manjupriya, R.; Roopan, S.M. Carbon Dots-Based Catalyst for Various Organic Transformations. *J. Mater. Sci.* **2021**, *56*, 17369–17410. [CrossRef]
46. Sui, G.; Li, J.; Du, L.; Zhuang, Y.; Zhang, Y.; Zou, Y.; Li, B. Preparation and Characterization of G-$C_3N_4$/Ag–$TiO_2$ Ternary Hollowsphere Nanoheterojunction Catalyst with High Visible Light Photocatalytic Performance. *J. Alloys Compd.* **2020**, *823*, 153851. [CrossRef]

47. Sun, H.; Zou, C.; Liao, Y.; Tang, W.; Huang, Y.; Chen, M. Modulating Charge Transport Behavior across the Interface via G-C$_3$N$_4$ Surface Discrete Modified BiOI and Bi$_2$MoO$_6$ for Efficient Photodegradation of Glyphosate. *J. Alloys Compd.* **2023**, *935*, 168208. [CrossRef]
48. Yoshinaga, T.; Iso, Y.; Isobe, T. Particulate, Structural, and Optical Properties of D-Glucose-Derived Carbon Dots Synthesized by Microwave-Assisted Hydrothermal Treatment. *ECS J. Solid State Sci. Technol.* **2018**, *7*, R3034–R3039. [CrossRef]
49. Alarfaj, N.; El-Tohamy, M.; Oraby, H. CA 19-9 Pancreatic Tumor Marker Fluorescence Immunosensing Detection via Immobilized Carbon Quantum Dots Conjugated Gold Nanocomposite. *Int. J. Mol. Sci.* **2018**, *19*, 1162. [CrossRef]
50. Geng, R.; Yin, J.; Zhou, J.; Jiao, T.; Feng, Y.; Zhang, L.; Chen, Y.; Bai, Z.; Peng, Q. In Situ Construction of Ag/TiO$_2$/g-C$_3$N$_4$ Heterojunction Nanocomposite Based on Hierarchical Co-Assembly with Sustainable Hydrogen Evolution. *Nanomaterials* **2019**, *10*, 1. [CrossRef]
51. Zhang, X.; Ren, B.; Li, X.; Liu, B.; Wang, S.; Yu, P.; Xu, Y.; Jiang, G. High-Efficiency Removal of Tetracycline by Carbon-Bridge-Doped g-C$_3$N$_4$/Fe$_3$O$_4$ Magnetic Heterogeneous Catalyst through Photo-Fenton Process. *J. Hazard. Mater.* **2021**, *418*, 126333. [CrossRef]
52. Anandan, S.; Wu, J.J.; Bahnemann, D.; Emeline, A.; Ashokkumar, M. Crumpled Cu$_2$O-g-C$_3$N$_4$ Nanosheets for Hydrogen Evolution Catalysis. *Colloids Surf. A Physicochem. Eng. Asp.* **2017**, *527*, 34–41. [CrossRef]
53. Xu, Q.; Jiang, C.; Cheng, B.; Yu, J. Enhanced Visible-Light Photocatalytic H$_2$-Generation Activity of Carbon/g-C$_3$N$_4$ Nanocomposites Prepared by Two-Step Thermal Treatment. *Dalt. Trans.* **2017**, *46*, 10611–10619. [CrossRef] [PubMed]
54. Zhu, Z.; Ma, C.; Yu, K.; Lu, Z.; Liu, Z.; Huo, P.; Tang, X.; Yan, Y. Synthesis Ce-Doped Biomass Carbon-Based g-C$_3$N$_4$ via Plant Growing Guide and Temperature-Programmed Technique for Degrading 2-Mercaptobenzothiazole. *Appl. Catal. B Environ.* **2020**, *268*, 118432. [CrossRef]
55. Yang, X.; Qian, F.; Zou, G.; Li, M.; Lu, J.; Li, Y.; Bao, M. Facile Fabrication of Acidified G-C$_3$N$_4$/g-C$_3$N$_4$ Hybrids with Enhanced Photocatalysis Performance under Visible Light Irradiation. *Appl. Catal. B Environ.* **2016**, *193*, 22–35. [CrossRef]
56. Kumar, A.; Raizada, P.; Singh, P.; Saini, R.V.; Saini, A.K.; Hosseini-Bandegharaei, A. Perspective and Status of Polymeric Graphitic Carbon Nitride Based Z-Scheme Photocatalytic Systems for Sustainable Photocatalytic Water Purification. *Chem. Eng. J.* **2020**, *391*, 123496. [CrossRef]
57. Rohilla, S.; Gupta, A.; Kumar, V.; Kumari, S.; Petru, M.; Amor, N.; Dalal, J. Excellent UV-light triggered photocatalytic performance of ZnO.SiO$_2$ nanocomposite for water pollutant compound methyl orange dye. *Nanomaterials* **2021**, *11*, 2548. [CrossRef]
58. Aditya, M.N.; Chellapandi, T.; Prasad, G.K.; Venkatesh, M.J.P.; Khan, M.M.R.; Madhumitha, G.; Roopan, S.M. Biosynthesis of Rod Shaped Gd$_2$O$_3$ on g-C$_3$N$_4$ as Nanocomposite for Visible Light Mediated Photocatalytic Degradation of Pollutants and RSM Optimization. *Diam. Relat. Mater.* **2022**, *121*, 108790. [CrossRef]
59. Li, Y.; Sun, H.; Peng, T.; Qing, X. Effect of Temperature on the Synthesis of G-C$_3$N$_4$/Montmorillonite and Its Visible-Light Photocatalytic Properties. *Clays Clay Miner.* **2022**, *70*, 555–565. [CrossRef]
60. Chellapandi, T.; Madhumitha, G. Facile Synthesis and Characterization of Carrisa Edulis Fruit Extract Capped NiO on MK30 Surface Material for Photocatalytic Behavior against Organic Pollutants. *Mater. Lett.* **2023**, *330*, 133215. [CrossRef]
61. Chellapandi, T.; Roopan, S.M.; Madhumitha, G. Interfacial Charge Transfer of Carrisa Edulis Fruit Extract Capped Co$_3$O$_4$ Nanoparticles on the Surface of MK30: An Efficient Photocatalytic Removal of Methylthioninium Chloride and Tetracycline Organic Pollutants. *Environ. Res.* **2022**, *219*, 115052. [CrossRef] [PubMed]
62. Shimi, A.K.; Parvathiraj, C.; Kumari, S.; Dalal, J.; Kumar, V.; Wabaidur, S.M.; Alothman, Z.A. Green synthesis of SrO nanoparticles using leaf extract of *Albizia julibrissin* and its recyclable photocatalytic activity: An eco-friendly approach for treatment of industrial wastewater. *Environ. Sci. Adv.* **2022**, *1*, 849–861. [CrossRef]
63. Ding, Y.; Lin, Z.; Deng, J.; Liu, Y.; Zhang, L.; Wang, K.; Xu, S.; Cao, S. Construction of Carbon Dots Modified Hollow G-C$_3$N$_4$ Spheres via in Situ Calcination of Cyanamide and Glucose for Highly Enhanced Visible Light Photocatalytic Hydrogen Evolution. *Int. J. Hydrog. Energy* **2022**, *47*, 1568–1578. [CrossRef]
64. Zhang, L.; Israel, E.M.; Yan, J.; Ritter, T. Copper-Mediated Etherification via Aryl Radicals Generated from Triplet States. *Nat. Synth.* **2022**, *1*, 376–381. [CrossRef]
65. Rosso, C.; Filippini, G.; Prato, M. Use of Nitrogen-Doped Carbon Nanodots for the Photocatalytic Fluoroalkylation of Organic Compounds. *Chem. Eur. J.* **2019**, *25*, 16032–16036. [CrossRef]
66. Meth-Cohn, O.; Narine, B.; Tarnowski, B. A Versatile New Synthesis of Quinolines and Related Fused Pyridines, Part 5. The Synthesis of 2-Chloroquinoline-3-Carbaldehydes. *J. Chem. Soc. Perkin Trans. 1* **1981**, 1520. [CrossRef]

**Disclaimer/Publisher's Note:** The statements, opinions and data contained in all publications are solely those of the individual author(s) and contributor(s) and not of MDPI and/or the editor(s). MDPI and/or the editor(s) disclaim responsibility for any injury to people or property resulting from any ideas, methods, instructions or products referred to in the content.

*Article*

# Highly Selective Nitrogen-Doped Graphene Quantum Dots/Eriochrome Cyanine Composite Photocatalyst for NADH Regeneration and Coupling of Benzylamine in Aerobic Condition under Solar Light

Ruchi Singh [1], Rajesh K. Yadav [1,*], Ravindra K. Shukla [1], Satyam Singh [1], Atul P. Singh [2], Dilip K. Dwivedi [3], Ahmad Umar [4,5,*,†] and Navneet K. Gupta [6]

[1] Department of Chemistry and Environmental Science, Madan Mohan Malviya University of Technology, Gorakhpur 273010, India
[2] Department of Chemistry, Chandigarh University Mohali, Mohali 140413, India
[3] Department of Physics and Materials Science, Madan Mohan Malaviya University of Technology, Gorakhpur 273010, India
[4] Department of Chemistry, Faculty of Science and Arts, and Promising Centre for Sensors and Electronic Devices (PCSED), Najran University, Najran 11001, Saudi Arabia
[5] Department of Materials Science and Engineering, The Ohio State University, Columbus, OH 43210, USA
[6] Centre for Sustainable Technologies, IISC Bangalore Gulmohar Marg, Banglaru 560012, India
* Correspondence: rajeshkr_yadav2003@yahoo.co.in (R.K.Y.); ahmadumar786@gmail.com (A.U.)
† Adjunct Professor at the Department of Materials Science and Engineering, The Ohio State University, Columbus, OH 43210, USA.

**Abstract:** Photocatalysis is an ecofriendly and sustainable pathway for utilizing solar energy to convert organic molecules. In this context, using solar light responsive graphene-based materials for C–N bond activation and coenzyme regeneration (nicotinamide adenine dinucleotide hydrogen; NADH) is one of the utmost important and challenging tasks in this century. Herein, we report the synthesis of nitrogen-doped graphene quantum dots (NGQDs)-eriochrome cyanine (EC) solar light active highly efficient "NGQDs@EC" composite photocatalyst for the conversion of 4-chloro benzylamine into 4-chloro benzylamine, accompanied by the regeneration of NADH from NAD+, respectively. The NGQDs@EC composite photocatalyst system is utilized in a highly efficient and stereospecific solar light responsive manner, leading to the conversion of imine (98.5%) and NADH regeneration (55%) in comparison to NGQDs. The present research work highlights the improvements in the use of NGQDs@EC composite photocatalyst for stereospecific NADH regeneration and conversion of imine under solar light.

**Keywords:** NADH regeneration; NGQDs@EC composite; Glucose; Graphene

## 1. Introduction

Using solar light for the conversion of sustainable resources into chemicals by mimicking natural photosynthesis is a challenge. Usages of solar light harvesting artificial photosynthesis are increasing. Due to amazing chemical, mechanical and physical properties, light-harvesting composites have attracted massive attention. These materials also show unique π-conjugated structures. Light harvesting composites are extremely useful in a variety of applications, including organic separation [1,2] gas storage [3] biocatalysis [4] etc. For the catalytic process, a range of solar light harvesting 2D graphene materials have been studied, including organic dye-based 2D graphene material [5], nitrogen-containing 2D graphene materials [6,7], and 2D graphene-based nanomaterials [8]. Among this, nitrogen-enriched graphene has attracted attention due to its solar light active photocatalytic ability [9], chemical stability, and extraordinary biocompatibility [10]. Furthermore, bulk nitrogen-enriched graphene has the problem of easy, fast recombination of solar light-irradiated electron–hole couples [11] and allegation of inadequate solar light absorption capacity-related lighting, which ruthlessly limits the catalytic performance in the presence of solar light for the regeneration of redox-active cofactors. To date, various efforts have been made to improve either of the two aspects in order to uplift the redox-active cofactors in photocatalytic phenomena. Therefore, to enhance the photocatalytic ability of nitrogen-enriched graphene quantum dots via π–π stacking of dyes [12], we report the synthesis of NGQDs-based eriochrome cyanine (EC) solar light highly efficient, i.e., NGQDs@EC composite photocatalyst for excellent conversion of NAD$^+$ (nicotinamide adenine dinucleotide) to NADH (nicotinamide adenine dinucleotide hydrogen) regeneration and coupling of benzylamine in aerobic conditions under solar lights, as shown in Scheme 1.

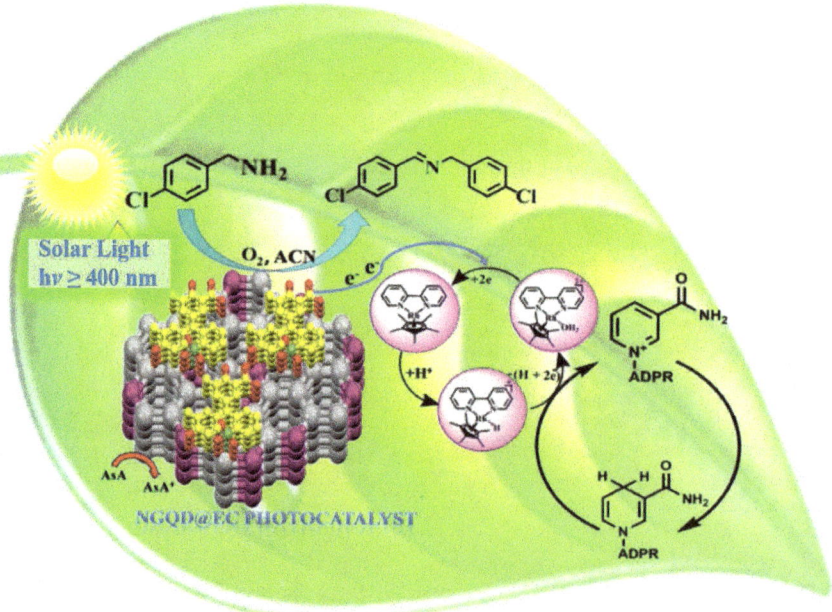

**Scheme 1.** Photocatalytic NADH regeneration and organic transformation.

## 2. Results and Discussion

The UV absorption studies of NGQDs and NGQDs@EC have been conducted in DMF. We evaluated the UV–visible spectra of the NGQDs composite to evaluate the optical properties of the newly generated highly selective NGQDs@EC, as shown in Figure 1a. In comparison to NGQDs, NGQDs@EC have a broad absorption peak at 576 nm in the visible region [13]. The optical band gap of NGQDs and NGQDs@EC photocatalyst are 2.32 eV and 2.15 eV, respectively. The optical band gap energy was calculated using the following equation: $E_g$ (eV) = 1240/λ (wavelength in nm). The covalent conjugation of melamine and glucose units (NGQDs) results in a significant increase in the absorption coefficient in the spectral window of 400 nm, allowing for a more efficient light-harvesting nature [14,15]. Figure 1b,c shows the direct and correct extrapolation approaches used to estimate the optical band gap of the synthesized photocatalysts, as shown in the tauc plots. From the tauc plots, the obtained band gap of the NGQDs@EC photocatalyst is 2.15 eV. Additionally, we compared the energy level or band gap position from the cyclic voltammetry [16]. The cyclic voltammetry (CV) experiment also confirms the same type of band gap (Figure 2a,b). As shown in Figure 2a, NGQDs@EC photocatalyst has oxidation and reduction potentials near +1.22 V and −0.88 V, respectively.

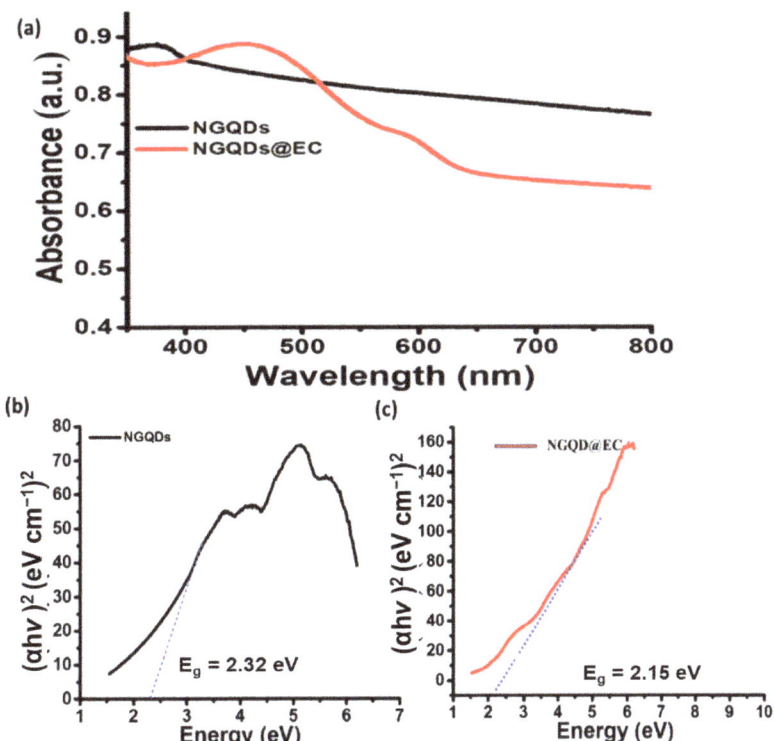

**Figure 1.** (a) UV–visible diffuse reflectance spectra (DRS) of NGQDs (black line) and NGQDs@EC (red line) photocatalyst, respectively, (b) Tauc plot of NGQDs, (c) Tauc plot of NGQDs@EC photocatalyst.

The collected redox potential data can be used to calculate the band gap [17]. Figure 2b shows the Latimer diagram derived from cyclic voltammetry, which confirmed the band gap of 2.10 eV, which is similar to the calculated band gap from Figure 1.

The Fourier-transform spectroscopy (FTIR) was used to identify the functional group. In Figure 3a, the FTIR spectrum of NGQDs has two main peaks: a peak centered at 1637 cm$^{-1}$ and a broad peak at 3402 cm$^{-1}$, both revealing O–H bonding. Along with these, peaks at 1255 cm$^{-1}$ and 1078 cm$^{-1}$ indicate the existence of C–H and C–O, respectively. After the π–π stacking of EC on NGQDs, the peaks were shifted to 3100 cm$^{-1}$ and 1092 cm$^{-1}$ due to C–H and C–O stretching and new peaks formed at 2164 cm$^{-1}$ [18–20]. These results clearly indicated that the EC chromophore was stacked on NGQDs via π–π stacking.

The thermal behavior of NGQDs and NGQDs@EC photocatalysts was investigated by differential scanning calorimetry (DSC, model: 2910) in Figure 3b at a heating rate of 5 °C/min under N$_2$ flow in the temperature range from 50 to 300 °C. The NGQDs@EC photocatalyst has strong water adsorption effects. Figure 3b clearly indicated that the synthesized NGQD@EC photocatalyst is stable up to 225 °C. As well the sublimation and thermal condensation of NGQDs was observed at 50–140 °C. So, the NGQD@EC photocatalyst is more stable and more efficient than NGQDs [21].

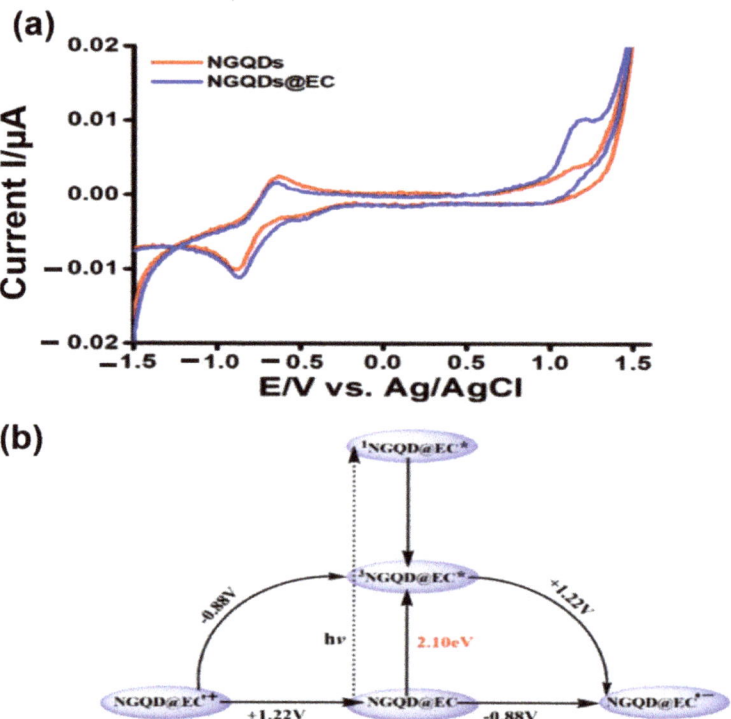

**Figure 2.** (**a**) Cyclic voltammetry (CV) of NGQDs (red line) and NGQDs@EC (blue line) photocatalyst, respectively, (**b**) Latimer diagram of NGQDs@EC photocatalyst.

The nature and the size of particles were investigated by X-ray diffraction pattern (XRD). Figure 4 shows the diffraction pattern of NGQDs@EC and NGQDs photocatalysts. The observed diffraction pattern exhibited few well-defined diffraction peak for NGQDs appeared at 2θ = 11.03°, 13.07°, 21.09°, and 27.08°. Interestingly, after the π–π stacking of EC on the NGQDS, the crystalline nature of the composite was increased with a significant shift in the diffraction peaks. Thus, the observed diffraction pattern for the NGQD@EC composite show various high-intensity peaks at 2θ = 13.1°, 20.4°, 24.32°, and 27.89° [22].

The zeta potential of NGQDs and NGQDs@EC were found to be −11.8 and −29.2 mV, as shown in Figure 5a,b respectively. In comparison to NGQDs, the NGQDs@EC photocatalyst has a higher negative zeta potential value, indicating that it is more stable. The higher chemical stability of the NGQDs@EC due to the creation of the C-N bond is likely to explain the greater negative zeta potential value when compared to the NGQDs [16,23].

### 2.1. The Enzymatically Active and Inactive 1,4-NADH Cofactor Regeneration

The goal of this research is to recover the enzymatically active 1,4-NADH (1,4-Nicotinamide adenine dinucleotide) cofactor from its oxidized versions, $NAD^+$. The $NAD^+$ undergoes an unselective protonation and radical coupling reaction, as shown in Scheme 2.

Because of this process, numerous NAD isomers, both active and inactive, are produced. Using an electron mediator, you can prevent the formation of enzymatically inactive isomers. Under sunlight irradiation, the rhodium complex mediator helps to regenerate enzymatically active 1,4-NADH isomer only. At room temperature in an inert atmosphere, photochemical regenerations of 1,4-NADH cofactors were carried out under artificial sunlight irradiation (λ > 420 nm) [15,16,24].

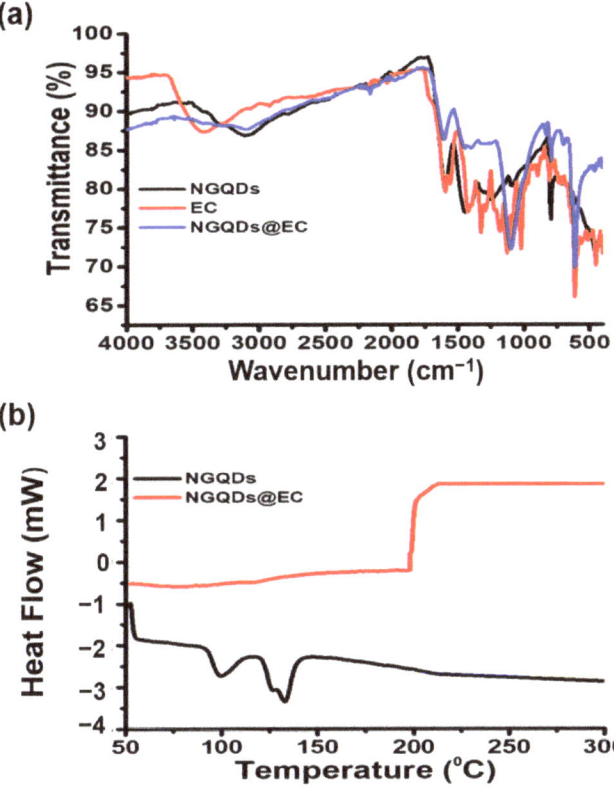

**Figure 3.** (**a**) FTIR spectra of EC (red line), NGQDs (black line), and NGQDs@EC (blue line) photocatalysts respectively, and (**b**) DSC of NGQDs@EC (red line) and NGQDs (black line) photocatalyst respectively.

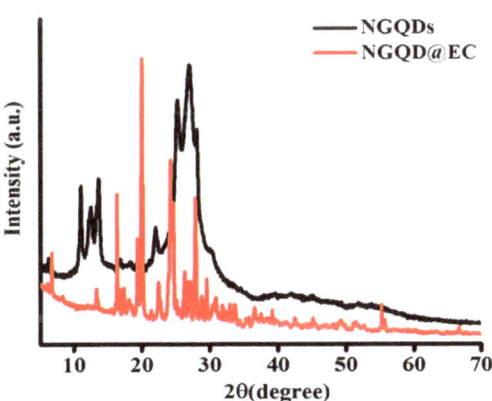

**Figure 4.** X-ray Diffraction pattern (XRD) of NGQDs and NGQD@EC.

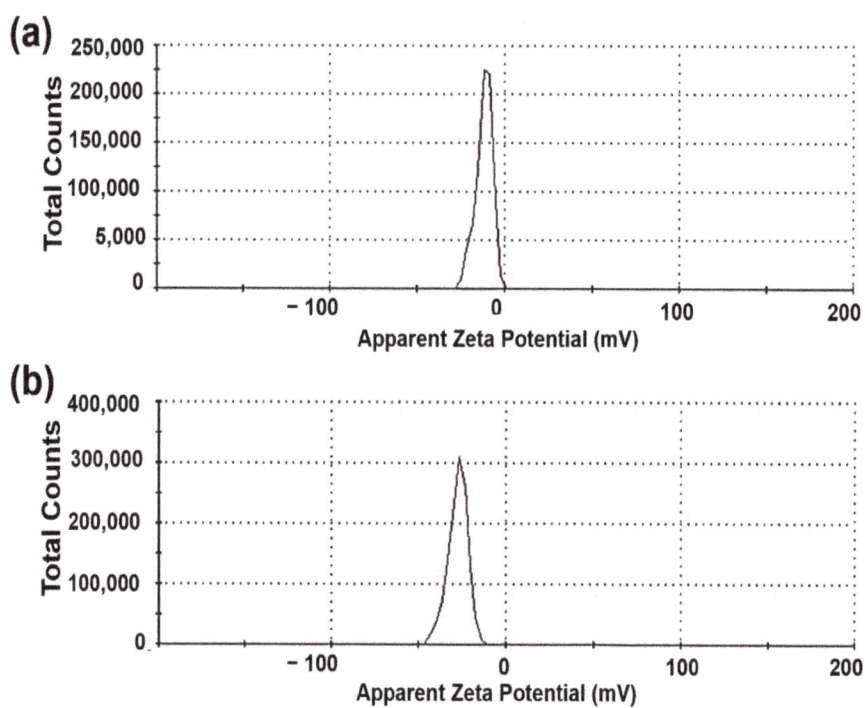

**Figure 5.** The negative potential graph of (**a**) NGQDs (−11.8 mV) and, (**b**) NGQDs@EC photocatalyst (−29.2 mV).

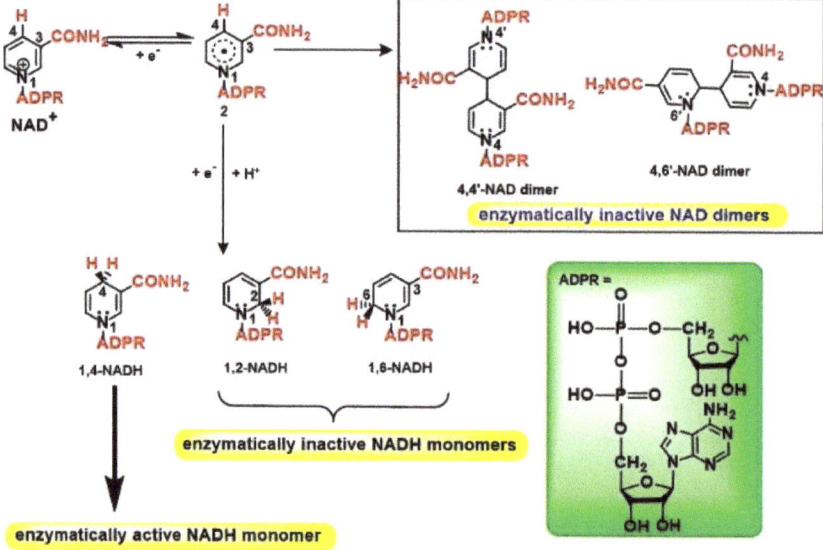

**Scheme 2.** Enzymatically active and inactive NADH isomer production via electrochemical reduction of $NAD^+$.

## 2.2. Mechanistic Pathway during the Regeneration of NADH Cofactors

Scheme 3 depicts a mechanistic route for the rebirth of NADH cofactors. The cationic form of Rh-complex, designated as A, can be readily hydrolyzed in an aqueous buffer medium to give the water-coordinated complex [Cp*Rh(bpy) (H$_2$O)]$^{2+}$, designated as B. The formate (HCOO$^-$) reacts with complex A via the hydride elimination process to produce complex B with the removal of CO$_2$ molecule [25]. The reduced form of complex D is formed after the supplying of charges to complex C by the photocatalyst. The rhodium hydride complex receives external electrons from the photocatalyst NGQDs@EC, resulting in the reduced intermediate D. NAD$^+$ can be coordinated with the D intermediate at this point, allowing hydride to be transferred to produce the regioselective NAD cofactors [26].

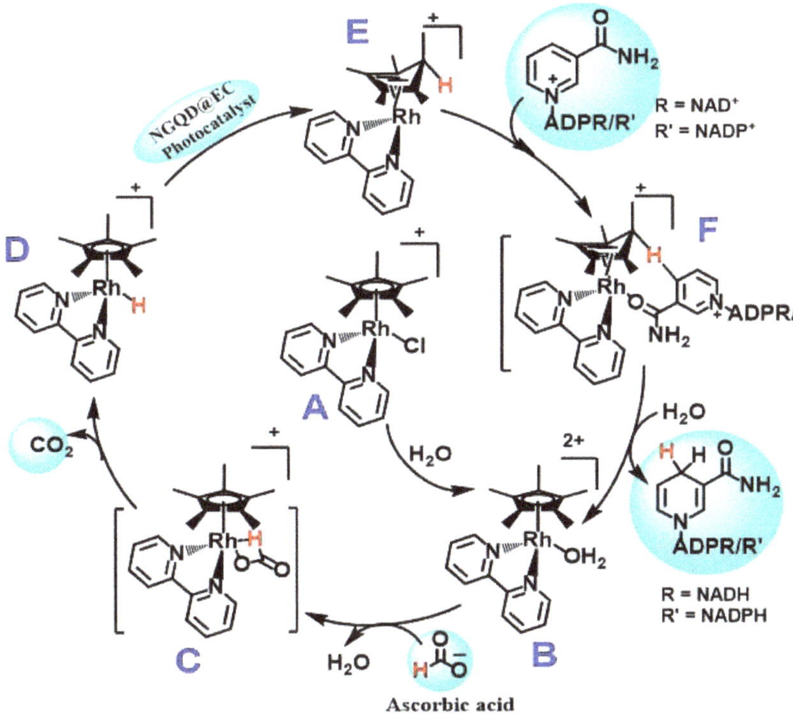

**Scheme 3.** The regeneration of NADH cofactors using NGQDs@EC photocatalyst.

## 2.3. Schematic Representation of Energy Level Diagram for Transfer of Photo-Excited Electron

The potential energy diagram is shown in Scheme 4. With the absorption of solar light by NGQDs@EC photocatalyst, an electron–hole pair is created in the valence band (VB). From the HOMO level (−5.72 eV) of NGQDs@EC photocatalyst to its conduction band (CB), photoexcited electrons are transferred via AsA. Thereafter, these form the LUMO (lowest unoccupied molecular orbital) level (−3.62 eV) of NGQDs@EC photocatalyst electrons transfers to NAD$^+$ (−4.20 eV) via rhodium complex (−3.96 eV) and lead to the regeneration of NADH. Hence, highly efficient regeneration of NADH cofactor occurs through the use of NGQDs@EC photocatalyst [24].

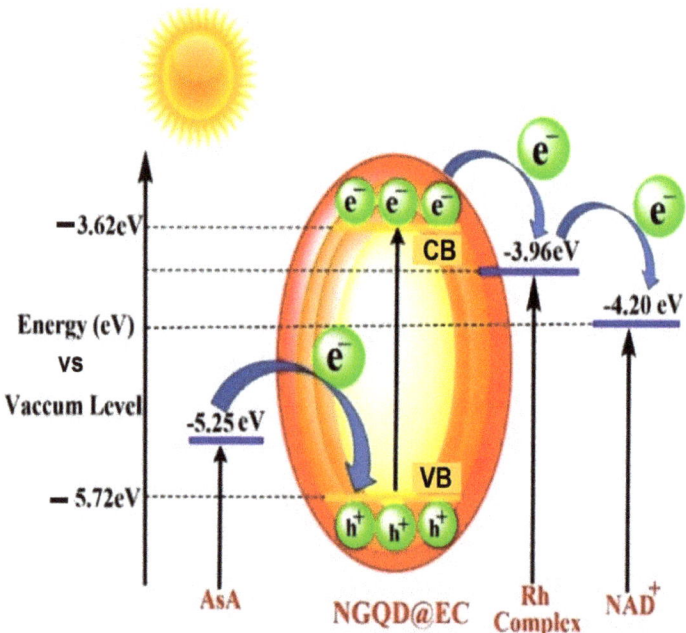

**Scheme 4.** The energy diagram shows electron generation and transfer of photoexcited electron.

### 2.4. Quantitative Analysis for Regeneration of NADH

As demonstrated in Figure 6, the yield (%) grew steadily in response to sun radiation. There was no yield obtained in the dark. In this experiment, product accumulation was faster, and the conversion of $NAD^+$ to NADH was 55% achieved in just 120 min from the NGQDs@EC photocatalyst. Therefore, the comparison of the photocatalytic performance of NGQDs and NGQDs@EC photocatalyst is important in this context. As a result, the ability of NGQDs and NGQDs@EC to photo-generate NADH under solar light was investigated, as shown in Figure 6, and an absolute increase in regeneration yield of about 55% was observed with NGQDs@EC photocatalyst compared to NGQDs under similar circumstances. As a result, a very promising production of NADH was observed with a yield percent of 55%, suggesting the huge potential of NGQDs@EC as a solar light harvesting photocatalyst [7].

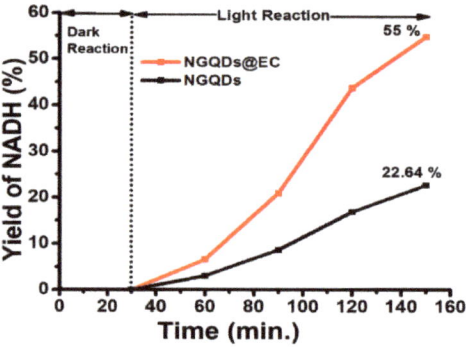

**Figure 6.** Regeneration of NADH from NGQDs (black line) and NGQDs@EC (red line) photocatalyst respectively.

The UV–visible performance for NADH concentration at 340 nm is studied as shown in Figure 7. Under photo-stationary circumstances, the constant rise in the absorption peak at 340 nm represents an increase in the transformation of $NAD^+$ to NADH over time. At 120 min, the highest conversion rate is achieved. In an artificial photosynthetic system, utilizing electron and proton transport channels, $NAD^+$ was reduced to NADH cofactors. [27].

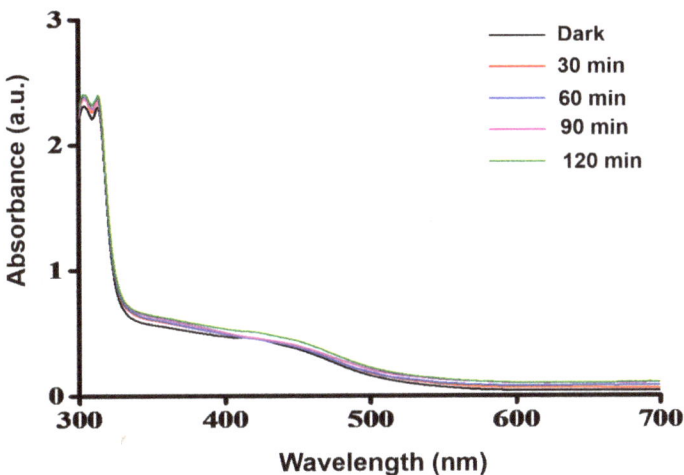

**Figure 7.** NGQDs@EC photocatalyst was used to record UV-visible spectroscopy for photocatalytic NADH regeneration at various time intervals.

Additionally, in Figure 8, the photostability of the NGQDs@EC photocatalyst was investigated under the same experimental conditions.

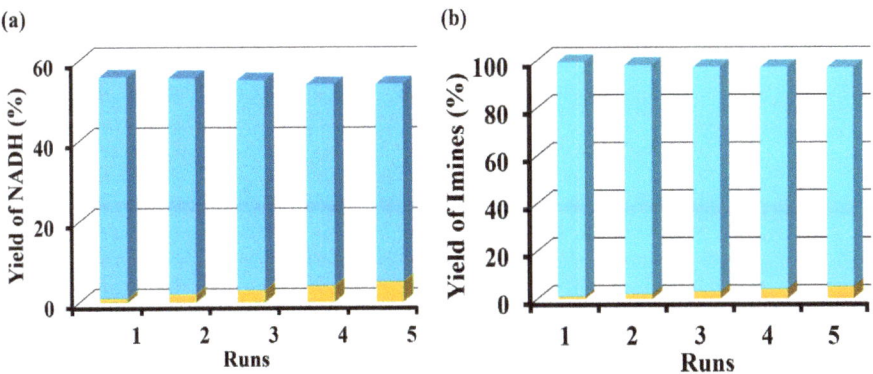

**Figure 8.** Photostability test of NGQDs@EC photocatalyst for (**a**) 1,4-NADH regeneration and, (**b**) Formation of Imine.

## 2.5. Photo-Chemically Coupling of Chlorobenzyl Amine in Presence of Oxygen

The oxidative coupling activity of the photocatalyst was screened in Table 1. When we optimized the reaction by choosing chlorobenzyl amine (125 μL) as a substrate, and NGQDs@EC (25 mg) as a photocatalyst in 10 mL ACN in the presence of $O_2$ under solar light, we obtained 99% yield and 99% selectivity of the product. In addition, standard reaction conditions using EC as a photocatalyst and NGQDs as a starting material provide a 33% and 28% yield of product, respectively. We have also screened the reaction in the absence of photocatalyst, solar light, and solvent; no product was received. The result confirms that a photocatalyst, sunlight, and a solvent (acetonitrile)are essential requirements for a photocatalytic oxidative coupling reaction (Scheme 5) [28].

**Table 1.** Results of screening experiments.

| S. No. | Photocatalyst | Solvent | Solar Light | Yield (%) |
|---|---|---|---|---|
| 1. | NGQD@EC | ACN | Yes | 98.5 |
| 2. | EC | ACN | Yes | 34 |
| 3. | NGQDs | ACN | Yes | 12 |
| 4. | Absence | ACN | Yes | 0 |
| 5. | NGQD@EC | ACN | No | 5 |

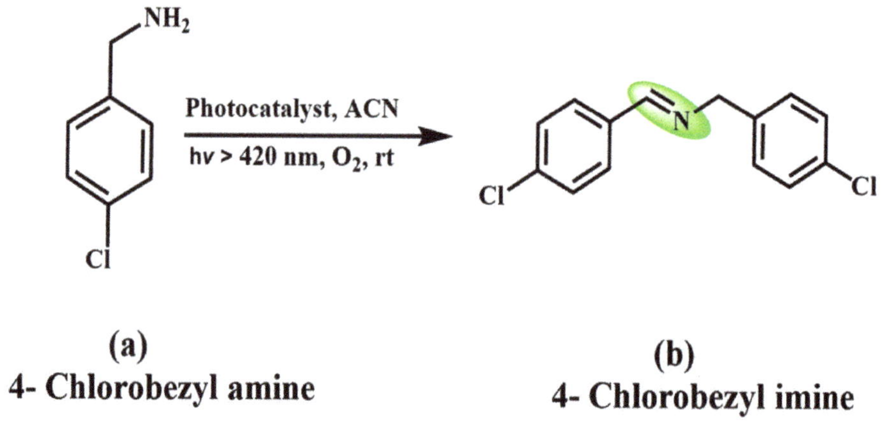

**(a)**
**4- Chlorobezyl amine**

**(b)**
**4- Chlorobezyl imine**

**Scheme 5.** Conversion of 4-Chlorobenzyl amine (a) to 4-Chlorobenzyl imine (b).

*2.6. Reaction Mechanism during the Photocatalytic Coupling Reaction*

As shown in Scheme 6, the reaction is given based on the previously reported literature [26,29]. Light irradiation caused charge separation in the photocatalyst, with photogenerated electrons created in the conduction band (CB) and photogenerated holes staying in the valence band (VB). Due to its 2D planar conjugated structure, the presence of NGQDs@EC promoted charge separation and supplied electron mobility on the surface of the NGQDs@EC photocatalyst [30]. These heated electrons recombined with $O_2$ molecules adsorbing on the surface of NGQDs@EC, forming the $O_2$ radical [30]. These $O_2$ radicals are extremely powerful oxidizing agents, capable of converting benzylamine to imine when exposed to solar light. At the same time, due to their great oxidizing properties, reactive holes in NGQDs@EC VB can directly oxidize benzylamine molecules. As a result, the electrons and holes that have been separated are fully participating in the photocatalytic process. The NGQDs@EC photocatalyst system is utilized in a highly efficient and stereospecific solar light active manner, leading to a higher conversion of imine (98.5%) in comparison to NGQDs.

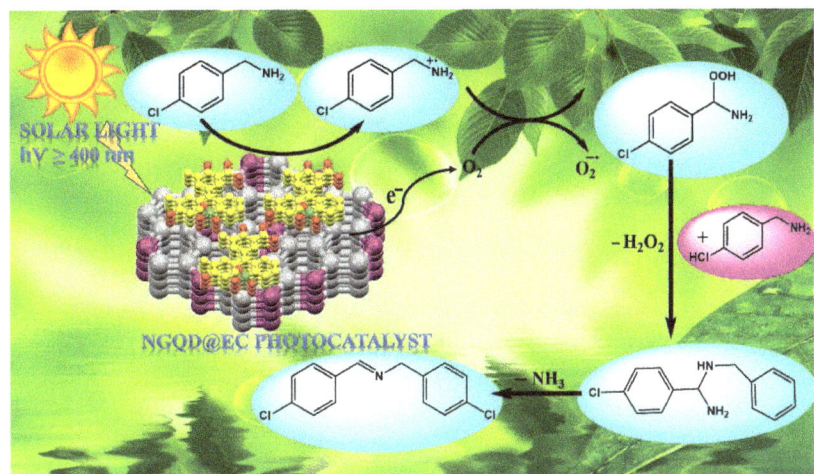

**Scheme 6.** The mechanistic route represents the coupling of benzylamine in the presence of oxygen and NGQDS@EC photocatalyst.

The green technology used in photocatalysis has, among its key benefits, the ability to purify water and clean the environment through solar light-induced photocatalysis. Numerous significant applications of photocatalysis exist, such as $CO_2$ reduction, organic pollutant degradation, removal of toxic ions and heavy metal ions, water splitting, antibacterial activity, self-cleaning process, and others. Lack of solar sensitivity and poorer efficiency are the key drawbacks of photocatalysis. The prepared NGQD@EC photocatalyst's performance was compared to that of a number of other photocatalysts that have already been published; intriguingly, it was observed that the studied NGQD@EC photocatalyst had greater photocatalytic performance compared to other published ones (Table 2). The prepared photocatalyst demonstrated outstanding photocatalytic performance, strong stability, reusability, and a highly light harvesting property.

**Table 2.** Comparative study of different photocatalyst for light reaction.

| S.No. | Photocatalyst | NADH Regeneration (%) | Conversion of Amine (%) | References |
|---|---|---|---|---|
| 1. | 5%Ag@rGO | — | 98% | [26] |
| 2. | CCG-BIODPY | 54.02% | 95% | [29] |
| 3. | CN/BW | — | 95% | [30] |
| 4. | CCGCMAQSP | 45.54% | — | [31] |
| 5. | NGQDs@EC | 55% | 98.5% | Our work |

## 3. Experimental Details

### 3.1. Chemicals and Materials

Graphite flakes, $NAD^+$ (nicotinamide adenine dinucleotide), (Pentamethylcyclopentadienyl) rhodium (III) chloride dimer, Glucose (G), melamine (M), eriochrome cyanine (EC), ethanol ($C_2H_5OH$), 4-chlorobenzylamine ($C_7H_8ClN$) and acetonitrile (MeCN) were purchased from Sigma Aldrich and were used as received.

### 3.2. Synthesis of Nitrogen-Doped Graphene Quantum Dot (NGQDs)

Nitrogen-doped graphene quantum dots (NGQDs) were synthesized (Scheme 7) in a one-step process reaction. In a typical method, glucose (1 g), and melamine (4 g) were mixed together and placed in the crucible. After, that crucible was placed in the muffle furnace for 2 h at 400 °C. The compound turned from a white color into a black color; finally, nitrogen-doped graphene quantum dots were obtained [22,32].

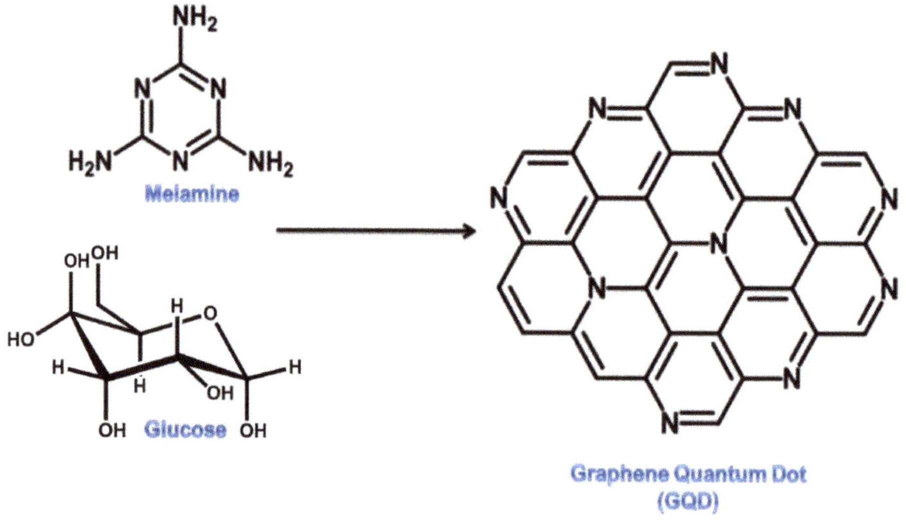

**Scheme 7.** Synthesis of NGQDs from Melamine and Glucose.

### 3.3. Synthesis of NGQDs@EC Photocatalyst

Eriochrome cyanine (750 mg) and previously prepared nitrogen-doped graphene quantum dots were dissolved in 50 mL of ethanol solution. The mixture was stirred at room temperature for 12 h and then centrifuged at 2000 rpm for 10 min. The filtrate was collected for additional analysis and photocatalysis. The compound NGQDs@EC photocatalyst was found to be 0.484 mg. The synthesis process is shown in Scheme 8 [33,34].

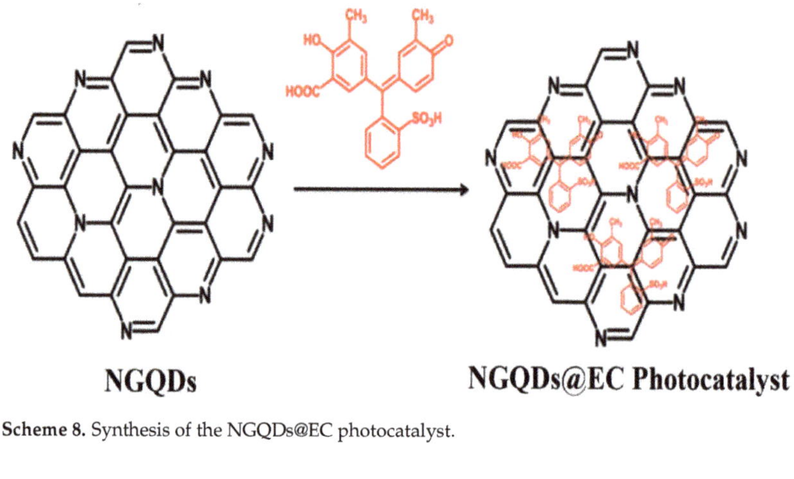

**Scheme 8.** Synthesis of the NGQDs@EC photocatalyst.

*3.4. Photocatalytic Studies*

Here, we report a self-assembled NGQDs@EC photocatalyst for 1,4-NADH regeneration, in which photoexcited electrons are rapidly pumped for NADH regeneration. Firstly, we prepared a reaction mixture for photocatalytic regeneration of 1,4-NADH. The reaction mixture contains 0.4 mM NAD$^+$ (248 µL), 0.2 mM electron mediator (124 µL) 0.1 M ascorbic acid (310 µL), and 10 µM NGQDs@EC photocatalyst (31 µL). Therefore, the reaction mixture was transferred to a quartz cuvette for solar light irradiation in a UV–visible spectrometer. All the ingredients were dissolved in a 0.1 M sodium phosphate buffer at pH 7.0. Then, 1,4-NADH regeneration was carried out at room temperature in a nitrogen environment. A UV–visible spectrometer was used to track the conversion in absorbance at 340 nm to quantify the photocatalytic 1,4-NADH regeneration [14,15,35].

*3.5. Formation of Imine in the Presence of Oxygen*

The synthesized NGQDs@EC photocatalyst was used to convert 4-cholrobenzylamine to imine due to the oxidative coupling of oxygen in the presence of solar light at room temperature. Acetonitrile was used as a solvent in the reaction medium. In the photocatalytic experiment, a 20 W white LED light was used as the light source along with a 20 mL quartz flask filled with acetonitrile (10 mL) and 4-chlorobenzylamine (125 µL), NGQDs@EC photocatalyst (25 mg). In the presence of an oxygen molecule, the resulting mixture was exposed to solar light while being stirred for 10 h. Thin layer chromatography (TLC) was used to track the progress of the reaction. The photocatalyst was separated through filtration after the reaction was completed, and the residue was concentrated under decreased pressure to get a crude product. Purification was performed using ethyl acetone column chromatography on silica gel with hexane as eluent, yielding the pure compound (shown in Scheme 9) [36].

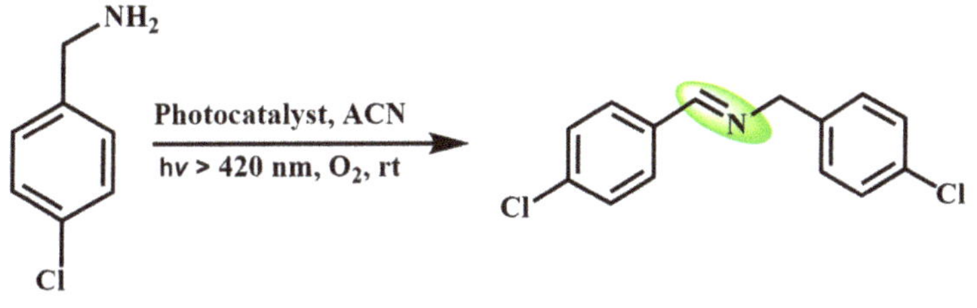

**(a)** 4- Chlorobezyl amine

**(b)** 4- Chlorobezyl imine

**Scheme 9.** Oxidative coupling of benzylamines under solar light.

### 4. Conclusions

We demonstrated an eco-friendly and sustainable pathway for the production and regeneration of imine and NADH from 4-chlorobenzyl amine and NAD$^+$ via highly selective nitrogen-doped graphene/eriochrome cyanine composite (NGQDs@EC) photocatalyst. The NGQD@EC photocatalyst was thoroughly studied for its photocatalytic performance using UV–visible spectroscopy, FTIR spectroscopy, DSC, Zeta potential, and cyclic voltametric studies, and important, influential factors were found. Additionally, five consecutive recycle stability tests were conducted and a comparison table for the conversion and regeneration of amines and NAD$^+$ was compiled. This study offers a simple technique for creating benign photocatalysts that are environmentally safe and have reasonably strong photocatalytic activity for the conversion and regeneration of industrial chemicals under visible light illumination.

**Author Contributions:** Conceptualization, R.S., R.K.Y., R.K.S., S.S., A.P.S. and A.U.; software, R.S., R.K.Y., R.K.S., S.S., A.P.S., A.U. and N.K.G.; validation, R.S., R.K.Y., R.K.S., S.S., A.P.S., A.U. and N.K.G.; formal analysis, R.S., R.K.Y., R.K.S., S.S., A.P.S., A.U. and N.K.G.; investigation, R.S., R.K.Y., R.K.S., S.S., A.P.S., A.U. and N.K.G.; writing—original draft, R.S., R.K.Y., R.K.S., S.S., A.P.S., D.K.D. and A.U.; writing—review & editing, R.S., R.K.Y., R.K.S., S.S., A.P.S., D.K.D., A.U. and N.K.G.; visualization, D.K.D.; project administration, R.K.Y. All authors have read and agreed to the published version of the manuscript.

**Funding:** The authors are thankful to the Deanship of Scientific Research at Najran University, Najran, Kingdom of Saudi Arabia for funding under the Research Group funding program grant no. NU/RG/SERC/11/1.

**Data Availability Statement:** No Supplementary data is available.

**Acknowledgments:** Authors are thankful to MMMUT, Gorakhpur-273010, U. P., India for providing the platform for research. The authors are thankful to the Deanship of Scientific Research at Najran University, Najran, Kingdom of Saudi Arabia for funding under the Research Group funding program grant no. NU/RG/SERC/11/1.

**Conflicts of Interest:** The authors declare no competing financial interests.

## References

1. Yasuhiro, T.; Vayssieres, L.; Durrant, J. Artificial photosynthesis for solar water-splitting. *Nat. Photonics* **2012**, *6*, 511–518.
2. Serena, B.; Drouet, S.; Francà, L.; Gimbert-Suriñach, C.; Guttentag, M.; Richmond, C.; Stoll, T.; Llobet, A. Molecular artificial photosynthesis. *Chem. Soc. Rev.* **2014**, *43*, 7501–7519.
3. Osterloh, F.E. Inorganic Materials as Catalysts for Photochemical Splitting of Water. *Chem. Mater.* **2008**, *20*, 35–54. [CrossRef]
4. Wang, X.; Saba, T.; Yiu, H.H.P.; Howe, R.; Anderson, J.; Shi, J. Cofactor NAD(P)H regeneration inspired by heterogeneous pathways. *Chem* **2017**, *2*, 621–654. [CrossRef]
5. Liu, J.; Antonietti, M. Bio-inspired NADH regeneration by carbon nitride photocatalysis using diatom templates. *Energy Environ. Sci.* **2013**, *6*, 1486–1493. [CrossRef]
6. Huang, J.; Antonietti, M.; Liu, J. Bio-inspired carbon nitride mesoporous spheres for artificial photosynthesis: Photocatalytic cofactor regeneration for sustainable enzymatic synthesis. *J. Mater. Chem. A* **2014**, *2*, 7686–7693. [CrossRef]
7. Gupta, S.K.; Yadav, R.; Gupta, A.; Yadav, B.; Singh, A.; Pande, B. Highly Efficient S-g-CN/Mo-368 Catalyst for Synergistically NADH Regeneration Under Solar Light. *Photochem. Photobiol.* **2021**, *97*, 1498–1506. [CrossRef]
8. Yang, D.; Zou, H.; Wu, Y.; Shi, J.; Zhang, S.; Wang, X.; Han, P.; Tong, Z.; Jiang, Z. Constructing quantum dots@flake g-$C_3N_4$ isotype heterojunctions for enhanced visible-light-driven NADH regeneration and enzymatic hydrogenation. *Ind. Eng. Chem. Res.* **2017**, *56*, 6247–6255. [CrossRef]
9. Wan, J.; Choi, W.; Kim, J.; Kuk, S.; Lee, S.; Park, C. Self-Assembled Peptide-Carbon Nitride Hydrogel as a Light-Responsive Scaffold Material. *Biomacromolecules* **2017**, *18*, 3551–3556.
10. Shifa, T.A.; Wang, F.; Liu, Y.; He, J. Heterostructures Based on 2D Materials: A Versatile Platform for Efficient Catalysis. *Adv. Mater.* **2019**, *31*, 1804828. [CrossRef]
11. Iyer, M.S.K.; Patil, S.; Singh, A. Flame Synthesis of Functional Carbon Nanoparticles. *Trans. Indian Natl. Acad. Eng.* **2022**, *7*, 787–807. [CrossRef]
12. Kang, H.; Liu, H.; Li, C.; Sun, L.; Zhang, C.; Gao, H.; Yin, J.; Yang, B.; You, Y.; Jiang, K.; et al. Polyanthraquinone-Triazine—A Promising Anode Material for High-Energy Lithium-Ion Batteries. *ACS Appl. Mater. Interfaces* **2018**, *10*, 37023–37030. [CrossRef] [PubMed]
13. Chaubey, S.; Yadav, R.; Tripathi, S.K.; Yadav, B.; Singh, S.; Kim, T.W. Covalent Triazine Framework as an Efficient Photocatalyst for Regeneration of NAD(P)H and Selective Oxidation of Organic Sulfide. *Photochem. Photobiol.* **2022**, *98*, 150–159. [CrossRef] [PubMed]
14. Singh, S.; Yadav, R.; Kim, T.; Singh, C.; Singh, P.; Singh, A.; Singh, A.; Singh, A.; Beag, J.; Gupta, S. Rational design of a graphitic carbon nitride catalytic–biocatalytic system as a photocatalytic platform for solar fine chemical production from $CO_2$. *React. Chem. Eng.* **2022**, *7*, 1566–1572. [CrossRef]
15. Singh, P.; Yadav, R.; Kim, T.; Yadav, T.; Gole, V.; Gupta, A.; Singh, K.; Kumar, K.; Yadav, B.; Dwivedi, D. Solar light active flexible activated carbon cloth-based photocatalyst for Markovnikov-selective radical-radical cross-coupling of S-nucleophiles to terminal alkyne and liquefied petroleum gas sensing. *J. Chin. Chem. Soc.* **2021**, *68*, 1435–1444. [CrossRef]
16. Pan, D.; Zhang, J.; Li, Z.; Wu, M. Hydrothermal Route for Cutting Graphene Sheets into Blue-Luminescent Graphene Quantum Dots. *Adv. Mater.* **2010**, *22*, 734–738. [CrossRef]
17. Li, Y.; Hu, Y.; Zhao, Y.; Shi, G.; Deng, L.; Hou, Y.; Qu, L. An electrochemical avenue to green-luminescent graphene quantum dots as potential electron-acceptors for photovoltaics. *Adv. Mater.* **2011**, *23*, 776–780. [CrossRef]
18. Chen, J.; Collier, C. Noncovalent Functionalization of Single-Walled Carbon Nanotubes with Water-Soluble Porphyrins. *J. Phys. Chem. B* **2005**, *109*, 7605–7609. [CrossRef]

19. Li, B.; Cao, H.; Yin, G.; Lu, Y.; Yin, J. Facile synthesis of silver@graphene oxide nanocomposites and their enhanced antibacterial properties. *J. Mater. Chem.* **2011**, *21*, 13765–13768. [CrossRef]
20. Silva, S.P.; Moraes, D.; Samios, D. Iron Oxide Nanoparticles Coated with Polymer Derived from Epoxidized Oleic Acid and Cis-1,2-Cyclohexanedicarboxylic Anhydride: Synthesis and Characterization. *J. Mater. Sci. Eng.* **2016**, *5*, 1000247. [CrossRef]
21. Wang, J.; Lu, C.; Chen, T.; Hu, L.; Du, Y.; Yao, Y.; Goh, M. Simply synthesized nitrogen-doped graphene quantum dot (NGQD)-modified electrode for the ultrasensitive photoelectrochemical detection of dopamine. *Nanophotonics* **2020**, *9*, 3831–3839. [CrossRef]
22. Chaubey, S.; Singh, P.; Singh, C.; Singh, S.; Shreya, S.; Yadav, R.; Mishra, S.; Jeong, Y.-J.; Biswas, B.; Kim, T. Ultra-efficient synthesis of bamboo-shape porphyrin framework for photocatalytic $CO_2$ reduction and consecutive C-S/C-N bonds formation. *J. CO2 Util.* **2022**, *59*, 101968. [CrossRef]
23. Cimino, P.; Troiani, A.; Pepi, F.; Garzoli, S.; Salvitti, C.; Di Rienzo, B.; Barone, V.; Ricci, A. From ascorbic acid to furan derivatives: The gas phase acid catalyzed degradation of vitamin C. *Phys. Chem. Chem. Phys.* **2018**, *20*, 17132–17140. [CrossRef]
24. Singh, C.; Yadav, R.; Kim, T.; Baeg, J.-O.; Singh, A. Greener One-step Synthesis of Novel In Situ Selenium-doped Framework Photocatalyst by Melem and Perylene Dianhydride for Enhanced Solar Fuel Production from $CO_2$. *Photochem. Photobiol.* **2022**, *98*, 998–1007. [CrossRef]
25. Yadav, R.K.; Oh, G.; Park, N.; Kumar, A.; Kong, K.; Baeg, J. Highly selective solar-driven methanol from $CO_2$ by a photocatalyst/biocatalyst integrated system. *J. Am. Chem. Soc.* **2014**, *136*, 16728–16731. [CrossRef]
26. Kumar, A.; Sadanandhana, A.M.; Jain, S.L. Silver doped reduced graphene oxide as promising plasmonic photocatalyst for oxidative coupling of benzylamines under visible light irradiation. *New J. Chem.* **2019**, *43*, 9116–9122. [CrossRef]
27. Bajorowicz, B.; Reszczyńska, J.; Lisowski, W.; Klimczuk, T.; Winiarski, M.; Słoma, M.; Zaleska-Medynska, A. Perovskite-type $KTaO_3$–reduced graphene oxide hybrid with improved visible light photocatalytic activity. *RSC Adv.* **2015**, *5*, 91315–91325. [CrossRef]
28. Feng, Y.; Wang, G.; Liao, J.; Li, W.; Chen, C.; Li, M.; Li, Z. Honeycomb-like ZnO Mesoporous Nanowall Arrays Modified with Ag Nanoparticles for Highly Efficient Photocatalytic Activity. *Sc. Rep.* **2017**, *7*, 11622. [CrossRef]
29. Yadav, R.K.; Baeg, J.-O.; Kumar, A.; Kong, K.; Oh, G.; Park, N.-J. Graphene–BODIPY as a photocatalyst in the photocatalytic–biocatalytic coupled system for solar fuel production from $CO_2$. *J. Mater. Chem.* **2014**, *2*, 5068–5076. [CrossRef]
30. Yuan, A.; Lei, H.; Wang, Z.; Dong, X. Improved photocatalytic performance for selective oxidation of amines to imines on graphitic carbon nitride/bismuth tungstate heterojunctions. *J. Colloid Interface Sci.* **2020**, *560*, 40–49. [CrossRef]
31. Yadav, R.K.; Baeg, J.-O.; Oh, G.; Park, N.-J.; Kong, K.; Kim, J.; Hwang, D.W.; Biswas, S.K. A Photocatalyst–Enzyme Coupled Artificial Photosynthesis System for Solar Energy in Production of Formic Acid from $CO_2$. *J. Am. Chem. Soc.* **2012**, *134*, 11455–11461. [CrossRef]
32. Liu, F.; Huang, K.; Ding, S.; Dai, S. One-step synthesis of nitrogen-doped graphene-like meso-macroporous carbons as highly efficient and selective adsorbents for $CO_2$ capture. *J. Mater. Chem. A* **2016**, *4*, 14567–14571. [CrossRef]
33. Zhang, K.; Li, H.; Shi, H.; Hong, W. Polyimide with enhanced $\pi$ stacking for efficient visible-light-driven photocatalysis. *Catal. Sci. Technol.* **2021**, *11*, 4889–4897. [CrossRef]
34. Mou, Z.; Dong, Y.; Li, S.; Du, Y.; Wang, X.; Yang, P.; Wang, S. Eosin Y functionalized graphene for photocatalytic hydrogen production from water. *Int. J. Hydrogen Energy* **2011**, *36*, 8885–8893. [CrossRef]
35. Singh, S.; Yadav, R.; Kim, T.; Singh, C.; Singh, P.; Chaubey, S.; Singh, A.; Beag, J.; Gupta, S.; Tiwary, D. Generation and Regeneration of the C($sp^3$)–F Bond and 1,4-NADH/NADPH via Newly Designed S-$gC_3N_4$@$Fe_2O_3$/LC Photocatalysts under Solar Light. *Energy Fuels* **2022**, *36*, 8402–8412. [CrossRef]
36. Kumar, A.; Hamdi, A.; Coffinier, Y.; Addad, A.; Roussel, P.; Boukherroub, R.; Jain, S. Visible light assisted oxidative coupling of benzylamines using heterostructured nanocomposite photocatalyst. *Chemistry* **2018**, *356*, 457–463. [CrossRef]

**Disclaimer/Publisher's Note:** The statements, opinions and data contained in all publications are solely those of the individual author(s) and contributor(s) and not of MDPI and/or the editor(s). MDPI and/or the editor(s) disclaim responsibility for any injury to people or property resulting from any ideas, methods, instructions or products referred to in the content.

*Article*

# Highly Efficient Self-Assembled Activated Carbon Cloth-Templated Photocatalyst for NADH Regeneration and Photocatalytic Reduction of 4-Nitro Benzyl Alcohol

Vaibhav Gupta [1], Rajesh K. Yadav [1,*], Ahmad Umar [2,3,4,*,†], Ahmed A. Ibrahim [2,3], Satyam Singh [1], Rehana Shahin [1], Ravindra K. Shukla [1], Dhanesh Tiwary [5], Dilip Kumar Dwivedi [6], Alok Kumar Singh [7], Atresh Kumar Singh [7] and Sotirios Baskoutas [8]

1. Department of Chemistry and Environmental Science, Madan Mohan Malviya University of Technology, Gorakhpur 273010, India
2. Department of Chemistry, Faculty of Science and Arts, Promising Centre for Sensors and Electronic Devices (PCSED), Najran University, Najran 11001, Saudi Arabia
3. Centre for Scientific and Engineering Research, Najran University, Najran 11001, Saudi Arabia
4. Department of Materials Science and Engineering, The Ohio State University, Columbus, OH 43210, USA
5. Department of Chemistry, Indian Institute of Technology (BHU), Varanasi 221005, India
6. Department of Physics and Materials Science, Madan Mohan Malaviya University of Technology, Gorakhpur 273010, India
7. Department of Chemistry, Deen Dayal Upadhyaya Gorakhpur University, Gorakhpur 273009, India
8. Department of Materials Science, University of Patras, 26504 Patras, Greece
* Correspondence: rajeshkr_yadav2003@yahoo.co.in (R.K.Y.); ahmadumar786@gmail.com (A.U.)
† Adjunct Professor at the Department of Materials Science and Engineering, The Ohio State University, Columbus, OH 43210, USA.

**Citation:** Gupta, V.; Yadav, R.K.; Umar, A.; Ibrahim, A.A.; Singh, S.; Shahin, R.; Shukla, R.K.; Tiwary, D.; Dwivedi, D.K.; Singh, A.K.; et al. Highly Efficient Self-Assembled Activated Carbon Cloth-Templated Photocatalyst for NADH Regeneration and Photocatalytic Reduction of 4-Nitro Benzyl Alcohol. *Catalysts* **2023**, *13*, 666. https://doi.org/10.3390/catal13040666

Academic Editors: Indra Neel Pulidindi, Archana Deokar and Aharon Gedanken

Received: 10 January 2023
Revised: 19 March 2023
Accepted: 23 March 2023
Published: 29 March 2023

**Copyright:** © 2023 by the authors. Licensee MDPI, Basel, Switzerland. This article is an open access article distributed under the terms and conditions of the Creative Commons Attribution (CC BY) license (https://creativecommons.org/licenses/by/4.0/).

**Abstract:** This manuscript emphasizes how structural assembling can facilitate the generation of solar chemicals and the synthesis of fine chemicals under solar light, which is a challenging task via a photocatalytic pathway. Solar energy utilization for pollution prevention through the reduction of organic chemicals is one of the most challenging tasks. In this field, a metal-based photocatalyst is an optional technique but has some drawbacks, such as low efficiency, a toxic nature, poor yield of photocatalytic products, and it is expensive. A metal-free activated carbon cloth (ACC)–templated photocatalyst is an alternative path to minimize these drawbacks. Herein, we design the synthesis and development of a metal-free self-assembled eriochrome cyanine R (EC-R) based ACC photocatalyst (EC-R@ACC), which has a higher molar extinction coefficient and an appropriate optical band gap in the visible region. The EC-R@ACC photocatalyst functions in a highly effective manner for the photocatalytic reduction of 4-nitro benzyl alcohol (4-NBA) into 4-amino benzyl alcohol (4-ABA) with a yield of 96% in 12 h. The synthesized EC-R@ACC photocatalyst also regenerates reduced forms of nicotinamide adenine dinucleotide (NADH) cofactor with a yield of 76.9% in 2 h. The calculated turnover number (TON) of the EC-R@ACC photocatalyst for the reduction of 4-nitrobenzyl alcohol is $1.769 \times 10^{19}$ molecules. The present research sets a new benchmark example in the area of organic transformation and artificial photocatalysis.

**Keywords:** EC-R@ACC photocatalyst; NADH regeneration; 4-nitro benzyl alcohol; solar light; photocatalytic

## 1. Introduction

Solar light has emerged as a sustainable and greener energy source for various solar chemical synthesis reactions. In the past few years, solar light-induced chemical transformations have been extensively achieved by eco-friendly processes [1,2]. In this context, the enlargement of an artificial substitute for this smart system continues to be an extraordinary challenge in the chemical society [1–6]. The recent research, therefore, involves synthesizing and designing a photoreactor system as a photocatalyst for the selective regeneration of fine chemicals and the reduction of aromatic compounds under solar light. As reported previously, we noted that about 4% and 46% of the overall solar light accessible on the planet falls in the UV and visible ranges, respectively [7,8]. Consequently, a solar light i.e., a visible light-responsive photocatalyst is significantly important for the synthesis of solar fine chemicals, such as nicotinamide adenine dinucleotide phosphate (NADPH) and nicotinamide adenine dinucleotide (NADH) cofactor and the photoreduction of organic compounds. Additionally, a significantly important feature to be noted is that in photocatalytic systems, NADH is important for the fixation of carbon dioxide ($CO_2$). Therefore, the regeneration of the NADH cofactor and the reduction of the organic compound via a highly efficient pathway is the only way to make it economically and industrially feasible [9–12].

In this context, green chemistry and a related photoreactor, solar light, as one of the reactants of chemical synthesis, is a rising research area [13,14]. Solar light sponsored NADPH, NADH cofactor, and the photoreduction of organic compounds by various solar light harvestings materials, such as graphitic carbon nitride, graphene composites, and titanium oxide $TiO_2$, were well discovered [15,16]. Besides the photoreduction and degradation in the presence of solar light, different chemical reactions, such as [2 + 2] cycloaddition [17,18], reductive dehalogenation [19], hydrogen formation from 1,4 asymmetric alkylation [20,21], and dihydropyridine [22], are also described in the reported literature where various solar light-responsive metal complexes are utilized as the solar light-assisted photocatalyst. In spite of the utmost expensive metal complexes, solar chemical conversion can also be attained by using energy harvesting materials as a solar light-responsive photocatalyst [23]. It is evident from the literature that photocatalytic NADH regeneration and organic transformation consuming solar light illumination are a rising research zone, which provides various possibilities for future work. In this addition, $NAD^+$ is a bio-enzyme that needs a steady flux of solar light to photocatalyze a solar chemical synthesis [24]. Thus, solar light is essential in the chemical transformation reaction catalyzed by the enzyme. In the nonexistence of solar light and a photocatalyst, the $NAD^+$ type enzyme remains completely inactive during the catalytic reaction. To date, various types of photo enzymes have been investigated, which are used as a photosystem for organic transformation and solar chemical regeneration [25]. It is supposed that most of the possible formerly existing solar light active enzyme derivatives were sorted out by progression and that nowadays, solar light active coenzymes are only the past survivors of this pathway [24]. The use of many metal-based compounds for solar chemical regeneration and environmental remediation has achieved the utmost attention in current times due to the active utilization of naturally existing solar energy and an effective solar light active system to terminate various types of unwanted materials [26,27]. Furthermore, photosynthetic pathways can increase the fast and complete conversion of solar energy into solar fine chemicals [28]. Expensive metals such as CdS, ZnO, and $TiO_2$ are expensive materials utilized as a photocatalyst in various fields due to their good photocatalytic ability [29]. However, the key weakness of such types of photocatalysts is captured only in the ultraviolet (UV) quota of the solar light spectrum [30]. A diversity of semiconductor materials has been broadly utilized as light-harvesting photocatalysts by engineering or tuning the energy gap position and for effective use in the solar light spectrum [31]. Over the decades, expensive metal-based semiconductor materials have played an important role in solar light photocatalysis due to their narrow band gap and ionic conductivity, etc. [32,33]. The metal orbital along with a lone pair is combined with supporting materials, such as graphene, carbon activated cloth, and graphitic carbon nitride to generate a shift valence band (VB) and conduction band (CB) that tends to create the suitable band gap [34]. Among the metal-based photocatalysts, graphene has lately been utilized for the photocatalytic NADH/NADPH regeneration and conversion of organic substrates in polar and non-polar solvents under solar light illumination [35]. Expensive metals may exist as different crystalline phases along with scheelite tetragonal zircon tetragonal and scheelite-monoclinic [36]. It is a fact that the properties of the photocatalyst always depend on the nature of the crystal structure [37].

The monoclinic structure of a few expensive metals exhibits stronger photocatalyst ability under ultraviolet light illumination due to a small energy gap compared to other phasic structures [38]. Additionally, pure expensive metal has some restrictions, such as the rate of low absorption capacity of incident light, the fast recombination ability of solar light-created electrons, and a lack of pores in the structure [39]. It is of utmost importance to strengthen its solar light or solar spectrum absorption range ability and confine the recombination of solar light-generated holes and electrons in order to improve the ability of the photocatalyst under solar light illumination [40]. In this regard, many approaches have been designed by different researchers [16], such as n and p-type doping, fabricating a heterostructure solar light active system, and combining bias energy [41]. The combination of a few expensive metals with different types of porous materials, such as silica, alumina, glass, zeolites, and activated carbon cloth (ACC) is conducted to enhance the performance of the solar light absorption ability and overwhelm the charge carrier's recombination rate [42,43]. Among these, activated carbon cloth (ACC) has excellent physical and chemical properties to construct solar light active materials. The porous properties, structural stability, strong solar spectrum adsorption efficiency, and larger surface area of ACC enable healthier adsorption of substrates, making it a promising material that supports a photocatalytic procedure [44,45]. Thus, it permits the photocatalyst to absorb a solar spectrum that further leads to solar fine chemical production and the conversion of organic substrates [46,47]. A number of alterations have been described by many researchers on the solar light active catalyst surface using different supporting materials by various pathways of synthesis, such as sol-gel, co-precipitation, hydrothermal, solvothermal, and microwave synthesis [48]. Among these, the hydrothermal pathway is very easy to use to prepare different types of light-harvesting composites for solar chemical regeneration and organic transformations.

The solar chemical regeneration and photoreduction of NADH and organic compounds using solar radiation are developing new disciplines for future research. For the synthesis of new organic chemicals, the mechanism for the reduction of organic functionality is critical. To reduce organic functional groups, numerous methods have been reported, including (i) metal/acid reduction, (ii) photocatalytic reduction, (iii) electrolytic reduction, and (iv) catalytic hydrogen transfer [49]. One of the most prominent pollution control and disposal processes is the photocatalytic reduction of aromatic nitro compounds. Nitrogen-containing compounds are commonly created as by-products in a variety of industries and factories, including agrichemicals and pharmaceuticals. Among the many nitrogen-containing compounds, 4-nitrobenzyl alcohol is one of the most common by-products that is harmful to the environment [50,51].

4-amino benzyl alcohol (4-ABA) is made in the pharmaceutical industry by reducing 4-nitro benzyl alcohol (4-NBA). 4-ABA is a necessary precursor for the manufacture of a variety of drugs, including paracetamol, phenacetin, and acetanilide [52]. Metal-based photocatalysts in acidic conditions are utilized to reduce 4-NBA to 4-ABA to the greatest extent possible. However, such a procedure produces toxic metal oxide sludge that is harmful to the environment [52].

To address the aforementioned difficulties, a self-assembled metal-free self-assembled eriochrome cyanine R (EC-R) based ACC photocatalyst (EC-R@ACC) for the regeneration of NADH cofactors and the conversion of 4-NBA to 4-ABA is created. The synthesized metal-free EC-r@ACC photocatalyst has received a lot of interest in photocatalytic reactions because of its outstanding physicochemical properties, such as a suitable band gap, high molar extinction coefficient, low rate of intersystem crossing, excellent photocatalytic ability, easier synthesis, and excellent chemical stability. When compared to the metal-based photocatalyst, the metal-free self-assembled EC-r@ACC composite demonstrated significantly higher efficiency for NADH cofactor regeneration and the production of 4-ABA via artificial photocatalysis. Due to the utilization of environmentally acceptable and sustainable solar energy, artificial photocatalysis has sparked great interest in the synthesis of solar chemicals (NADH) and the reduction of 4-NBA to 4-ABA. A schematic representation of the photocatalytic reduction of 4-NBA to 4-ABA and NADH regeneration under a solar light spectrum is represented in Scheme 1.

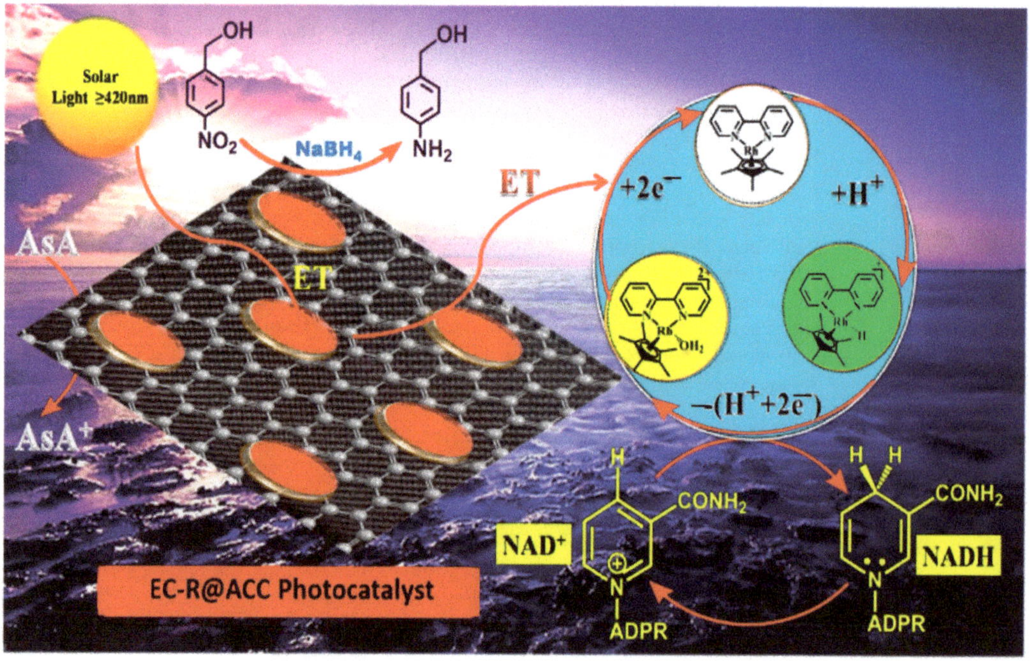

**Scheme 1.** Schematic representation of NADH regeneration and photocatalytic reduction of 4-NBA under solar light.

## 2. Results and Discussion

We introduced our study utilizing 4-NBA (a) as the substrate in the open air. The optimization of the photocatalytic reaction was performed in different solvents (Table 1). For photocatalytic optimization under different solvents, a low-cost, environmentally friendly EC-R@ACC (0.010 g) and 4-NBA (0.045 g) were utilized as a solar light spectrum harvester photocatalyst and starting material model substrate, respectively. The intended 4-ABA product was achieved with 50% conversion when the reaction was optimized in $C_2H_5OH$ (49% yield) in 12 h. We screened the same reaction in different solvents: $C_2H_5OH$, PEG (Polyethylene glycol), $CH_2Cl_2$, and DMF under the same reaction conditions. DMF has the highest polar nature among the solvents, so it provides an excellent yield in 12 h.

We found that when the reaction was carried out with 4-NBA (0.045 g), the EC-R@ACC photocatalyst (0.010 g), and DMF (30 mL) at room temperature under a solar light spectrum for 12 h, the highest conversion (97%) and yield (96%) of 4-ABA was achieved. In contrast, the conversion and yield were reduced in different organic solvents as the reaction time increased, indicating that DMF is the most efficient at promoting the organic transformation reaction [53].

### 2.1. Mechanistic Pathway for the Photoreduction of 4-NBA

Based on the outcomes and the fiction, it is clear that the current photocatalytic reduction pathway includes several critical steps, involving: (I) the adsorption of the reacting molecules on the catalyst's surface, (II) the excitation of the EC-R@ACC photocatalyst to its triplet state and the transfer of electrons easily from sodium borohydride to the newly designed EC-R@ACC photocatalyst, (III) electron transfer from the EC-R@ACC photocatalyst to 4-NBA, (IV) the transfer of hydrogen from the $BH_4$/solvent to 4-NBA, and (V) the desorption of the products from the edges of the newly designed photocatalyst. Scheme 2 depicts the likely mechanistic routes of the current photocatalytic conversion of 4-NBA to 4-ABA.

## 2.2. Presence of Conformational Isomers during 1,4-NADH Synthesis

As shown in Scheme 3, NAD$^+$ is reduced directly for radical coupling and unselective protonation reaction. During this phase, many NAD isomers are formed, which have a presence in both enzymatically inactive and active states. To avoid the generation of enzymatically inactive isomers, an electron mediator must be utilized. Only when exposed to sun rays, the Rh-complex electron mediator supports the creation of the enzymatically solar light spectrum active 1,4-NADH isomer [54].

The reaction buffer medium for NADH regeneration contained 248 µL NAD$^+$ solution, 124 µL Rh-complex, 310 µL ascorbic acid (AsA), 2387 µL phosphate buffer, and 15 mg of the photocatalyst EC-R@ACC photocatalyst. Under continuous solar light irradiation cut by a 420 nm band-pass filter, the reaction was performed in a quartz cuvette as a reactor with a magnetic stir.

**Table 1.** Optimization of photocatalytic reduction of 4-NBA.

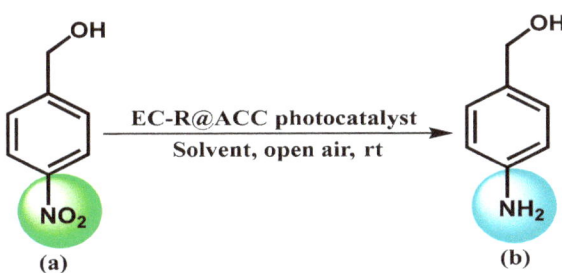

| Entry | Solar Light | Photocatalyst | Solvent | Time | Conversion (%) | Yield (%) |
|---|---|---|---|---|---|---|
| 1. | Yes | EC-R@ACC | C$_2$H$_5$OH | 12 | 50 | 49 |
| 2. | Yes | EC-R@ACC | C$_2$H$_5$OH | 6 | 47 | 48 |
| 3. | Yes | EC-R@ACC | PEG | 12 | 55 | 56 |
| 4. | Yes | EC-R@ACC | PEG | 6 | 51 | 49 |
| 5. | Yes | EC-R@ACC | CH$_2$Cl$_2$ | 12 | 60 | 58 |
| 6. | Yes | EC-R@ACC | CH$_2$Cl$_2$ | 6 | 58 | 57 |
| 7. | Yes | EC-R@ACC | DMF | 12 | 97 | 96 |
| 8. | Yes | EC-R@ACC | DMF | 6 | 79 | 78 |
| 9. | Yes | EC-R | DMF | 6 | 46 | 45 |
| 10. | No | EC-R@ACC | DMF | 12 | 05 | 05 |
| 11. | Yes | EC-R@ACC | Absent | 12 | 10 | 10 |
| 12. | Yes | Absent | DMF | 12 | 05 | 05 |

Reaction conditions: EC-R@ACC photocatalyst (0.010 g), 'a' (0.045 g), and NaBH$_4$ (5 mg/L, various solvents (30 mL)) illuminated under a solar light spectrum for 12 h in an inert atmosphere at room temperature. (a) and (b) represent the reactant and product.

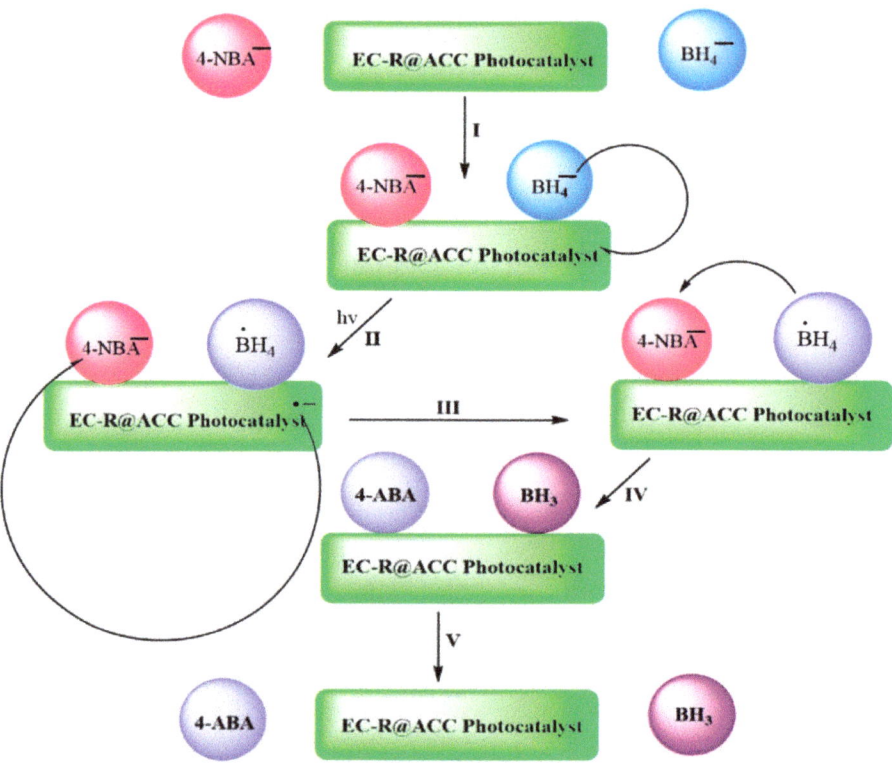

**Scheme 2.** Mechanistic studies of the photocatalytic reduction of 4-NBA to 4-ABA.

The cofactor of 1, 4-NADH rejuvenation was carried out in an inert environment at ambient temperature under the effect of sunlight (>420 nm). An FG@ACC photocatalyst (31 µL), AsA (310 µL), electron mediator (124 µL), and β –NAD$^+$ (248 µL) were dissolved in 2387 µL of sodium phosphate buffer at neutral pH 7.0. First, the process was run in the absence of a solar light spectrum for 30 min, and no cofactor of NADH regeneration was achieved. The cofactor of 1,4-NADH regeneration was achieved in presence of solar light spectrum illumination, as illustrated in Figure 1. It was discovered that as the reaction time increased, so did the yield of 1,4-NADH. The absorbance at 340 nm in the UV-visible spectrum was used to calculate the amount of 1,4-NADH produced. As per a previous report [54], the molar absorption/extinction coefficient of the cofactor of 1,4-NADH is 6.22 mM$^{-1}$ cm$^{-1}$ [54]. We achieved 76.9% catalytic efficiency of 1,4-NADH in two hours (2 h) utilizing a highly stable and solar light active newly designed EC-R@ACC photocatalyst (shown in Figure 1). Because of the π-π interaction, the photocatalytic ability of the solar light active newly designed EC-R@ACC photocatalyst is greater than that of its precursor EC-R [40,54].

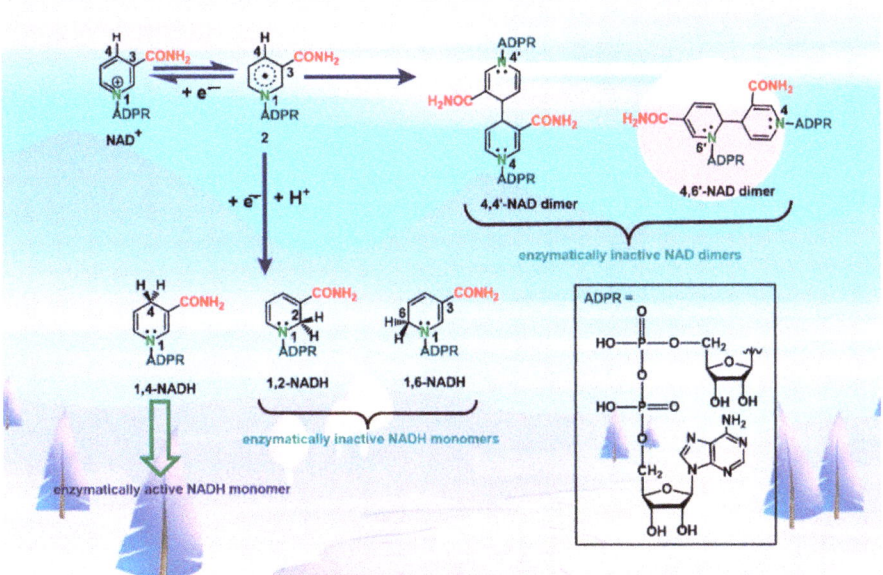

**Scheme 3.** The enzymatically active 1,4 NADH cofactor regeneration after NAD+ reduction.

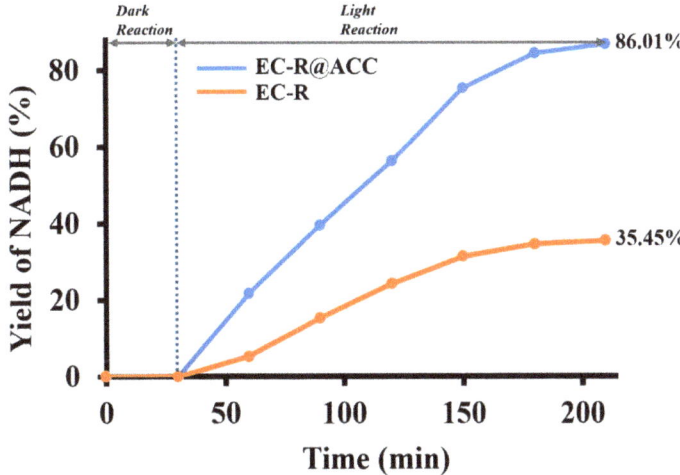

**Figure 1.** The photocatalytic activity of the EC-R@ACC photocatalyst and EC-R for NADH regeneration under solar light.

*2.3. Reaction Mechanism of Photocatalytic NADH Regeneration*

Scheme 4 demonstrates a possible approach for photocatalytic regeneration of NADH using the EC-R@ACC photocatalyst. During the photocatalytic activity, the cationic type of the electron mediator (A) hydrolyzes, yielding a water-coordinated complex (B), symbolized as $[Cp*Rh(bpy)(H_2O)]^{2+}$. The complex (B) interacts with the formate ($HCOO^-$) during the hydride removal procedure [54]. This reaction generates the Rh hydride complex (C), i.e., $[Cp*Rh(bpy)(H)]^+$, and the release of $CO_2$. When the EC-R@ACC photocatalyst contributes electrons to the complex of Rh, the reduced intermediate complex (D) is generated

(C). NAD$^+$ interacts with complex D via the activity of amide and transfer of hydride, due to which the NADH cofactor's region-selectivity regenerates.

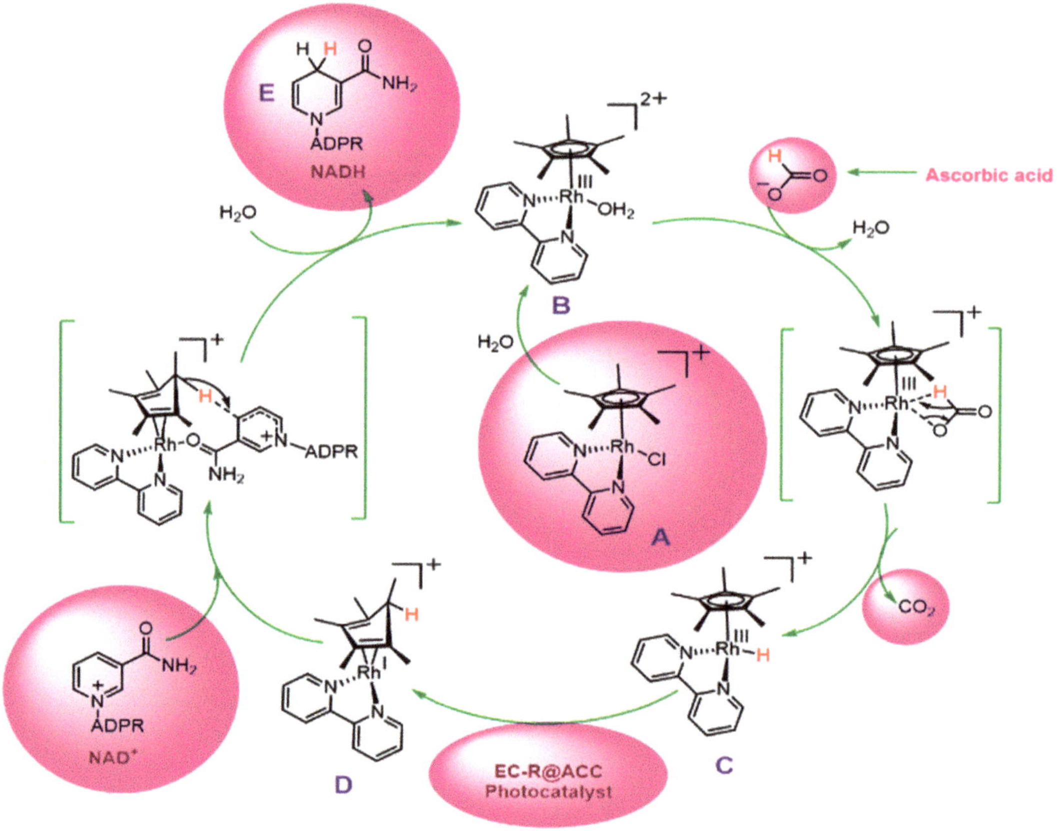

**Scheme 4.** Photocatalytic NADH regeneration utilizing an EC-R@ACC photocatalyst mechanism.

### 2.4. Solar Light-Induced Catalytic 1,4 NADH Regeneration

We exclusively focused on regenerating the enzymatically active form of 1,4-NADH from the oxidized form of NAD$^+$. As shown in Scheme 5, an electron mediator was used to prevent the transformation of undesirable isomeric forms, resulting in the artificial photocatalytic transformation of 1,4 NADH under sunlight irradiation. A neutral solution (pH 7.0) of phosphate-buffered solution (NaH$_2$PO$_4$–Na$_2$HPO$_4$, 0.1 M) and an NAD+ cofactor along with scavenger agents were used to regenerate NADH. In addition, combined with the recently synthesized EC-R@ACC photocatalyst, [Cp*Rh(bpy)Cl]Cl was introduced to the reaction media.

Recycling experiments for NADH synthesis in the presence of EC-R@ACC photocatalyst were conducted by recycling the newly designed same photocatalyst several times under identical conditions to investigate the utility and sustainability of the EC-R@ACC photocatalyst under the same experimental conditions. During the reusability test, a nearly constant conversion yield was observed with no appreciable decline in efficiency, suggesting that the EC-R@ACC photocatalyst has strong catalytic strength [55]. Furthermore, the extra experiments were carried out under sunlight in the absence of NAD$^+$. No absorbance peak was observed at a wavelength of 340 nm in this experiment, which suggests that NADH cannot be produced in the absence of NAD$^+$ (Figure 1). Generally, every component of the artificial photosynthetic machinery is important, including solar light, EC-R@ACC, and NAD$^+$. It should be noted that to eliminate the photo-saturation during the measurement of UV-Visible spectra, the concentration of the reaction media, which included AsA, NAD$^+$, EC-R@ACC, and Rhodium complex, was kept quite low [54,55].

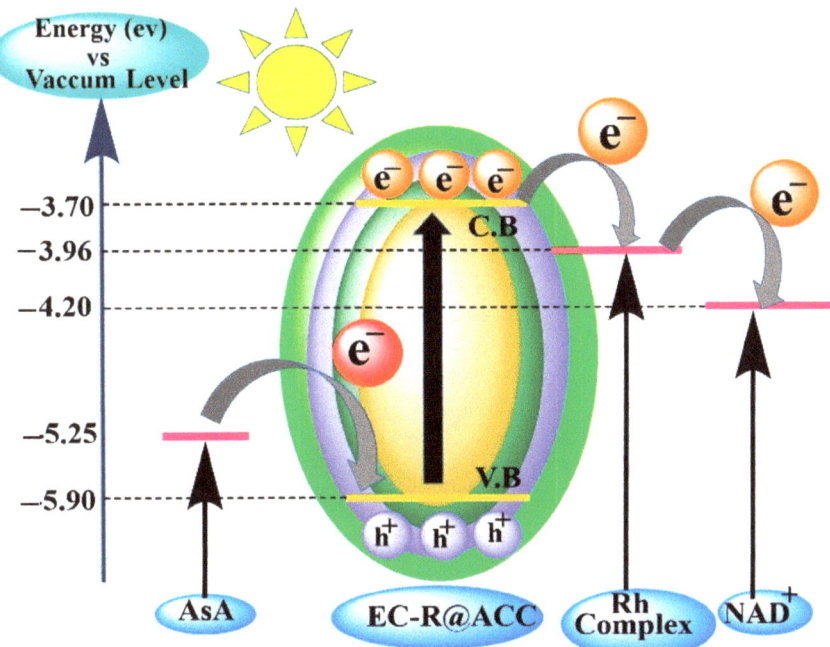

**Scheme 5.** A simplified potential energy diagram showing carrier generation and its migration in the photocatalytic system.

Scheme 5 shows the energy-labeled diagram, which depicts the pathway of the induced charge carriers in the photocatalytic system. Initially, on the irradiation of solar light, ascorbic acid becomes oxidized, and the electron of the EC-R@ACC photocatalyst is transferred from the valence band/highest occupied molecular orbital (HOMO) to the conduction band. Subsequently, the electron jumps from the conduction band/lowest unoccupied molecular orbital (LUMO) of the EC-R@ACC photocatalyst to the conduction band Rh-complex due to the lower band gap of the Rh-complex. Thus, the Rh-complex acts as an electron mediator. After the Rh-complex electron, the electron follows the same manner and easily jumps to the conduction band of NAD$^+$. Here, NAD$^+$ is reduced in NADH (Nicotinamide Dinucleotide Adenine Hydrate). In this photocatalytic process, after the reduction of NADH, the photocatalytic system is able to follow the Calvin cycle and mimic the natural photosynthetic route [54].

*2.5. Study of UV-Visible Spectra of Newly Designed Solar Light Spectrum Responsive EC-R@ACC Photocatalyst*

UV-Visible spectroscopy (UV-Visible-1900i, Shimadzu, Japan) was utilized to study the absorption spectrum of the ACC and EC-R@ACC in DMF (Figure 2). The UV-Visible spectra of the ACC were observed at about 250 nm [56], whereas the absorption band of the EC-R@ACC was observed at 545 nm. We estimated the optical band gap using the Scherrer equation (1240/λ) and found it to be 4.96 eV and 2.29 eV, respectively, indicating that it can operate as an active catalyst. The predicted optical band gap (2.29 eV) validates redshift and boosts solar-driven activation. The results showed that the EC-R and ACC absorb a lot of visible light. The absorption spectra of the newly designed EC-R@ACC photocatalyst in the bathochromic shift are most significant and are responsible for cofactor 1,4-NADH regeneration and organic transformation, which improves its solar light harvesting abilities/capabilities in the solar light spectrum region.

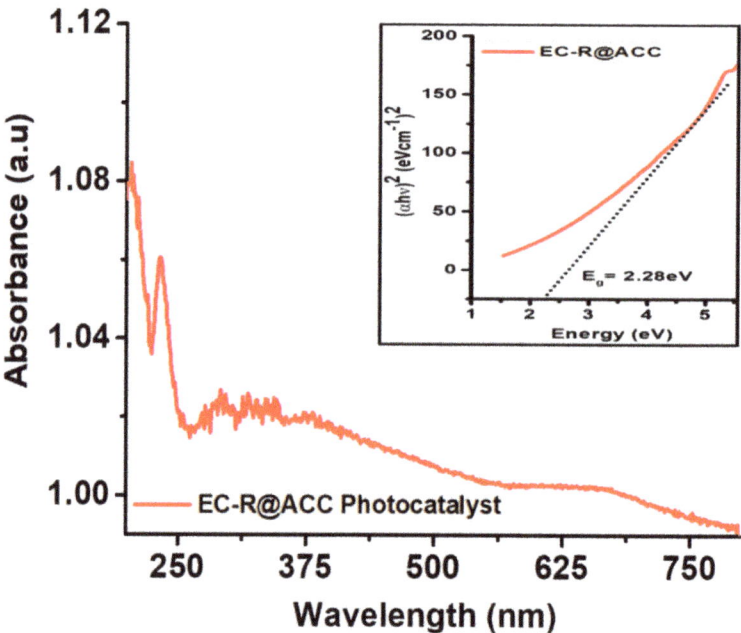

**Figure 2.** UV-Visible absorption spectra of the photocatalyst.

The optical band gap of the EC-R@ACC photocatalyst was computed using the Scherrer equation (1240/λ) [57], and it is close to 2.28 eV at about 540 nm. The cyclic voltammetry (CV, K-lyte electrochemical station,) measurement supports the computed optical band gap by the Scherrer method with the value of 2.20 eV [57]. The reduction and oxidation energy potential values of the newly designed EC-R@ACC photocatalyst were achieved by the CV measurement technique (Figure 3). The EC-R@ACC photocatalyst oxidation and reduction potential values were +1.10 V and −1.10 V, respectively. The collected reduction and oxidation energy potential data were utilized to calculate the energy gap/band gap.

The CV experiment authorizes the energy gap/band gap calculation (see Figures 3 and 4) [57]. A CV experiment was used to measure the reduction and oxidation energy potential values of the EC-R@ACC photocatalyst. The reduction and oxidation energy potentials were measured as +1.10 V and −1.10 V, respectively. The reduction and oxidation values gathered can be utilized to compute the band gap using the Latimer diagram (Figure 4), which verifies the optical band gap. Bathochromic shifts in the absorption spectra of the EC-R@ACC photocatalyst were detected, which boosts its ability to harvest sunlight.

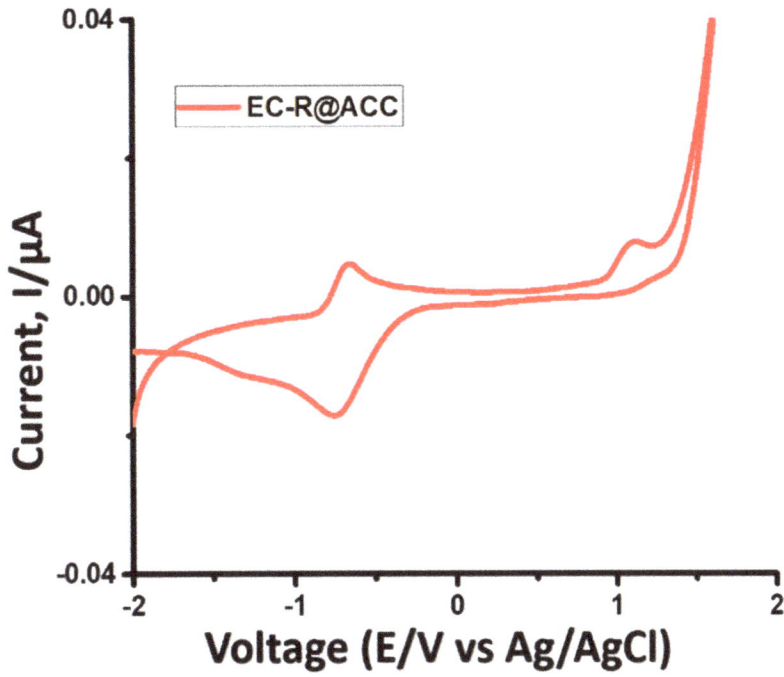

**Figure 3.** Cyclic voltammetry of the EC-R@ACC photocatalyst.

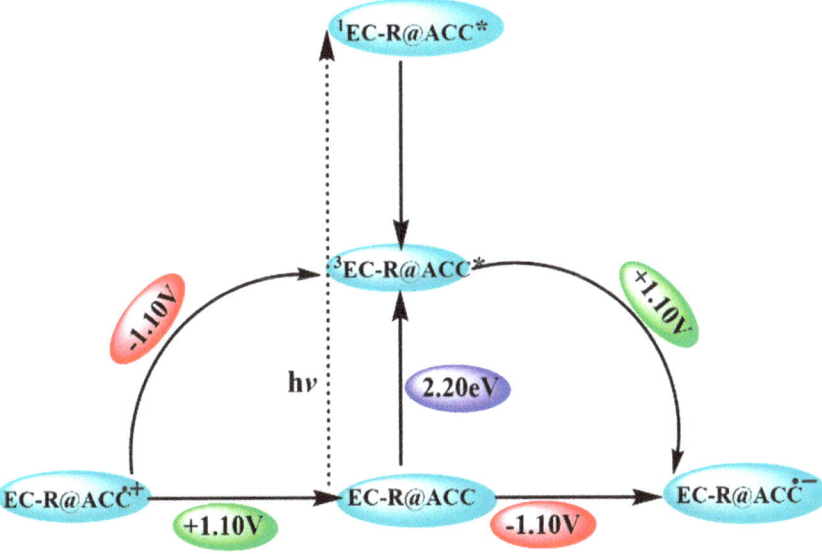

**Figure 4.** Latimer Diagram displaying the photo-redox property of the EC-R@ACC photocatalyst; (* represent the excited state.)

*2.6. Study of Zeta Potential of ACC and EC-R@ACC Photocatalyst*

The zeta potential (Malvern Panalytical, Nano-zetasizer (NZS90), Malvern, UK) of the EC-R@ACC photocatalyst was observed and showed an additional negative value of

−40.1 mV, while the ACC showed a value of −23.7 mV (Figure 5) [58]. It is illustrated that the synthesis of the EC-R@ACC composite provides the more negatively charged fractions as it has a high content of EC-R. Additionally, the more negative zeta potential value for the EC-R@ACC photocatalyst proves that the interaction between the ACC and EC-R is quite good [59].

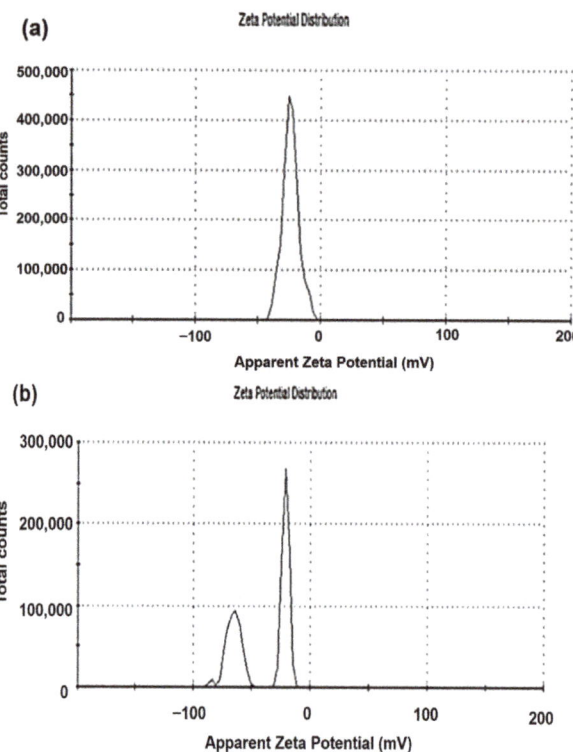

**Figure 5.** Studies of the (**a**) ACC (−23.7 mV) and (**b**) EC-R@ACC solar light spectrum photocatalyst (−40.1 mV) by Zeta potential pathway.

The FTIR spectra (Shimadzu, IRspirit FTIR-8000, Anan, Japan) of the EC-R@ACC photocatalyst, as well as the ACC and EC-R in Figure 6, demonstrated the occurrence of an interaction in the EC-R@ACC. Figure 6 indicates that the FTIR spectrum of the EC-R@ACC displays a stretching peak of approximately 3450 cm$^{-1}$, confirming the existence of the -OH group [60]. The stretching peak of $SO_3^-$ is also found at about 1250 cm$^{-1}$ [61]. The stretching peak of –COONa is also found at about 950 cm$^{-1}$ [62]. The stretching peak of –CO is also found at about 1050 cm$^{-1}$ [40]. The stretching peak of –CH$_3$ is also found at about 2850 cm$^{-1}$; however, it is completely absent in the ACC FTIR spectra [61]. The results show that the interaction in the EC-R@ACC photocatalyst was formed successfully. Additionally, we recycled the EC-R@ACC photocatalyst for more than four consecutive runs (i.e., four reuses) under the same reaction circumstances (Figure 7). It was perceived that the photocatalytic cofactor 1,4-NADH regeneration is almost constant in all the recycles, confirming the highest solar light spectrum harvesting stability of the EC-R and EC-R@ACC photocatalyst, respectively. In addition, the observed results revealed that the EC-R@ACC photocatalyst possesses higher stability (Figure 7a) compared to the EC-R photocatalyst (Figure 7b), which clearly revealed that the EC-R@ACC is superior to the EC-R photocatalyst. The turnover number (TON) of the EC-R photocatalyst for the reduction of 4-nitrobenzyl alcohol is calculated from the below-mentioned equation [63]:

TON = No. of substrate molecules converted into the product by 1 g of photocatalyst

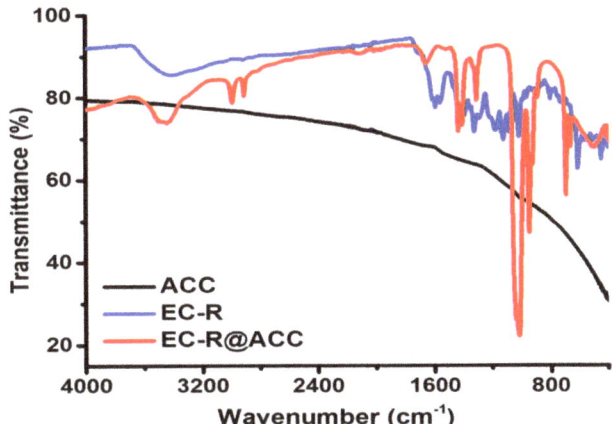

**Figure 6.** Studies of the ACC, EC-R, and EC-R@ACC photocatalyst by FTIR technique.

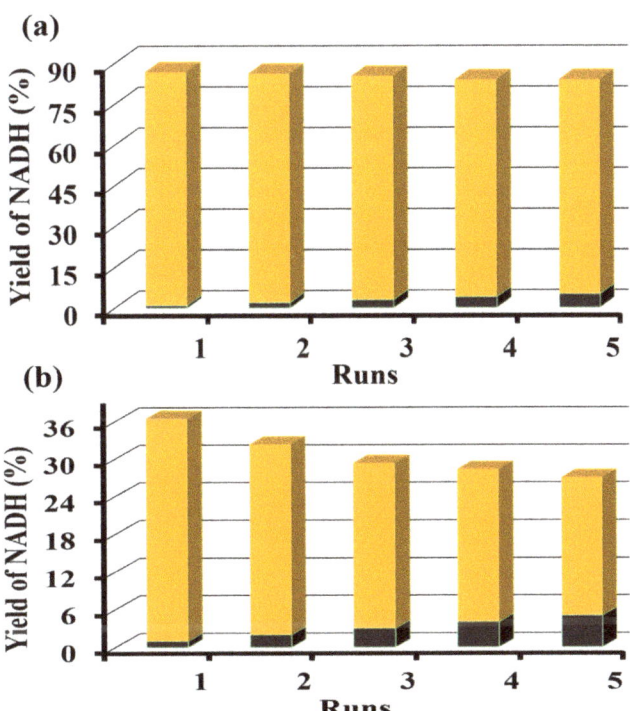

**Figure 7.** The recycle stability for NADH regeneration by the (**a**) EC-R@ACC photocatalyst and (**b**) EC-R photocatalyst.

So, the calculated TON of the EC-R@ACC photocatalyst for the reduction of 4-nitrobenzyl alcohol is $1.769 \times 10^{19}$ molecules.

## 3. Experimental Details

### 3.1. Materials and Chemicals

The ACC, sodium borohydride (NaBH$_4$), EC-R, N, N-dimethyl formamide (DMF), 4-nitro benzyl alcohol (4-NBA), sodium phosphate monobasic dihydrate (NaH$_2$PO$_4$·2H$_2$O), sodium phosphate dibasic dihydrate (Na$_2$HPO$_4$·2H$_2$O), nicotinamide adenine dinucleotide (NAD$^+$), and 2,2 bipyridine (pentamethylcyclopentadienyl) rhodium (III) chloride dimer were purchased from Sigma Aldrich (Munich, Germany) and TCI (Portland, OR, USA).

### 3.2. Synthesis of ACC

Activated Carbon Cloth (ACC) was synthesized in the reported way. Carbon fabric (1 cm × 1 cm) was initially washed many times with acetone and distilled water. Following multiple washes, the carbon fabric was cured with conc. HNO$_3$ at more than 90 °C for roughly 4 h. Following the acid treatment, the carbon fabric was thoroughly cleaned with distilled water and acetone. After washing, the freshly produced activated carbon cloth (ACC) was dried in a 70 °C oven [64].

### 3.3. Synthesis of EC-R@ACC Photocatalyst

Typically, 350 mg of carbon powder (graphene) and 150 mg of EC-R were mixed in 20 mL DMF and stirred for 2 h to ensure complete mixing. Then, the solution was autoclaved at 150 °C for 12 h (Figure 8). Furthermore, the solution was cooled to room temperature. Then, the solvent in the solution was evaporated at its boiling point. The obtained compound was thoroughly washed with distilled water 2–3 times. Finally, the newly designed EC-R@ACC photocatalyst was dried in the oven overnight at 100 °C. The amount of EC-R@ACC achieved was 203 mg [65].

### 3.4. Synthesis of Rh-Complex

The Rh-complex [Cp*Rh(bpy)Cl]$^+$ was prepared using a well-standard technique. In 5 mL distilled methanol, 0.025 g of rhodium compound ([Rh(C$_5$Me$_5$)Cl$_2$]$_2$) was dissolved in an N$_2$-purged environment. The methanol solution was then mixed in a dark-incubated environment at room temperature with 0.013 g of 2,2'-bipyridyl (2 eq.) [15].

As soon as diethyl ether was added, a yellow precipitate formed. In an N$_2$-purged environment, the complete product was received by the filtration method and dried at room temperature.

### 3.5. Synthesis of 4-ABA

The mixture of the EC-R@ACC photocatalyst (0.010 g), 4-NBA (0.045 g), and NaBH$_4$ (5 mg/L) was prepared in 30 mL DMF in a glass vial and mixed with a magnetic stirrer. The reaction mixture was stirred at room temperature for 12 h in the presence of air under continuous high solar irradiation. The reaction mixture was then examined using TLC after it was completed. After filtering, the mixture was thoroughly washed with 50 mL of distilled water. The filtrate was concentrated using a rotary evaporator to abstract the final product. The compound's yield was 97.61% [52].

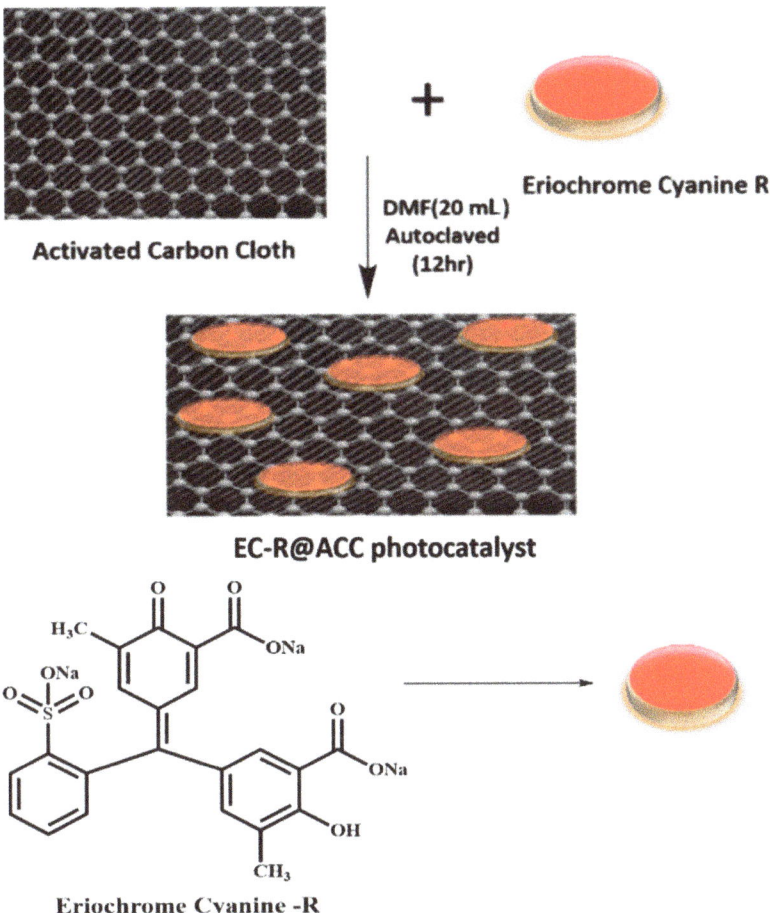

**Figure 8.** The schematic diagram for the synthesis of the EC-R@ACC photocatalyst.

## 4. Conclusions

Overall, with this support, we have explained that the newly designed photocatalyst is feasible for the regeneration of NADH and organic transformation under solar light. In this context, photochemically under solar light irradiation, the regeneration of NADH and the reduction of 4-nitrophenol with $NaBH_4$ can be carried out using a metal-free ACC templated EC-R@ACC photocatalyst. The regeneration of NADH, as well as the reduction of 4-NBA into 4-ABA, was accomplished using an EC-R doped ACC photocatalyst (EC-R@ACC) in conjunction with artificial photosynthetic machinery. The EC-R@ACC photocatalyst demonstrated good maintenance of catalytic effectiveness during numerous cycles of photocatalytic reaction due to its great thermal and chemical stability. Most importantly, the EC-R@ACC, under continuous solar light irradiation, permits the effective regeneration of NADH cofactors with a yield of 76.9%. This research suggests that solar light could be used to produce more effective and cost-effective NADH regeneration along with photocatalytic reduction of 4-NBA and many more reductive processes.

**Author Contributions:** Conceptualization, V.G., R.K.Y., A.U. and R.K.S.; software, V.G., R.K.Y., A.U., A.A.I., S.S., R.S., R.K.S., D.T., D.K.D., A.K.S. (Alok Kumar Singh), A.K.S. (Atresh Kumar Singh) and S.B., validation, V.G., R.K.Y., A.U., A.A.I., S.S., R.S., R.K.S., D.T., D.K.D., A.K.S. (Alok Kumar Singh), A.K.S. (Atresh Kumar Singh) and S.B.; formal analysis and investigation, V.G., R.K.Y., A.U., A.A.I., S.S., R.S., R.K.S., D.T., D.K.D., A.K.S. (Alok Kumar Singh), A.K.S. (Atresh Kumar Singh) and S.B.; writing—original draft, V.G., R.K.Y., A.U. and R.K.S.; writing—review and editing, V.G., R.K.Y., A.U., A.A.I., S.S., R.S., R.K.S., D.T., D.K.D., A.K.S. (Alok Kumar Singh), A.K.S. (Atresh Kumar Singh) and S.B.; visualization, D.K.D. and S.B.; project administration, R.K.Y. All authors have read and agreed to the published version of the manuscript.

**Funding:** The authors are thankful to the Deanship of Scientific Research and supervision of the Centre for Scientific and Engineering Research at Najran University, Najran, Kingdom of Saudi Arabia for funding under the Research Centers funding program Grant No. NU/RCP/SERC/12/6.

**Data Availability Statement:** Not applicable.

**Acknowledgments:** The authors are thankful to the Deanship of Scientific Research and supervision of the Centre for Scientific and Engineering Research at Najran University, Najran, Kingdom of Saudi Arabia for funding under the Research Centers funding program Grant No. NU/RCP/SERC/12/6.

**Conflicts of Interest:** The authors declare no conflict of interest.

## References

1. Horvath, I.T.; Anastus, P.T. Innovations and Green Chemistry. *Chem. Rev.* **2007**, *107*, 2169–2173. [CrossRef] [PubMed]
2. Jamali, A.A.; Solangi, A.R.; Memon, N.; Nizamani, S.M.; Khaskheli, A.A.; Hussain, M.; Mahmoud, M.H.; Fouad, H.; Akhtar, M.S. Abiotic Degradation of Imidacloprid Pesticide with L-Threonine Capped Nickel Nanoparticles. *Sci. Adv. Mater.* **2021**, *13*, 2043–2048. [CrossRef]
3. Izzudin, N.M.; Jalil, A.A.; Aziz, F.F.; Azami, A.; Ali, M.S.; Hassan, M.W.; Rahman, A.F.A.; Fauzi, A.A.; Vo, N.S. Simultaneous remediation of hexavalent chromium and organic pollutants in wastewater using period 4 transition metal oxide-based photocatalysts: A review. *Environ. Chem. Lett.* **2021**, *19*, 4489–4517. [CrossRef]
4. Lv, Y.; Zhan, Q.; Yu, X. Microbial-Induced Mineralization of Zinc Ions Based on the Degradation of Toluene and Its Characterization. *Sci. Adv. Mater.* **2021**, *13*, 656–661. [CrossRef]
5. Lang, X.; Chen, X.; Zhao, J. Rational Design of Hybrid Nanostructures for Advanced Photocatalysis. *Chem. Soc. Rev.* **2014**, *43*, 473–486. [CrossRef] [PubMed]
6. Tan, X.; Zheng, Z.; Peng, B.; Wu, X.; Huang, X.; Chen, X. Simultaneous Degradation of p-Nitrophenol and Recovery of Copper from Wastewater in Electrochemical Reactor under High Salinity. *Sci. Adv. Mater.* **2021**, *13*, 2450–2459. [CrossRef]
7. Dalle, K.E.; Julien, W.; Jane, J.L.; Bertrand, R.; Isabell, S.K.; Erwin, R. Electro- and Solar-Driven Fuel Synthesis with First Row Transition Metal Complexes. *Chem. Rev.* **2019**, *119*, 2752. [CrossRef]
8. Yadav, R.K.; Oh, G.; Park, N.-J.; Kumar, A.; Kong, K.-J.; Baeg, J.-O. Highly selective solar-driven methanol from $CO_2$ by a photocatalyst/biocatalyst integrated system. *J. Am. Chem. Soc.* **2014**, *136*, 16728. [CrossRef]
9. Yoon, S.K.; Choban, E.R.; Kane, C.; Tzedakis, T.; Kenis, P.J.A. Active control of the depletion boundary layers in microfluidic electrochemical reactors. *J. Am. Chem. Soc.* **2005**, *127*, 10546. [CrossRef]
10. Liu, J.; Antoniettia, M. Bio-inspired NADH regeneration by carbon nitride photocatalysis using diatom templates. *Energy Environ. Sci.* **2013**, *6*, 1486–1493. [CrossRef]
11. Maenaka, Y.; Suenobu, T.; Fukuzumi, S. Efficient Catalytic Interconversion between NADH and $NAD^+$ Accompanied by Generation and Consumption of Hydrogen with a Water-Soluble Iridium Complex at Ambient Pressure and Temperature. *J. Am. Chem. Soc.* **2012**, *134*, 367–374. [CrossRef] [PubMed]
12. Okamoto, Y.; Kohler, V.; Paul, C.E.; Hollmann, F.; Ward, T.R. Efficient In Situ Regeneration of NADH Mimics by an Artificial Metalloenzyme. *ACS Catal.* **2016**, *6*, 3553. [CrossRef]
13. Mathew, S.C.; Mohlmann, L.; Antonietti, M.; Wang, X.; Blechert, S. Aerobic oxidative coupling of amines by carbon nitride photocatalysis with visible light. *Angew. Chem. Int. Ed.* **2011**, *50*, 657–660.
14. Su, F.; Mathew, S.C.; Lipner, G.; Fu, X.; Antonietti, M.; Blechert, S.; Wang, X. mpg-$C_3N_4$-Catalyzed Selective Oxidation of Alcohols Using $O_2$ and Visible Light. *J. Am. Chem. Soc.* **2010**, *132*, 16299–16301. [CrossRef]
15. Singh, S.; Yadav, R.; Kim, T.; Singh, C.; Singh, P.; Chaubey, S.; Singh, A.; Beag, J.; Gupta, S.; Tiwary, D. Generation and Regeneration of the C($sp^3$)–F Bond and 1,4-NADH/NADPH via Newly Designed S-g$C_3N_4$@$Fe_2O_3$/LC Photocatalysts under Solar Light. *Energy Fuels* **2022**, *36*, 8402–8412. [CrossRef]
16. Sawunyama, P.; Fujishima, A.; Hashimoto, K. Titanium dioxide photocatalysis. *Langmuir* **1999**, *15*, 3551–3556. [CrossRef]
17. Ischay, M.A.; Anzovino, M.E.; Du, J.; Yoon, T.P. Efficient Visible Light Photocatalysis of [2+2] Enone Cycloadditions. *J. Am. Chem. Soc.* **2008**, *130*, 12886–12887. [CrossRef]
18. Zeitler, K. Photoredox Catalysis with Visible Light. *Angew. Chem. Int. Ed.* **2009**, *48*, 9785–9789. [CrossRef]

19. Narayanam, J.M.R.; Tucker, J.W.; Stephenson, C.R.J. Electron-Transfer Photoredox Catalysis: Development of a Tin-Free Reductive Dehalogenation Reaction. *J. Am. Chem. Soc.* **2009**, *131*, 8756–8757. [CrossRef]
20. Nicewicz, D.A.; MacMillan, D.W.C. Merging photoredox catalysis with organocatalysis: The direct asymmetric alkylation of aldehydes. *Science* **2008**, *322*, 77–80. [CrossRef]
21. Shih, H.W.; Vander, M.N.; Wal, R.L.; MacMillan, D.W.C.; Grange, R.L. Enantioselective α-Benzylation of Aldehydes via Photoredox Organocatalysis. *J. Am. Chem. Soc.* **2010**, *132*, 13600–13603. [CrossRef] [PubMed]
22. Zhang, D.; Wu, L.Z.; Zhou, L.; Han, X.; Yang, Q.Z.; Zhang, L.P.; Tung, C.H. Photoresponsive hydrogen-bonded supramolecular polymers based on a stiff stilbene. *Unit. J. Am. Chem. Soc.* **2004**, *126*, 3440–3441.
23. Haag, B.A.; Mosrin, M.; Ila, H.; Malakhov, V.; Knochel, P. Regio- and Chemoselective Metalation of Arenes and Heteroarenes Using Hindered Metal Amide Bases. *Angew. Chem. Int. Ed.* **2011**, *50*, 9511–9954. [CrossRef] [PubMed]
24. Schmermund, L.; Jurkas, V.; Özgen, F.F.; Barone, G.D.; Grimm, H.C.; Winkler, C.K.; Schmidt, S.; Kourist, R.; Krouti, W. Photo-Biocatalysis: Biotransformations in the Presence of Light. *ACS Catal.* **2019**, *9*, 4115–4144. [CrossRef]
25. Zhao, Y.; Liu, H.; Wu, C.; Zhang, Z.; Pan, Q.; Hu, F.; Wang, R.; Li, P.; Huang, X.; Li, Z. Fully Conjugated Two-Dimensional sp2-Carbon Covalent Organic Frameworks as Artificial Photosystem I with High Efficiency. *Angew. Chem.* **2019**, *58*, 5376–5381. [CrossRef]
26. Rajeshwar, K. Solar Energy Conversion and Environmental Remediation Using Inorganic Semiconductor–Liquid Interfaces: The Road Traveled and the Way Forward. *J. Phys. Chem. Lett.* **2011**, *2*, 1301–1309. [CrossRef]
27. Yohannes, A.; Su, Y.; Yao, S. Emerging Applications of Metal−Organic Frameworks for Environmental Remediation. *ACS Symp. Ser.* **2021**, *1395*, 1–23.
28. Roy, A.; Sharma, A.; Yadav, S.; Jule, L.T.; Krishnaraj, R. Nanomaterials for Remediation of Environmental Pollutants. *Bioinorg Chem. Appl.* **2021**, *28*, 1764647. [CrossRef]
29. Gopinath, K.; Panchamoorthy, M.; Nagarajan, V.; Krishnan, A.; Malolan, R.; Rangarajan, G. Present applications of titanium dioxide for the photocatalytic removal of pollutants from water: A review. *J. Environ. Manag.* **2020**, *270*, 110906. [CrossRef]
30. Molinari, R.; Lavorato, C.; Argurio, P. Visible-Light Photocatalysts and Their Perspectives for Building Photocatalytic Membrane Reactors for Various Liquid Phase Chemical Conversions. *Catalyst* **2020**, *10*, 1334. [CrossRef]
31. Mohsin, M.; Ishaq, T.; Bhatti, I.A.; Maryam; Jilani, A.; Melaibari, A.A.; Abu-Hamdeh, N.H. Semiconductor Nanomaterial Photocatalysts for Water-Splitting Hydrogen Production: The Holy Grail of Converting Solar Energy to Fuel. *Nanomaterials* **2023**, *13*, 546. [CrossRef] [PubMed]
32. Chen, P.; Liu, H.; Cui, W.; Lee, S.C.; Wang, L.; Dong, F. Bi-based photocatalysts for light-driven environmental and energy applications: Structural tuning, reaction mechanisms, and challenges. *EcoMat* **2020**, *2*, e12047. [CrossRef]
33. Samadi, M.; Zirak, M.; Naseri, A.; Khorashadizade, E.; Moshfegh, A.Z. Recent progress on doped ZnO nanostructures for visible-light photocatalysis. *Thin Solid Films* **2016**, *605*, 2–19. [CrossRef]
34. Long, X.; Feng, C.; Yang, S.; Ding, D.; Feng, J.; Liu, M.; Chen, Y.; Tan, J.; Peng, X.; Shi, J.; et al. Oxygen doped graphitic carbon nitride with regulatable local electron density and band structure for improved photocatalytic degradation of bisphenol A. *J. Chem. Eng.* **2022**, *435*, 1385–8947. [CrossRef]
35. Choudhury, S.; Baeg, J.-O.; Park, N.-J.; Yadav, R.K. A Photocatalyst/Enzyme Couple That Uses Solar Energy in the Asymmetric Reduction of Acetophenones. *Angew. Chem.* **2012**, *51*, 11624–11628. [CrossRef] [PubMed]
36. Sakhare, P.A.; Pawar, S.S.; Bhat, T.S.; Yadav, S.D.; Patil, G.R.; Patil, P.S.; Sheikh, A.D. Magnetically Recoverable $BiVO_4/NiFe_2O_4$ Nanocomposite Photocatalyst For Efficient Detoxification of Polluted Water Under Collected Sunlight. *Mater. Res. Bull.* **2020**, *129*, 110908. [CrossRef]
37. Armaković, S.J.; Savanović, M.M.; Armaković, S. Titanium Dioxide as the Most Used Photocatalyst for Water Purification: An Overview. *Catalysts* **2023**, *13*, 26. [CrossRef]
38. Ibhadon, A.O.; Fitzpatrick, P. Heterogeneous Photocatalysis: Recent Advances and Applications. *Catalysts* **2013**, *3*, 189–218. [CrossRef]
39. Fang, B.; Xing, Z.; Sun, D.; Li, Z.; Zhou, W. Hollow semiconductor photocatalysts for solar energy conversion. *APM* **2022**, *1*, 100021. [CrossRef]
40. Yadav, R.K.; Baeg, J.-O.; Oh, G.; Park, N.-J.; Kong, K.; Kim, J.; Hwang, D.W.; Biswas, S.K. A Photocatalyst–Enzyme Coupled Artificial Photosynthesis System for Solar Energy in Production of Formic Acid from $CO_2$. *J. Am. Chem. Soc.* **2012**, *134*, 11455–11461. [CrossRef]
41. Yuan, J.; Li, H.; Gao, S.; Lin, Y.; Li, H. A facile route to n-type $TiO_2$-nanotube/p-type boron-doped-diamond heterojunction for highly efficient photocatalysts. *Chem. Comm.* **2010**, *46*, 3119. [CrossRef]
42. Natarajan, T.S.; Mozhiarasi, V.; Tayade, R.J. Nitrogen-Doped Titanium Dioxide (N-$TiO_2$): Synopsis of Synthesis Methodologies, Doping Mechanisms, Property Evaluation and Visible Light Photocatalytic Applications. *Photochemical* **2021**, *1*, 371–410. [CrossRef]
43. Tsang, C.H.A.; Li, K.; Zeng, Y.; Zhao, W.; Zhang, T.; Zhan, Y.; Xie, R.; Leung, D.Y.C.; Huang, H. Titanium oxide based photocatalytic materials development and their role of in the air pollutants degradation: Overview and forecast. *Environ. Int.* **2019**, *125*, 200–228. [CrossRef] [PubMed]
44. Plakas, K.V.; Taxintari, A.; Karabelas, A.J. Enhanced Photo-Catalytic Performance of Activated Carbon Fibers for Water Treatment. *Water* **2019**, *11*, 1794. [CrossRef]
45. Lawtae, P.; Tangsathitkulchai, C. The Use of High Surface Area Mesoporous-Activated Carbon from Longan Seed Biomass for Increasing Capacity and Kinetics of Methylene Blue Adsorption from Aqueous Solution. *Molecules* **2021**, *26*, 6521. [CrossRef] [PubMed]

46. Singh, C.; Yadav, R.K.; Kim, T.W.; Upare, P.P.; Gupta, A.K.; Singh, A.P.; Yadav, B.C.; Dwivedi, D.K. In Situ Prepared Solar Light-Driven Flexible Actuated Carbon Cloth-Based Nanorod Photocatalyst for Selective Radical-Radical Coupling to Vinyl Sulfides. *Photochem. Photobiol.* **2021**, *97*, 955–962. [CrossRef]
47. Xu, Y.; Guo, Z.; Wang, J.; Chen, Z.; Yin, J.; Zhang, Z.; Huang, J.; Qian, J.; Wang, X. Harvesting Solar Energy by Flowerlike Carbon Cloth Nanocomposites for Simultaneous Generation of Clean Water and Electricity. *ACS Appl. Mater.* **2021**, *13*, 27129–27139. [CrossRef]
48. Ndlwana, L.; Raleie, N.; Dimpe, K.M.; Ogutu, H.F.; Oseghe, E.O.; Motsa, M.M.; Msagati, T.A.M.; Mamba, B.B. Sustainable Hydrothermal and Solvothermal Synthesis of Advanced Carbon Materials in Multidimensional Applications: A Review. *Materials* **2021**, *14*, 5094. [CrossRef]
49. Carmen, Z.; Daniela, S. Rijeka: Textile Organic Dyes–Characteristics, Polluting Effects and Separation/Elimination Procedures from Industrial Effluents–A Critical Overview. *IntechOpen* **2012**, *3*, 55–86.
50. Sahiner, N.; Ozay, H.; Ozay, O.; Aktas, N. A soft hydrogel reactor for cobalt nanoparticle preparation and use in the reduction of nitrophenols. *Appl. Catal. B Environ.* **2010**, *101*, 137–143. [CrossRef]
51. Yang, Y.; Zhang, C.; Hu, Z. Impact of metallic and metal oxide nanoparticles on wastewater treatment and anaerobic digestion. *Environ. Sci. Process. Impacts* **2013**, *15*, 39–48. [CrossRef] [PubMed]
52. Gazi, S.; Ananthakrishnan, R. Metal-free-photocatalytic reduction of 4-nitrophenol by resin-supported dye under the visible irradiation. *Appl. Catal. B Environ.* **2011**, *105*, 317–325. [CrossRef]
53. Krishna, M.B.M.; Venkatramaiah, N.; Venkatesan, R.; Rao, D.N. Synthesis and structural, spectroscopic and nonlinear optical measurements of graphene oxide and its composites with metal and metal free porphyrins. *J. Mater. Chem.* **2012**, *22*, 3059–3068. [CrossRef]
54. Chaubey, S.; Singh, P.; Singh, C.; Singh, S.; Shreya, S.; Yadav, R.K.; Mishra, S.; Jeong, Y.J.; Biswas, B.K.; Kim, T.W. Ultra-efficient synthesis of bamboo-shape porphyrin framework for photocatalytic $CO_2$ reduction and consecutive C-S/C-N bonds formation. *J. CO2 Util.* **2022**, *59*, 101968. [CrossRef]
55. Yadav, S.N.; Kumar, B.; Yadav, R.K.; Singh, P.; Gupta, S.K.; Singh, S.; Singh, A.P. Synthesis of highly efficient selenium oxide hybridized g-$C_3N_4$ photocatalyst for NADH/NADPH regeneration to facilitate solar-to-chemical reaction. *Main Group Chem.* **2022**, *21*, 1077–1089. [CrossRef]
56. Kumar, K.S.; Huerta, G.V.; Castellanos, A.R.; Varaldo, H.M.P.; Feria, O.S. Microwave Assisted Synthesis and Characterizations of Decorated Activated Carbon. *Int. J. Electrochem. Sci.* **2012**, *7*, 5484–5494.
57. Chaubey, S.; Yadav, R.K.; Kim, T.W.; Singh, A.P.; Kumar, K.; Yadav, B.C. Ultrahigh sun-light-responsive/not responsive integrated catalyst for C-S arylation/humidity sensing. *Vietnam J. Chem.* **2021**, *59*, 500–510.
58. Singh, C.; Kim, T.W.; Yadav, R.K.; Baeg, J.O.; Gole, V.; Singh, A.P. Flexible Covalent Porphyrin Framework Film: An Emerged Platform for Photocatalytic C-H Bond Activation. *App. Surface Sci.* **2021**, *544*, 148938. [CrossRef]
59. Singh, P.; Yadav, R.K.; Kim, T.W.; Kumar, A.; Dwivedi, D.K. Chitosan-based fluorescein isothiocyanate film as a highly efficient metal-free photocatalyst for solar-light-mediated direct C-H arylation. *Int. J. Energy Res.* **2021**, *45*, 5964–5973. [CrossRef]
60. Lacroix, M.R.; Gao, X.Y.; Liu, S.H. Strauss, Unusually sharp FTIR ν(OH) bands and very weak O single bondH···F hydrogen bonds in $M_2(H_2O)_{1,2}B_{12}F_{12}$ hydrates (Mdouble bond Nasingle bond Cs). *J. Fluorine Chem.* **2019**, *217*, 105–108. [CrossRef]
61. Yen, C.Y.; Lee, C.H.; Lin, Y.F.; Lin, H.L.; Hsiao, Y.H.; Liao, S.H.; Chuang, C.Y.; Ma, C.C.M. Proton conductivity of Nafion/ex situ Stöber silica nanocomposite membranes as a function of silica particle size and temperature. *J. Power Sources* **2007**, *173*, 36–44. [CrossRef]
62. Thompson, W.R.; Cai, M.; Ho, M.; Pemberton, J.E. Hydrolysis and Condensation of Self-Assembled Monolayers of (3-Mercaptopropyl) trimethoxysilane on Ag and Au Surfaces. *Langmuir* **1997**, *13*, 2291–2302. [CrossRef]
63. Bano, M.; Ahirwar, D.; Thomas, M.; Naikoo, G.A.; Sheikh, M.U.-D.; Khan, F. Hierarchical synthesis of silver monoliths and their efficient catalytic activity for the reduction of 4-nitrophenol to 4-aminophenol. *New J. Chem.* **2016**, *40*, 6787–6795. [CrossRef]
64. Singh, S.; Yadav, R.; Kim, T.; Singh, C.; Singh, P.; Singh, A.; Singh, A.; Singh, A.; Beag, J.; Gupta, S. Rational design of a graphitic carbon nitride catalytic–biocatalytic system as a photocatalytic platform for solar fine chemical production from $CO_2$. *React. Chem. Eng.* **2022**, *7*, 1566–1572. [CrossRef]
65. Kumar, A.; Yadav, R.K.; Park, N.J.; Baeg, J.O. Facile One-Pot Two-Step Synthesis of Novel in Situ Selenium-Doped Carbon Nitride Nanosheet Photocatalysts for Highly Enhanced Solar Fuel Production from $CO_2$. *ACS Appl. Nano Mater.* **2017**, *1*, 47–54. [CrossRef]

**Disclaimer/Publisher's Note:** The statements, opinions and data contained in all publications are solely those of the individual author(s) and contributor(s) and not of MDPI and/or the editor(s). MDPI and/or the editor(s) disclaim responsibility for any injury to people or property resulting from any ideas, methods, instructions or products referred to in the content.

*Article*

# One-Pot Synthesis of Green-Emitting Nitrogen-Doped Carbon Dots from Xylose

**Gabriela Rodríguez-Carballo [1,*], Cristina García-Sancho [1], Manuel Algarra [2], Eulogio Castro [3] and Ramón Moreno-Tost [1,*]**

[1] Department of Inorganic Chemistry, Faculty of Science, Campus de Teatinos s/n, University of Málaga, 29071 Malaga, Spain; cristinags@uma.es
[2] Department of Sciences, INAMAT2-Institute for Advanced Materials and Mathematics, Public University of Navarre, Campus de Arrosadía, 31006 Pamplona, Spain; manuel.algarra@unavarra.es
[3] Department Chemical, Environmental and Materials Engineering, Center for Advanced Studies in Earth Sciences, Energy and Environment (CEACTEMA), Universidad de Jaén, Campus las Lagunillas, 23071 Jaén, Spain; ecastro@ujaen.es
\* Correspondence: gabrielarc@uma.es (G.R.-C.); rmtost@uma.es (R.M.-T.)

**Abstract:** Carbon dots (CDs) are interesting carbon nanomaterials that exhibit great photoluminescent features, low cytotoxicity, and excellent water stability and solubility. For these reasons, many fields are starting to integrate their use for a variety of purposes. The catalytic performance of $VOPO_4$ has been evaluated in the synthesis of nitrogen-doped carbon dots (N-CDs). The synthesis reaction was carried out at 180 °C using $VOPO_4$ as a heterogeneous catalyst for 2 to 4 h of reaction time. After reaction, the N-CDs were purified using a novel method for the protection of the functional groups over the surfaces of the N-CDs. The morphological, superficial, and photoelectronic properties of the N-CDs were thoroughly studied by means of TEM, HRTEM, XPS, and photoluminescence measurements. The conversion of the carbon precursor was followed by HPLC. After three catalytic runs, the catalyst was still active while ensuring the quality of the N-CDs obtained. After the third cycle, the catalyst was regenerated, and it recovered its full activity. The obtained N-CDs showed a great degree of oxidized groups in their surfaces that translated into high photoluminescence when irradiated under different lasers. Due to the observed photoelectronic properties, they were then assayed in the photocatalytic degradation of methyl orange.

**Keywords:** carbon dots; doping; $VOPO_4$; heterogeneous catalysis; xylose; acetic acid; hydrothermal method; photoluminescence; photocatalysis

## 1. Introduction

Carbon dots (CDs) are zero-dimensional particles normally smaller than 10 nm in diameter [1]. Their morphology is quasi-spherical and comprises a graphite-like core composed mainly of carbon, which can present inclusions of different adatoms such as N and S, and of a surface decorated by different organic functional groups. This superficial functionalization allows CDs to develop different surface trap states, lowering the energy band gap [2]. This is coupled with the fact that graphitic $sp^2$ structures favor the projection of the recombination of electrons and hole [3], thus promoting a possible transition HOMO-LUMO and explaining their outstanding electronic properties. As for their optical properties, these nanoparticles are widely recognized as highly photoluminescent systems, with tunable and up-conversion photoluminescence properties [3]. CDs' optical properties show enormous variation with size. An increment in size directly translates into a decrease in the energy gap between HOMO and LUMO. Therefore, a bathochromic shift to higher-wavelength emissions is observed. However, $sp^2$ clusters can act as auxochromes, decreasing the energy gaps and exhibiting a completely opposite effect to the formerly described size-gap relation [4]. As CDs are functionalized by a great variety of groups, they exhibit surface state photoluminescence, considering that the functional groups on their surface have multiple energy levels that can interact and result in emissive traps, thus dominating the emissions. The surface state is a synergetic contribution of all groups conforming to the surface of the carbon dot that hybridize with the outer graphitic core [5]. The presence of different functional groups on the surfaces of the CDs not only control their electronic properties, but are the cause of their water stability and solubility. Since they are carbon-based materials, their cytotoxicity has been proven to be very low and more environmentally friendly than their analogues, metallic quantum dots. Due to this phenomenon, known as quantum confinement, along with their low cytotoxicity and excellent optical properties, CDs have numerous practical applications: bioimaging [6], light-harvesting [7], photocatalysts [8] and photosensitizers [9], bio- and chemosensors [10], and drug delivery [11].

Their outstanding optical features are very sensitive to minimal structural changes, including heteroatomic inclusions [12]. For that reason, one of the main methods of tuning the photoluminescent properties of CDs consists of doping the CDs with either one kind of element, such as N, P, or S, in different concentrations, or combining them to synthesize codoped CDs [13]. The nature and quantity of the dopant are key parameters for the improvement of the photophysical properties of CDs, as their electronic structure is modified by the presence of adatoms originating from $n$ or $p$ carriers [14]. Nitrogen is frequently used for doping CDs, as the variety of precursors that can be used is wide, from organic to inorganic compounds. These dopants can be added during synthesis or post-synthesis, and its presence can highly impact the quantum yield (QY) of CDs [15], increasing it to values as high as 26% when it acts as a codopant along with sulfur atoms [16] in water, and up to nearly 100% depending on the emission wavelength when using solvents different from water [17]. Doping [18], along with surface functionalization [19] and size [20,21], are the three main factors controlling the emission wavelengths of CDs. A high amount of N as the dopant [22] reduces the bandgap between valence and conduction bans, causing a bathochromic shift. The presence of several organic groups on the surfaces of CDs can be observed, as different contributions appear in the photoluminescence spectra. Each emission can be assigned to a different group [23,24]. CDs energy gaps are also strongly affected by size. As size increases, the bandgap decreases, red-shifting the emission [25].

Generally, for the synthesis of CDs, whether the carbon precursor is a green precursor [26] or a commercial one and whether the dopant is added during the synthesis or post-synthesis, the preferred method is hydrothermal synthesis. Although hydrothermal synthesis is, in general terms, a green synthetic process, many authors still rely on strong mineral acids for the production of CDs [27,28]. These mineral acids are corrosive, contaminant, and cannot be recovered or reused after reaction, generating streams that must be neutralized before disposal. This is why greener alternatives based on the use of solid catalysts are starting to be considered [26,29]. However, heterogeneous catalysis also faces a major issue. In terms of acidity, they present much lower values of Brønsted acidity, which is necessary for the major transformation of the precursors into CDs. In this sense, catalysts based on V, Nb, Sn, and Zr are widely used in processes that require strong Brønsted acidity [30–32].

The aim of this work is to synthesize CDs doped with N adatoms by means of heterogeneous catalysis. Thus, in this work, vanadyl phosphate (VOPO$_4$) is proposed as a bifunctional catalyst that will promote the dehydration and aggregation of the carbon precursor while oxidizing the functional groups on the surfaces of N-CDs. With this approach to CD synthesis, we present a mineral acid-free hydrothermal alternative that replicates the conditions of real biomass-derived monosaccharides without compromising the purification, quality, or quantity of N-CDs, nor their photophysical or catalytic properties, and ensuring the recyclability of the catalyst.

## 2. Results and Discussion

### 2.1. Characterization of the CDs

#### 2.1.1. Morphological and Superficial Study

The synthesis of CDs was carried out following a bottom-up approach, using xylose as a model biomass molecule, $VOPO_4$ as a catalyst, and nitrogen as the doping adatom of the surface. Moreover, considering that the fractionation of lignocellulosic biomass yields a liquor of hemicellulosic sugars with an acidic pH, basically originating from acetic acid, this acid was added to the xylose solution.

It is important to note that the acetic acid was not sufficient to produce the CDs; therefore, the catalytic activity was only due to the presence of the solid catalyst. The morphology and size distribution (Figure S1) were studied by TEM (Figure 1) and HRTEM (Figure S2). The resulting images show that the N-CDs were effectively obtaining using $VOPO_4$ as the catalyst. There was a substantial difference in the yield of CDs when reaction time was doubled, as in Figure 1a, the reaction was stopped after 2 h, and in Figure 1b, it was kept for 4 h. The reaction time did not affect the quality of the nanoparticles in any way, as the shape remained quasi-spherical after 4 h and the average size ranged between 3.5 and 4 nm in diameter for both reaction times (Figure 1), meaning that a great degree of size homogeneity was achieved. It was thus decided to carry out the synthesis of CDs for 4 h. For further confirmation of the presence of N-CDs in the solution after reaction, the graphite spacing was identified by means of HRTEM and is presented in the Supplementary Materials (Figure S2). When measured, the spacing value was 0.287 nm on average. The obtained XRD pattern is also presented as further confirmation of the graphitic nature of the cores of N-CDs (Figure S3), as a broad peak at 21° was detected, corresponding to the (111) plane of graphite [33].

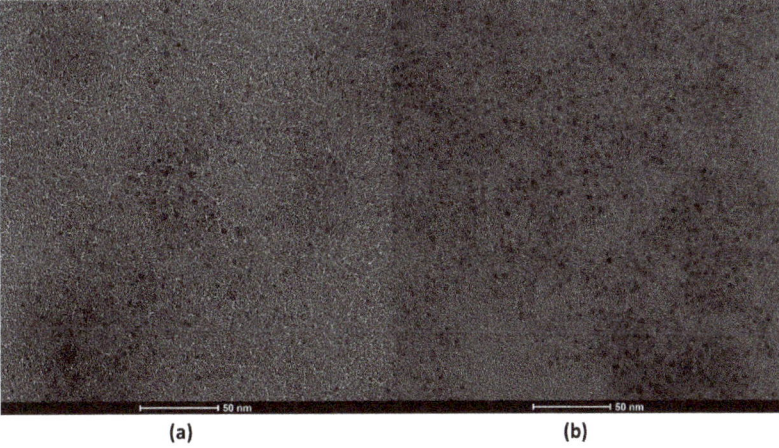

**Figure 1.** TEM images of N-CDs solutions synthesized using 100 mg $VOPO_4$ as the catalyst, 0.75 M of xylose as the C precursor, $NH_4Cl$ as the dopant, and 17 g/L of $CH_3COOH$ at 180 °C after (**a**) 2 h of reaction and (**b**) 4 h of reaction. Both samples were dialyzed prior to the analysis.

The acid properties of the catalyst were evaluated by means of the adsorption of pyridine coupled with FTIR spectroscopy (Figure S4). The TEM images confirm that the acidic properties of the catalyst are strong enough for the synthesis of the nanoparticles to be carried out. Due to the noise of the spectra and the fact that the catalyst was diluted in KBr, since its yellow coloration was blocking the IR beam, the concentrations of both the Brønsted and Lewis acid sites were not calculated. Thus, the spectra are provided as a qualitative approach to understand the nature of the acid sites of VOPO4, due to its importance regarding the acidity of the medium needed to carry out the synthesis of the

CDs. Despite the noise that hindered the interpretation of the spectra of VOPO$_4$, a band attributed to a strong Brønsted acidity site could be identified. A main band regarding the adsorption of pyridine on the catalyst was located at 1677 cm$^{-1}$. This band was ascribed to the presence of strong Brønsted acid sites [34] and corresponded to the ν8a vibration mode of pyridinium species [35].

As the temperature rose, the band maintained a similar value of absorbance until, drastically, it almost disappeared at 125 °C. The Lewis acid sites were not detected, indicating that this catalyst can be considered as a Brønsted solid. Additional characterization for the VOPO$_4$ catalyst (XRD pattern and Raman spectrum) is presented in the Supplementary Materials of this publication (Figures S5 and S6).

The surfaces of the N-CDs were analyzed by means of XPS analysis. We found that 45.6% of the surface was C (Table 1), as was expected due to the graphitic nature of the core. As can be observed in the deconvolution of the C1s core level (Figure 2a), there was a major contribution at 284.7 eV (Table 2) that corresponded to the C-C bond [36]. The existence of highly oxidized groups was doubly confirmed, on the one hand, by the high percentage of O present on the surface (47.7%), and on the other hand, by the next two bands in the deconvolution at 286.3 eV and 288.6 eV, which were associated with the C–N/C–O bond and the O=C–O type of bond, respectively [37]. The N1s core level was analyzed to determine the nature of the species included on the surfaces of the N-CDs (Figure 2b). The band at 401.3 eV (Table 2) made the greatest contribution. This band is usually attributed to graphitic N. The other band (399.4 eV) present in the deconvolution spectrum was related to a small amount of amine nitrogen [38,39].

**Table 1.** XPS mass concentration table of the surface of N-CDs synthesized using 100 mg VOPO$_4$ as the catalyst, 0.75 M of xylose as the C precursor, NH$_4$Cl as the dopant, and 17 g/L of CH$_3$COOH at 180 °C after 4 h of reaction.

| N-CDs | C (1s) | O (1s) | N (1s) | Cl (2p) |
|---|---|---|---|---|
| Mass concentration (%) | 45.6 | 47.7 | 5.0 | 1.6 |

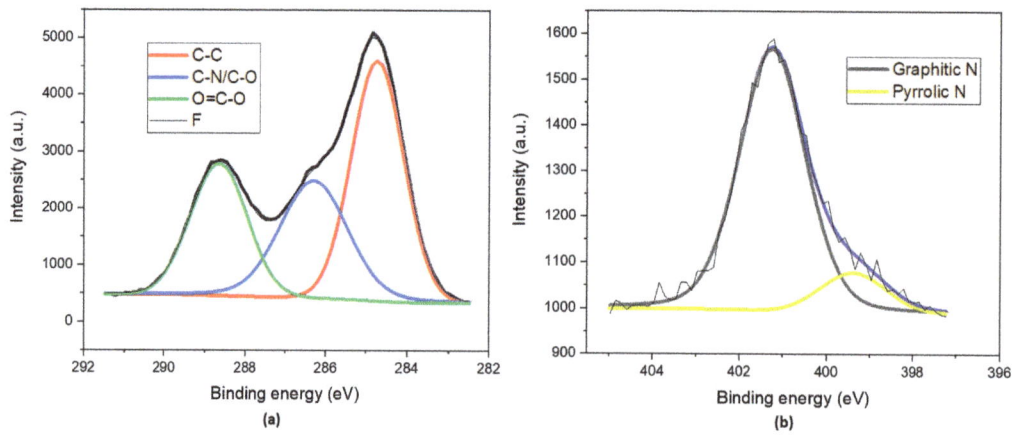

**Figure 2.** XPS deconvoluted spectra of (**a**) the C1s core level and (**b**) the N1s core level of dialyzed and lyophilized N-CDs synthesized from 100 mg VOPO$_4$ as the catalyst, 0.75 M of xylose as the precursor, NH$_4$Cl as the dopant, and 17 g/L of CH$_3$COOH at 180 °C after 4 h of reaction.

Table 2. XPS energy binding deconvoluted bands of the C1s core level and N1s core level of N-CDs synthesized using 100 mg VOPO4 as the catalyst, 0.75 M of xylose as the C precursor, NH$_4$Cl as the dopant, and 17 g/L of CH$_3$COOH at 180 °C after 4 h of reaction.

| C1s (eV) | N1s (eV) |
|---|---|
| 284.7 (43.76%) | 399.4 (12.68%) |
| 286.3 (29.97%) | 401.3 (87.32%) |
| 288.6 (26.27%) | |

In order to support the XPS results and to confirm the presence of N inclusions and the carboxylic functionalization of the surface of CDs, the Fourier-Transform Infrared (FTIR) spectrum of the sample was recorded (Figure 3).

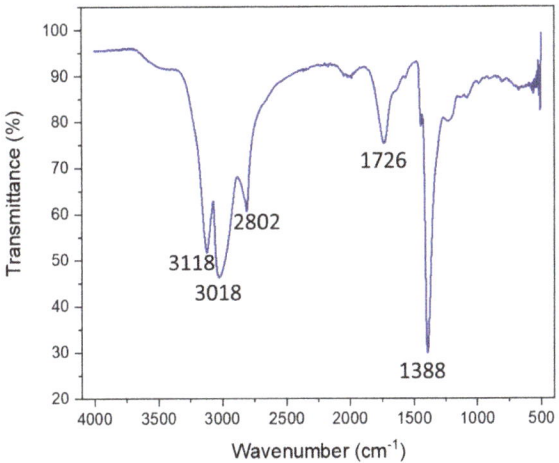

Figure 3. FTIR of dialyzed and lyophilized N-CDs synthesized using 100 mg VOPO$_4$ as the catalyst, 0.75 M of xylose as the C precursor, NH$_4$Cl as the dopant, and 17 g/L of CH$_3$COOH at 180 °C after 4 h of reaction. Before measurement, the N-CDs were dialyzed, purified, and lyophilized.

The IR spectrum presented five main bands, located at 3118 cm$^{-1}$, 3018 cm$^{-1}$, 2802 cm$^{-1}$, 1726 cm$^{-1}$, and 1388 cm$^{-1}$. The bands that appeared at 3118 cm$^{-1}$ and 3018 cm$^{-1}$ corresponded to the N–H vibration bands [40–42] of the two N species present on the CDs, amine/pyrrolic N and graphitic N, respectively. On the other hand, the 2800 cm$^{-1}$ band could be assigned to the vibration of the C–H bond and the 1388 cm$^{-1}$ band to its corresponding bending band. The signal at 1726 cm$^{-1}$ was associated with the C=O bond vibration band from carboxylic acids. Thus, this corroborated the XPS characterization data, as N was successfully included in the structure and functional groups, and the surfaces of the CDs were fully oxidized to carboxylic species.

2.1.2. Optical Properties of the CDs

The photoluminescent properties were studied at two different wavelengths, 325 nm and 473 nm, in order to study the possibility of tuning their emission. There was a clear difference between the photoluminescence of the solution of N-CDs that was left for 2 h and that that was left for 4 h. The intensity of the spectrum doubled as time passed, indicating a higher concentration of species emissions [43]. Two maximums can be identified in the spectrum of N-CDs irradiated under 325 nm (Figure 4a) after 4 h at 484 and 563 nm; these can be associated with C–N-emitting species and O–C=O-emitting species, respectively [44].

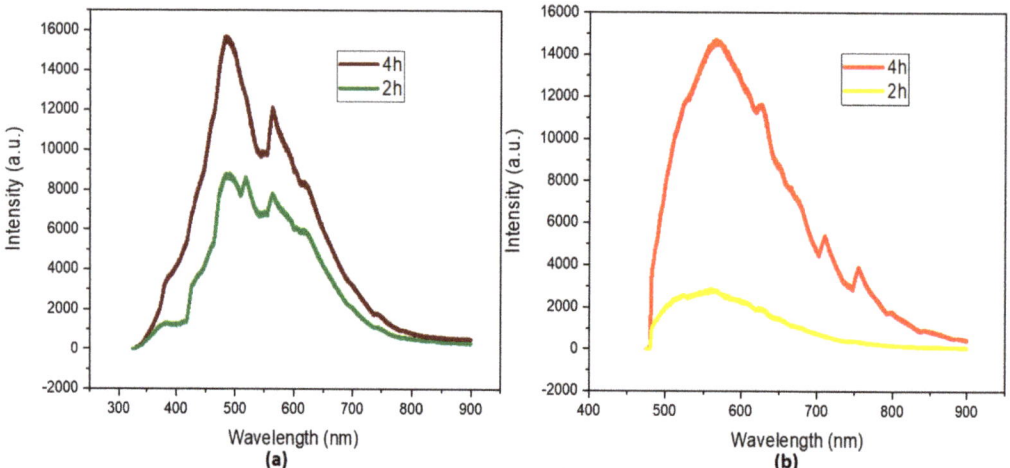

**Figure 4.** Photoluminescence spectra under (**a**) 325 nm (**b**) 473 nm irradiation of dialyzed solutions of N–CDs synthesized from 100 mg VOPO$_4$ as the catalyst, 0.75 M of xylose as the precursor, NH$_4$Cl as the dopant, and 17 g/L of CH$_3$COOH at 180 °C after 4 h of reaction. No dilution was performed prior to the analysis.

A third emission at 518 nm can be observed for the spectrum of the N-CDs after 2 h of reaction. This emission can be attributed to an intermediate species of C–O/C=O; after another 2 h, it fully oxidized into O–C=O.

When the N-CDs were irradiated under 473 nm (Figure 4b), the intensity of the spectra was lower, as a less energetic laser was used. two major contributions can be identified, at 566 nm and 625 nm, which can be associated with the formerly mentioned species. They suffered bathochromic shifts of 82 and 62 nm, respectively, proving the tunable photoluminescence properties of N-CDs [45–47].

The QY and fluorescence lifetime were also measured for both N-CDs, resulting in QYs of 2.3% when the synthesis lasted 2 h and 6.2% when it was left for 4 h. Fluorescence lifetimes of 2.59 ns for N-CDs and 3.04 ns for N-CDs were also observed when the reaction time was set at 2 h and when the reaction was maintained for 4 h.

2.1.3. Photostability Tests

Since the designated application for CDs is to work as photocatalysts, it becomes clear that photostability is a key parameter that confirms whether N-CDs are suitable for a photocatalysis reaction. The photostability tests were carried out in the same conditions as the photoluminescence measurements, changing only the exposure time. Samples for photoluminescence typically undergo a 2 min exposure time in order for the spectra to be recorded. During photostability assays, the solutions were irradiated for 28 min, and spectra were recorded every 2 min. During this time, there was a slight variation in intensity (Figure 5) in the spectra of both lasers, but it was not significant enough to consider the samples unstable in the selected conditions for the photocatalytic tests.

2.2. *Catalyst Recovery, Recycling, and Regeneration*

After 4 h of reaction, the catalyst was recovered, calcined to eliminate the organic matter over its surface, and re-assayed for several catalytic runs. The catalyst also underwent a regeneration step, at which point it was decided that its catalytic performance had excessively decreased. The changes in the active surface of the catalyst were followed by XPS (Figures S7–S10, Tables S1 and S2).

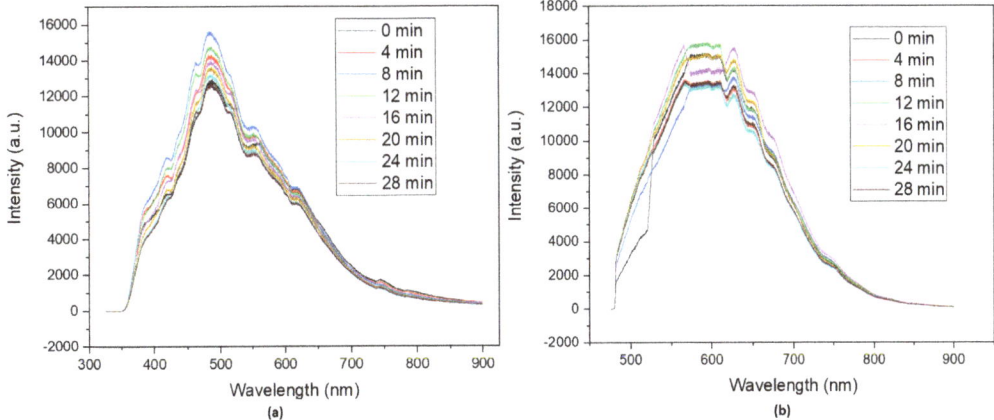

**Figure 5.** Photostability tests of dialyzed solutions of N–CDs using 100 mg of VOPO$_4$ as the catalyst, 0.75 M of xylose as the precursor, NH$_4$Cl as the dopant, and 17 g/L of CH$_3$COOH at 180 °C after 4 h of reaction under (**a**) a 325 nm irradiation laser and (**b**) a 473 nm irradiation laser. No dilution was performed prior to the analysis.

2.2.1. Recycling Tests

After three reaction cycles, the conversion did not suffer an alarming decrease in the conversion of xylose (Figure 6a). However, the presence of particles per squared micrometer dropped substantially with every cycle, suggesting that the catalyst's selectivity to N-CDs diminished as it was used in the reaction (Figure 6b). A possible reason for the decrease in the number of CD particles could be attributed to the presence of carbonaceous species adsorbed on the catalyst surface (Table S1), as shown by the increase in the atomic concentration of C after the reaction. Since no reactivation procedure was conducted between each catalytic run, these species may have been blocking the acid sites necessary for the dehydration reactions. Additionally, a partial reduction in V(V) was observed after the reaction (Table S2), indicating a potential decrease in the acidity of the catalyst. This reduction in acidity can be attributed to the fact that V(V) is more acidic than V(IV), providing a clue to the mechanism behind the observed changes.

In the bottom-up mechanism of CD synthesis, the role of the VOPO$_4$ catalyst was firstly to promote the dehydration of the xylose to furfural, which afterwards aggregated into higher structures that carbonized into N-CDs [43,48]. Secondly, the catalyst was able to oxidize the surfaces of the N-CDs. However, both the xylose and acetic acid dissolved in the reaction medium can act as reducing agents, promoting the rapid reduction of V(V) to V(IV), as confirmed via XPS (Table S2, before reaction). This partial reduction of vanadium could lead to leaching of the vanadium species, as it is a much more soluble species, especially in acidic media. Therefore, in order to avoid any contamination of the N-CDs with vanadium species, a purification step of the N-CDs was carried out. Thus, the pH was lowered to 2 after a reaction using a citrate/citric acid to ensure the protonation of the acidic groups on the surfaces of the N-CDs so they would not interact in any way with vanadium species such as VO$_2^+$ or VO$^{2+}$. After the purification of the N-CDs by means of dialysis, the presence of vanadium in the sample inside the dialysis tube was analyzed via ICP-MS (Table S3), confirming, along with the absence of vanadium in the XPS survey spectra of the N-CDs (Figure S11), that whether the reaction lasted 2 h or 4 h, the purification method was effective, as no significant quantities of V were detected on the samples of the N-CDs solutions.

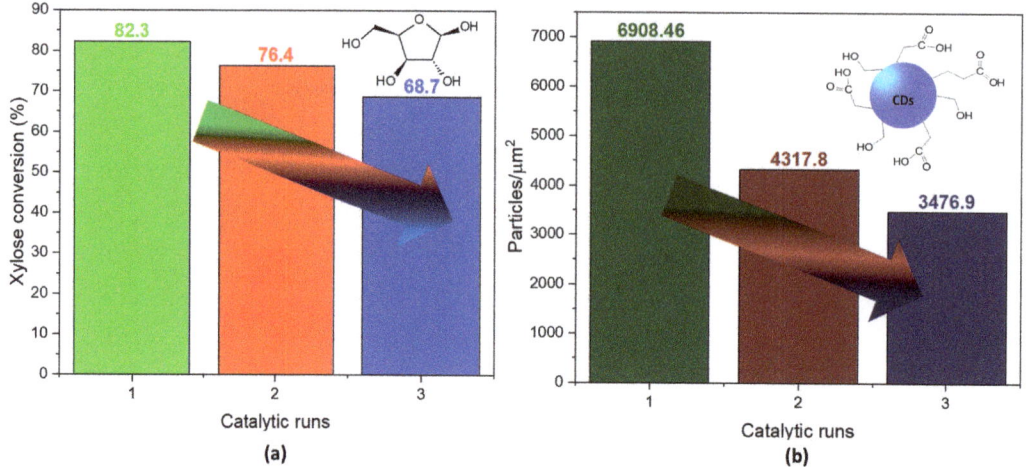

**Figure 6.** (a) Conversion of xylose in N-CDs solution samples after three catalytic runs of the VOPO$_4$ catalyst. (b) Particle density per squared micrometer in N-CDs dialyzed solutions after 4 h of reaction time. No dilution was performed prior to the analysis.

### 2.2.2. Regeneration Tests

After heating the used catalyst at 500 °C for the regeneration procedure, it was assayed for another catalytic run (Figure 7). The conversion of xylose attained in this new cycle was nearly the same as that obtained during the first catalytic run of the catalyst (blue bar). Nevertheless, this did not translate exactly into the same catalytic activity, as the ratio of particles per micrometer (red bar) achieved after the regeneration was higher than that obtained after the first recycling, but lower than the ratio reached when the catalyst was added fresh (Figure 7b). This can be attributed, again, to a partial blockage of the acidic sites due to residual carbonaceous species. Since, after regeneration, there is a reoxidation of V(IV) to V(V) a change in conversion or activity would not be due to a lack of V(V) species.

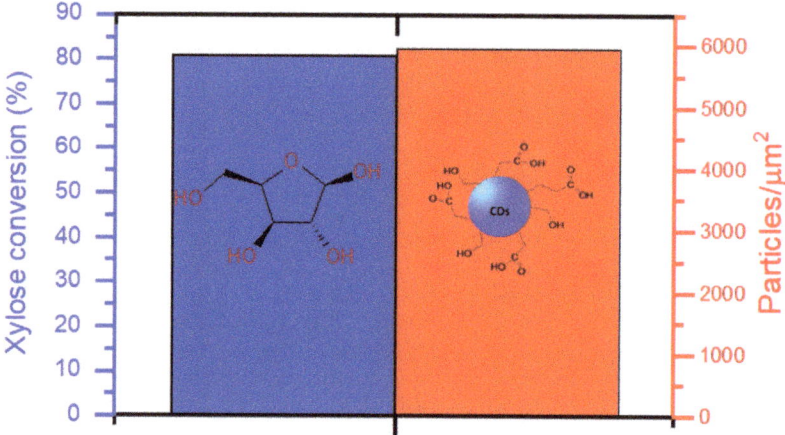

**Figure 7.** Conversion of xylose after the regeneration of the VOPO$_4$ catalyst and particle density per squared micrometer in N-CDs dialyzed solutions after 4 h of reaction time. No dilution was performed prior to the analysis.

## 2.3. Photocatalytic Assay

Since the photoluminescence measurements revealed that the synthesized N-CDs had good photoelectronic potential, meaning that their surface properties would be optimal for electronic transfer and movement, it was decided to test their activity as catalysts in the photodegradation of methyl orange (MO) (Figure 8). The assay was performed under visible light. During the first 20 min of the reaction, the degradation was moderate, but after 30 min, the absorbance observed in the UV-Vis had greatly diminished. By the time the hour of reaction was reached, nearly no absorbance was detected, meaning that the CDs had effectively promoted the degradation of the colorant. The UV-vis absorbance spectrum of N-CDs was measured in order to ensure that no secondary absorbance to MO absorbance was taking place (Figure S12).

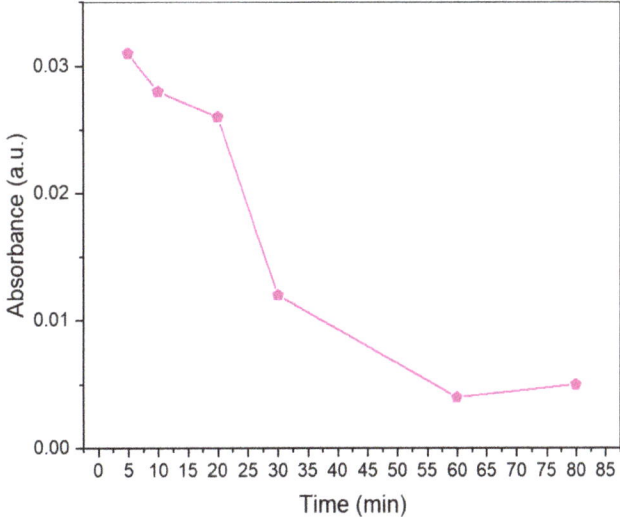

**Figure 8.** Photodegradation of methyl orange over time using 50 mg of N-CDs as the catalyst on an aqueous solution of 5 ppm of methyl orange.

## 3. Materials and Methods

### 3.1. Materials

Vanadium (V) oxide ($V_2O_5$) (>99.6%) and $NH_4Cl$ (>99.5%) were purchased from Sigma-Aldrich. Orthophosphoric ($H_3PO_4$) (>85%) and nitric acid ($HNO_3$) (>65%) were purchased from Panreac (Barcelone, Spain) and VWR (Radnor, PA, USA). Xylose (>98%) was purchased from Millipore (Burlington, MA, USA).

### 3.2. VOPO$_4$ Catalyst Synthesis

Briefly, the VOPO$_4$ preparation was based on a previous existing method [49], in which 1.93 g of $V_2O_5$ is magnetically stirred along with 10.5 mL of $H_3PO_4$, 22 mL of water, and 2 mL of concentrated $HNO_3$. The resulting suspension is kept for 2 h at 105 °C in reflux until the yellow precipitate is completely formed. Then, the vibrant yellow solid is filtered and left to dry in the stove overnight at 60 °C.

### 3.3. CDs Preparation

CDs were prepared following a hydrothermal procedure (Figure 9) that consisted of the addition of 100 mg of VOPO$_4$ as the catalyst, 1.62 mL of $CH_3COOH$, and 25 mL of 0.75 M xylose solution into a Teflon-lined steel hydrothermal reactor (Parr, Moline, IL, USA). As the ultimate target was to optimize the production for large-scale biomass transformation of olive pits, in order to replicate the conditions of acidity of a biomass-derived xylose solution, the expected amount of acetic acid that would be produced as a by-product when treating the biomass was added into reaction media. With the aim of doping the surfaces of the CDs to enhance their photocatalytic and photoluminescent properties, 5 g of NH$_4$Cl was added into the hydrothermal reactor along with the rest of the components for the synthesis. The reaction temperature was set at 180 °C inside a muffle furnace in all cases, while the reaction time varied between 2 h and 4 h. Since the obtained N-CDs showed a high photoluminescence yield, it was decided that the synthesized N-CDs would be assayed in the photocatalytic degradation of methyl orange.

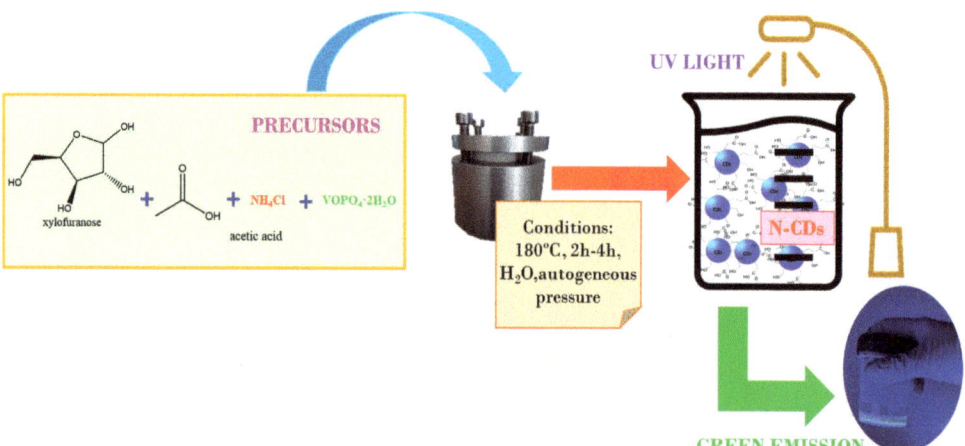

**Figure 9.** Scheme of the hydrothermal one-pot method for the synthesis of N-CDs from xylose as the carbon precursor, VOPO$_4$ as the catalyst, and NH$_4$Cl as the doping agent.

*3.4. CDs Purification*

After the reaction, the purification method involved a centrifugation step to separate the remaining carbonization solids and the majority of the catalyst; then, there was a second centrifugation step, along with 10 mL pH = 2 citrate-citric acid buffer. After centrifugation, the suspended solid particles of the catalyst were separated by filtrating the solution over 0.45 μm syringe filters. Before continuing with the purification process, after this filtering step, the photoluminescent emission of every sample was checked under a UV lamp (Electro DH, Barcelone, Spain) as a rapid way to confirm the success of the synthesis. Filtered solutions were then poured into dialysis membranes (Pur-A-LyzerTM 1 kDa, Sigma-Aldrich (St. Luis, MO, USA)), along with citric/citrate buffer solution. After 48 h, the samples were removed from the dialysis tubes, while keeping the rinsing water. To obtain solid N-CDs for XPS analysis and for their use in the photodegradation of methyl orange as catalysts, the samples were lyophilized after dialysis using Scanvac® Coolsave™ (Bjarkesvej, Denmark) lyophilizer equipment. After lyophilization, 10 mg of solid N-CD was recovered each time the procedure was carried out, so the average N-CD mass yield of the synthetic process rose to 0.36%.

*3.5. Catalyst Recovering*

The separated carbonaceous solid underwent a thermal treatment in order to remove all the organic residues masking the catalyst. The calcination of this solid was performed at 550 °C for 6 h, at a heating rate of 5 °C/min.

*3.6. Catalyst Recycling*

The calcined solid was then reused into further catalytic cycles in the same reaction conditions. As some of the catalyst was inevitably lost during the process of recovery and calcination, the recovered catalyst from two identical catalytic runs was evenly mixed for reuse.

*3.7. Catalyst Regeneration*

After three catalytic runs, the catalyst followed a regeneration step based on a previously published procedure [50], which involved the addition of $H_3PO_4$ along with $HNO_3$ to a suspension in the rinsing water used for the dialysis of the catalyst after the reaction.

*3.8. Photocatalytic Assay*

For the photocatalytic assay, 50 mg of CD was added into 0.5 L of an aqueous solution consisting of 5 ppm of methyl orange dye. A photoreactor (Luzchem (Gloucester, ON, Canada) Model CCP-4V 220 V 50 Hz 3 A) was used to irradiate the samples with visible light, employing the fourteen white visible lamps inside the reactor. After 5, 10, 20, 30, 60, and 80 min, an aliquot of the solution was taken to control methyl orange (MO) photodegradation by determining its remaining concentration. After 5, 10, 20, 30, 60, and 80 min, an aliquot of the solution was taken to control the methyl orange (MO) photodegradation by determining its remaining concentration. Prior to the measurement of the sample, a calibration curve was performed using five standards of methyl orange of 1, 2, 3, 4, and 5 ppm.

*3.9. Characterization Conditions*

Transmission electron microscopy (TEM) was carried out in a FEITalos F200X (Thermo Fisher Scientific, Waltham, MA, USA) equipped with an FEG 200 kV electron gun, four STEM detectors, and four FEG detectors. A Thermo Scientific-FEI Tecnai G2 20 Twin Transmission Electron Microscope equipped with LaB6 filament, tomography software, a cryo-transmission system, and an EDS/EDX (energy-dispersive X-ray spectroscopy) elemental analyzer (Thermo Fisher Scientific, Waltham, MA, USA) was used to evaluate the characteristics of the surfaces of the CDs. TEM and HRTEM images were processed using ImageJ v. 1.53k software. The photophysical characteristics of the CDs were studied by photoluminescence measurements using a LabRAM Odyssey PL microscope (Horiba, Kyoto, Japan). Samples were irradiated under two lasers: (1) 325 nm, ×40 lens, 100-hole aperture, 5% ND filter, 28 mW power. (2) 473 nm, ×40 lens, 100-hole aperture, 5% ND filter, 100 mW power. Photostability was measured using the same equipment and irradiation conditions over time, following a previously published study [51]. QY and the fluorescence lifetime were measured using an Edinburgh Instruments FLS920 fluorimeter coupled with a 1-M-1 integrating sphere for calculating the QY, and using the ultra-rapid F-G05 detector. XPS analysis was utilized to characterize the superficial composition of the CDs and the surface of the catalyst, $VOPO_4$. It was performed by means of a Physical Electronics PHI5700 spectrometer using monochromatic Al K$\alpha$ of 15 kV and 1486.6 eV, with a dual charge beam and a hemispheric multichannel detector for the $VOPO_4$ and N-CDs spectra. The analysis zone comprised an area 100 µm in diameter when AlK$\alpha$ radiation was used. The constant pass energy mode was set at 29.35 eV. The obtained spectrum was processed using MultiPak v.9.3 software. The XPS analysis was carried out for the lyophilized samples in the case of N-CDs. All values were referenced to adventitious carbon (C 1s at 284.8 eV). FTIR was performed for N-CDs in a Bruker Vertex70 coupled with a Golden Gate Single Reflection Diamond ATR System (Bruker, Billerica, MA, USA). The spectral resolution was set at 4 cm$^{-1}$ in a spectral range of 4000–500 cm$^{-1}$. The Raman spectrum of $VOPO_4$ was measured using a FT-Raman RFT-6000-JASCO spectrometer (JASCO, Tokyo, Japan) with a 1064 nm laser at 150 mW of power. The XRD patterns were collected on a PANanalytical EMPYREAN (Malvern Panalytical, Malvern, UK) automated diffractometer. The PIXcel 3D detector was set at a step size of 0.017° (2θ). The diffractograms were recorded between 4 and 70 in 2θ. The remaining xylose in the reaction samples was controlled using an HPLC instrument (JASCO, Tokyo, Japan) equipped with an autoinjector (AS-2055), which injected 6 µL of the sample into a Phenomenex (Torrance, CA, USA) Rezex ROA-Organic Acid H$^+$ (8%) (300 mm × 7.8 mm) column. The mobile phase (0.0025 M $H_2SO_4$) was pumped by a quaternary gradient pump (PU-2089) at a 0.35 mL/min$^{-1}$ flow rate to the column, heated at 40 °C. The possible remains of vanadium species after the reaction were studied by means of ICP-MS in a Nexon 300D at a flow rate of 0.8 L/min of nebulizer gas, 18 L/min of gas for the plasma, and 1.2 L/min of auxiliary gas at a potential RF of 1600 W. The Brönsted–Lewis acidic sites were determined via pyridine adsorption measurements. The catalyst was shaped into wafers along with KBr and saturated in pyridine for 10 min at room temperature; then, it was gradually heated in a tubular oven until 125 °C, recording the spectrum each 20 to 30 °C. The adsorption and desorption spectra, at different temperatures, were recorded by a SHIMADZU (Kyoto, Japan) FTIR-8300 infrared spectrophotometer at a fixed irradiation wavelength and power of 632.8 nm and 0.5 mW, respectively. The photodegradation of MO was followed by UV-vis spectrophotometry (UV 1800 SHIMADZU (Kyoto, Japan) UV). The UV-vis spectrum of N-CDs was recorded using the same equipment in a 300–900 nm range.

## 4. Conclusions

Green-emitting N-CDs were successfully obtained using heterogeneous catalysis via a hydrothermal synthetic procedure. Nevertheless, there is still work to be done as this procedure could be improved in terms of production yield. $VOPO_4$ played a bifunctional role in the synthesis, while Brønsted acidity promoted the dehydration and condensation of the carbon precursor. Its oxidizing properties provoked a complete oxidation of the organic groups functionalizing the surface of the N-CDs, as was deduced from the XPS results and confirmed by photoluminescent emissions. The addition of a certain amount of $CH_3COOH$ in order to replicate the composition of a real biomass liquor positively affected the synthesis, as it enhanced the oxidizing properties of the catalyst. The $VOPO_4$ catalyst was successfully recovered, recycled, and regenerated with no further negative effect on its performance in terms of the dehydration, aggregation, and oxidation of the surfaces of the N-CDs. CDs were doped with N atoms using $NH_4Cl$, and the doping was confirmed by XPS. The doping of CDs to N-CDs greatly improved the photophysical properties of pristine CDs, as was observed in the photoluminescent measurements that affected the photocatalytic performance of the CDs in a positive way, as methyl orange was easily degraded at a high rate after 30 min of reaction. After 1 h of reaction, the degradation was considered to have been completed.

**Supplementary Materials:** The following supporting information can be downloaded at: https://www.mdpi.com/article/10.3390/catal13101358/s1, Figure S1: Size histograms of N-CDs synthesized using 100 mg $VOPO_4$ as catalyst, 0.75 M of xylose as C precursor, $NH_4Cl$ as dopant, and 17 g/L of $CH_3COOH$ at 180 °C after 4 h of reaction calculated from TEM images; Figure S2: Graphitic spacing of N-CDs dialyzed solution samples synthesized using 100 mg $VOPO_4$ as catalyst, 0.75 M of xylose as C precursor, $NH_4Cl$ as dopant, and 17 g/L of $CH_3COOH$ at 180 °C after 4 h of reaction; Figure S3: XRD pattern of dialyzed and lyophilized N-CDs synthesized using 100 mg $VOPO_4$ as catalyst, 0.75 M of xylose as C precursor, $NH_4Cl$ as dopant, and 17 g/L of $CH_3COOH$ at 180 °C after 4 h of reaction; Figure S4: Pyridine adsorption and desorption on $VOPO_4$ followed by FTIR; Figure S5: XRD pattern of freshly synthesized $VOPO_4$; Figure S6: Raman spectrum of freshly synthesized $VOPO_4$; Figure S7: XPS spectrum of V 2p core level binding energy of freshly synthesized $VOPO_4$; Figure S8: XPS spectra of V 2p core level binding energy of (a) $VOPO_4$ before reaction* and (b) $VOPO_4$ after reaction; Figure S9: Survey XPS spectrum of freshly synthesized $VOPO_4$; Figure S10: Survey spectra of (a) $VOPO_4$ before reaction* and (b) $VOPO_4$ after reaction; Table S1: XPS atomic concentration table of $VOPO_4$; Table S2: XPS energy binding deconvoluted bands of $VOPO_4$; Table S3: ICP-MS measurements of vanadium species in N-CDs-dialyzed solutions; Figure S11: XPS survey spectrum of dialyzed and lyophilized N-CDs synthesized using 100 mg $VOPO_4$ as catalyst, 0.75 M of xylose as C precursor, $NH_4Cl$ as dopant, and 17 g/L of $CH_3COOH$ at 180 °C after 4 h of reaction; Figure S12:

UV-vis spectrum of N-CDs dialyzed synthesized using 100 mg VOPO$_4$ as catalyst, 0.75 M of xylose as C precursor, NH$_4$Cl as dopant, and 17 g/L of CH$_3$COOH at 180 °C after 4 h of reaction solution from 300 nm to 900 nm. No dilution was performed prior to the analysis. References [52–57] are cited in the Supplementary Materials.

**Author Contributions:** Conceptualization, R.M.-T., C.G.-S. and M.A.; methodology, G.R.-C. and M.A.; software, G.R.-C.; validation, M.A., C.G.-S. and R.M.-T.; formal analysis, G.R.-C.; investigation, G.R.-C.; resources, E.C.; data curation, G.R.-C.; writing—original draft preparation, G.R.-C.; writing—review and editing, G.R.-C., R.M.-T. and M.A.; visualization, R.M.-T. and C.G.-S.; supervision, R.M.-T. and C.G.-S.; project administration, M.A. and R.M.-T.; funding acquisition, R.M.-T. and M.A. All authors have read and agreed to the published version of the manuscript.

**Funding:** This research was funded by the Spanish Ministry of Science and Innovation (PID2021-122736OB-C42, PID2021-122613OB-I00) and FEDER (European Union) funds (PID2021-122736OB-C42, P20-00375, UMA20-FEDERJA88).

**Data Availability Statement:** No data is available.

**Conflicts of Interest:** The authors declare no conflict of interest. The funders had no role in the design of the study; in the collection, analyses, or interpretation of data; in the writing of the manuscript; or in the decision to publish the results.

# References

1. Moustafa, R.M.; Talaat, W.; Youssef, R.M.; Kamal, M.F. Carbon dots as fluorescent nanoprobes for assay of some non-fluorophoric nitrogenous compounds of high pharmaceutical interest. *Beni-Suef Univ. J. Basic Appl. Sci.* **2023**, *12*, 8. [CrossRef] [PubMed]
2. Lagos, K.J.; García, D.; Cuadrado, C.F.; de Souza, L.M.; Mezzacappo, N.F.; da Silva, A.P.; Inada, N.; Bagnato, V.; Romero, M.P. Carbon dots: Types, preparation, and their boosted antibacterial activity by photoactivation. Current status and future perspectives. *WIREs Nanomed. Nanobiotechnol.* **2023**, *15*, e1887. [CrossRef]
3. Khan, M.E.; Mohammad, A.; Yoon, T. State-of-the-art developments in carbon quantum dots (CQDs): Photo-catalysis, bio-imaging, and bio-sensing applications. *Chemosphere* **2022**, *302*, 134815. [CrossRef] [PubMed]
4. Macairan, J.-R.; de Medeiros, T.V.; Gazzetto, M.; Villanueva, F.Y.; Cannizzo, A.; Naccache, R. Elucidating the mechanism of dual-fluorescence in carbon dots. *J. Colloid Interface Sci.* **2022**, *606*, 67–76. [CrossRef] [PubMed]
5. Zhu, S.; Song, Y.; Zhao, X.; Shao, J.; Zhang, J.; Yang, B. The photoluminescence mechanism in carbon dots (graphene quantum dots, carbon nanodots, and polymer dots): Current state and future perspective. *Nano Res.* **2015**, *8*, 355–381. [CrossRef]
6. Gedda, G.; Sankaranarayanan, S.A.; Putta, C.L.; Gudimella, K.K.; Rengan, A.K.; Girma, W.M. Green synthesis of multi-functional carbon dots from medicinal plant leaves for antimicrobial, antioxidant, and bioimaging applications. *Sci. Rep.* **2023**, *13*, 6371. [CrossRef]
7. Lan, X.; Wang, Y.; Chen, X.; Liu, P.; Liu, C.; Xu, J.; Liu, C.; Jiang, F. Dual-action carbon quantum dots with light assist in enhancing the thermoelectric performance of polymers. *J. Mater. Chem. C* **2022**, *10*, 15906–15912. [CrossRef]
8. Madrid, A.; Martín-Pardillos, A.; Bonet-Aleta, J.; Sancho-Albero, M.; Martinez, G.; Calzada-Funes, J.; Martin-Duque, P.; Santamaria, J.; Hueso, J.L. Nitrogen-doped carbon nanodots deposited on titania nanoparticles: Unconventional near-infrared active photocatalysts for cancer therapy. *Catal. Today* **2023**, *419*, 114154. [CrossRef]
9. Han, W.; Li, D.; Hu, X.; Qin, W.; Sun, H.; Wang, S.; Duan, X. Photocatalytic activation of peroxymonosulfate by carbon quantum dots: Rational regulation of surface functionality and computational insights. *Mater. Today Chem.* **2023**, *30*, 101546. [CrossRef]
10. Jeong, G.; Park, C.H.; Yi, D.; Yang, H. Green synthesis of carbon dots from spent coffee grounds via ball-milling: Application in fluorescent chemosensors. *J. Clean. Prod.* **2023**, *392*, 136250. [CrossRef]
11. Liu, Y.; Sun, K.; Shi, N.; Li, R.; Zhang, J.; Zhao, J.; Geng, L.; Lei, Y. Dual functions of nitrogen and phosphorus co-doped carbon dots for drug-targeted delivery and two-photon cell imaging. *Arab. J. Chem.* **2023**, *16*, 104671. [CrossRef]
12. Ding, P.; Song, H.; Chang, J.; Lu, S. N-doped carbon dots coupled NiFe-LDH hybrids for robust electrocatalytic alkaline water and seawater oxidation. *Nano Res.* **2022**, *15*, 7063–7070. [CrossRef]
13. Zhou, P.; Xu, J.; Hou, X.; Dai, L.; Zhang, J.; Xiao, X.; Huo, K. Heteroatom-engineered multicolor lignin carbon dots enabling bimodal fluorescent off-on detection of metal-ions and glutathione. *Int. J. Biol. Macromol.* **2023**, *253*, 126714. [CrossRef]
14. Miao, S.; Liang, K.; Zhu, J.; Yang, B.; Zhao, D.; Kong, B. Hetero-atom-doped carbon dots: Doping strategies, properties and applications. *Nano Today* **2020**, *33*, 100879. [CrossRef]
15. Yang, X.; Li, X.; Wang, B.; Ai, L.; Li, G.; Yang, B.; Lu, S. Advances, opportunities, and challenge for full-color emissive carbon dots. *Chin. Chem. Lett.* **2021**, *33*, 613–625. [CrossRef]
16. Wareing, T.C.; Gentile, P.; Phan, A.N. Biomass-Based Carbon Dots: Current Development and Future Perspectives. *ACS Nano* **2021**, *15*, 15471–15501. [CrossRef] [PubMed]
17. Wang, B.; Lu, S. The light of carbon dots: From mechanism to applications. *Matter* **2022**, *5*, 110–149. [CrossRef]
18. Liao, L.; Lin, X.; Zhang, J.; Hu, Z.; Wu, F. Facile preparation of carbon dots with multicolor emission for fluorescence detection of ascorbic acid, glutathione and moisture content. *J. Lumin.* **2023**, *264*, 120169. [CrossRef]

19. Gong, P.; Sun, L.; Wang, F.; Liu, X.; Yan, Z.; Wang, M.; Zhang, L.; Tian, Z.; Liu, Z.; You, J. Highly fluorescent N-doped carbon dots with two-photon emission for ultrasensitive detection of tumor marker and visual monitor anticancer drug loading and delivery. *Chem. Eng. J.* **2018**, *356*, 994–1002. [CrossRef]
20. Zhu, X.; Han, L.; Liu, H.; Sun, B. A smartphone-based ratiometric fluorescent sensing system for on-site detection of pyrethroids by using blue-green dual-emission carbon dots. *Food Chem.* **2022**, *379*, 132154. [CrossRef] [PubMed]
21. Nguyen, K.G.; Baragau, I.-A.; Gromicova, R.; Nicolaev, A.; Thomson, S.A.J.; Rennie, A.; Power, N.P.; Sajjad, M.T.; Kellici, S. Investigating the effect of N-doping on carbon quantum dots structure, optical properties and metal ion screening. *Sci. Rep.* **2022**, *12*, 13806. [CrossRef] [PubMed]
22. Cao, M.; Zhao, X.; Gong, X. Ionic Liquid-Assisted Fast Synthesis of Carbon Dots with Strong Fluorescence and Their Tunable Multicolor Emission. *Small* **2022**, *18*, 2106683. [CrossRef]
23. Mattinzoli, D.; Cacioppo, M.; Ikehata, M.; Armelloni, S.; Alfieri, C.M.; Castellano, G.; Barilani, M.; Arcudi, F.; Messa, P.; Prato, M. Carbon dots conjugated to SN38 for improved colorectal anticancer therapy. *Mater. Today Bio* **2022**, *16*, 100286. [CrossRef] [PubMed]
24. Dai, R.; Chen, X.; Ouyang, N.; Hu, Y. A pH-controlled synthetic route to violet, green, and orange fluorescent carbon dots for multicolor light-emitting diodes. *Chem. Eng. J.* **2022**, *431*, 134172. [CrossRef]
25. Wang, B.; Yu, J.; Sui, L.; Zhu, S.; Tang, Z.; Yang, B.; Lu, S. Rational Design of Multi-Color-Emissive Carbon Dots in a Single Reaction System by Hydrothermal. *Adv. Sci.* **2020**, *8*, 2001453. [CrossRef] [PubMed]
26. Algarra, M.; Dos Orfãos, L.; Alves, C.S.; Moreno-Tost, R.; Pino-González, M.S.; Jiménez-Jiménez, J.; Rodríguez-Castellón, E.; Eliche-Quesada, D.; Castro, E.; Luque, R. Sustainable Production of Carbon Nanoparticles from Olive Pit Biomass: Understanding Proton Transfer in the Excited State on Carbon Dots. *ACS Sustain. Chem. Eng.* **2019**, *7*, 10493–10500. [CrossRef]
27. Wang, T.; Peng, L.; Wu, D.; Chen, B.; Jia, B. Crude fiber and protein rich cottonseed meal derived carbon quantum dots composite porous carbon for supercapacitor. *J. Alloys Compd.* **2023**, *947*, 169499. [CrossRef]
28. Hallaji, Z.; Bagheri, Z.; Ranjbar, B. One-Step Solvothermal Synthesis of Red Chiral Carbon Dots for Multioptical Detection of Water in Organic Solvents. *ACS Appl. Nano Mater.* **2023**, *6*, 3202–3210. [CrossRef]
29. Yin, C.-L.; An, B.-L.; Li, J.; Wang, X.-H.; Zhang, J.-M.; Xu, J.-Q. High-efficient synthesis of bright yellow carbon quantum dots catalyzed by $SnO_2$ NPs. *J. Lumin.* **2021**, *233*, 117850. [CrossRef]
30. Sarpiri, J.N.; Chermahini, A.N.; Saraji, M.; Shahvar, A. Dehydration of carbohydrates into 5-hydroxymethylfurfural over vanadyl pyrophosphate catalysts. *Renew. Energy* **2020**, *164*, 11–22. [CrossRef]
31. Mérida-Morales, S.; García-Sancho, C.; Oregui-Bengoechea, M.; Ginés-Molina, M.; Cecilia, J.; Arias, P.; Moreno-Tost, R.; Maireles-Torres, P. Influence of morphology of zirconium-doped mesoporous silicas on 5-hydroxymethylfurfural production from mono-, di- and polysaccharides. *Catal. Today* **2020**, *367*, 297–309. [CrossRef]
32. Faria, V.W.; Santos, K.M.A.; Calazans, A.M.; Fraga, M.A. Hydrolysis of Furfuryl Alcohol to Angelica Lactones and Levulinic Acid over Nb-based Catalysts. *ChemCatChem* **2023**, *15*, e202300447. [CrossRef]
33. Yan, J.; Zhou, Y.; Shen, J.; Zhang, N.; Liu, X. Facile synthesis of S, N-co-doped carbon dots for bio-imaging, $Fe^{3+}$ detection and DFT calculation. *Spectrochim. Acta Part A Mol. Biomol. Spectrosc.* **2023**, *302*, 123105. [CrossRef]
34. Parry, E. An infrared study of pyridine adsorbed on acidic solids. Characterization of surface acidity. *J. Catal.* **1963**, *2*, 371–379. [CrossRef]
35. Glazunov, V.; Odinokov, S. Infrared spectra of pyridinium salts in solution—II. Fermi resonance and structure of νNH bands. *Spectrochim. Acta Part A Mol. Spectrosc.* **1982**, *38*, 409–415. [CrossRef]
36. Zhao, Y.; Yu, L.; Deng, Y.; Peng, K.; Yu, Y.; Zeng, X. A multi-color carbon quantum dots based on the coordinated effect of quantum size and surface defects with green synthesis. *Ceram. Int.* **2023**, *49*, 16647–16651. [CrossRef]
37. Crispi, S.; Nocito, G.; Nastasi, F.; Condorelli, G.; Ricciardulli, A.; Samorì, P.; Conoci, S.; Neri, G. Development of a novel C-dots conductometric sensor for NO sensing. *Sens. Actuators B Chem.* **2023**, *390*, 133957. [CrossRef]
38. Wang, X.; Liuye, S.; Ma, X.; Cui, S.; Pu, S. Construction of a solid-state fluorescent switching with carbon dots and diarylethene. *Dye. Pigment.* **2023**, *216*, 111318. [CrossRef]
39. Li, R.; Shi, N.; Sun, K.; Fang, M.; Zhang, Z.; Geng, L.; Zhang, J. Nitrogen-doped carbon dots for doxorubicin-targeted delivery and two-photon cell imaging. *Arab. J. Chem.* **2023**, *16*, 105067. [CrossRef]
40. Matter, E.A.; El-Naggar, G.A.; Nasr, F.; Ahmed, G.H.G. Facile synthesis of N-doped carbon dots (N-CDs) for effective corrosion inhibition of mild steel in 1 M HCl solution. *J. Appl. Electrochem.* **2023**, *53*, 2057–2075. [CrossRef]
41. Gorji, Z.E.; Khodadadi, A.A.; Riahi, S.; Repo, T.; Mortazavi, Y.; Kemell, M. Functionalization of nitrogen-doped graphene quantum dot: A sustainable carbon-based catalyst for the production of cyclic carbonate from epoxide and $CO_2$. *J. Environ. Sci.* **2023**, *126*, 408–422. [CrossRef] [PubMed]
42. Ezati, P.; Rhim, J.-W.; Molaei, R.; Priyadarshi, R.; Roy, S.; Min, S.; Kim, Y.H.; Lee, S.-G.; Han, S. Preparation and characterization of B, S, and N-doped glucose carbon dots: Antibacterial, antifungal, and antioxidant activity. *Sustain. Mater. Technol.* **2022**, *32*, e00397. [CrossRef]
43. Rodríguez-Padrón, D.; Algarra, M.; Tarelho, L.A.C.; Frade, J.R.; Franco, A.; de Miguel, G.; Jiménez, J.; Rodríguez-Castellón, E.; Luque, R. Catalyzed Microwave-Assisted Preparation of Carbon Quantum Dots from Lignocellulosic Residues. *ACS Sustain. Chem. Eng.* **2018**, *6*, 7200–7205. [CrossRef]

44. Van Dam, B.; Nie, H.; Ju, B.; Marino, E.; Paulusse, J.M.J.; Schall, P.; Li, M.; Dohnalová, K. Excitation-Dependent Photoluminescence from Single-Carbon Dots. *Small* **2017**, *13*, 1702098. [CrossRef] [PubMed]
45. Anthony, A.M.; Pandurangan, P.; Abbas, S. Ligand engineering with heterocyclic aromatic thiol doped carbon quantum dots. *Carbon* **2023**, *211*, 118086. [CrossRef]
46. Liang, C.; Shi, Q.; Zhang, Y.; Xie, X. Water-soluble carbonized polymer dots with tunable solid- and dispersion-state fluorescence for multicolor films, anti-counterfeiting, and fungal imaging. *Mater. Today Nano* **2023**, *22*, 100324. [CrossRef]
47. Panigrahi, A.; Behera, R.K.; Mishra, L.; Dubey, P.; Dutta, S.; Sarangi, M.K. Regulating optoelectronics of carbon dots with redox-active dopamine. *Talanta Open* **2023**, *7*, 100198. [CrossRef]
48. Xiong, H.-F.; An, B.-L.; Zhang, J.-M.; Yin, C.-L.; Wang, X.-H.; Wang, J.-H.; Xu, J.-Q. Efficient one step synthesis of green carbon quantum dots catalyzed by tin oxide. *Mater. Today Commun.* **2020**, *26*, 101762. [CrossRef]
49. Luo, Z.; Liu, E.; Hu, T.; Li, Z.; Liu, T. Effect of synthetic methods on electrochemical performances of $VOPO_4 \cdot 2H_2O$ supercapacitor. *Ionics* **2014**, *21*, 289–294. [CrossRef]
50. Zazhigalov, V.; Kharlamov, A.; Haber, J.; Stoch, J.; Yaremenko, V.; Bacherikova, I.; Bogutskaya, L. Regeneration of VPMeO Catalysts for n-butane oxidation by means of mechanochemical and barothermal treatments. *Prep. Catal. V Sci. Bases Prep. Heterog. Catal. Proc. Fifth Int. Symp.* **1997**, *111*, 207–212. [CrossRef]
51. Ji, Y.; Wang, M.; Yang, Z.; Qiu, H.; Ji, S.; Dou, J.; Gaponenko, N.V. Highly stable Na: $CsPb(Br,I)_3@Al_2O_3$ nanocomposites prepared by a pre-protection strategy. *Nanoscale* **2020**, *12*, 6403–6410. [CrossRef] [PubMed]
52. He, Y.; Yang, X.; Bai, Y.; Zhang, J.; Kang, L.; Lei, Z.; Liu, Z.-H. Vanadyl phosphate/reduced graphene oxide nanosheet hybrid material and its capacitance. *Electrochim. Acta* **2015**, *178*, 312–320. [CrossRef]
53. Trchová, M.; Čapková, P.; Matějka, P.; Melánová, K.; Beneš, L.; Uhlířová, E. Intercalation of Water into Anhydrous Vanadyl Phosphate Studied by the Infrared and Raman Spectroscopies. *J. Solid State Chem.* **1999**, *148*, 197–204. [CrossRef]
54. De Luna, Y.; Bensalah, N. Synthesis, Characterization and Electrochemical Evaluation of Layered Vanadium Phosphates as Cathode Material for Aqueous Rechargeable Zn-ion Batteries. *Front. Mater.* **2021**, *8*, 645915. [CrossRef]
55. Griesel, L.; Bartley, J.K.; Wells, R.P.K.; Hutchings, G.J. Preparation of vanadium phosphate catalysts from $VOPO_4 \cdot 2H_2O$: Effect of $VOPO_4 \cdot 2H_2O$ preparation on catalyst performance. *J. Mol. Catal. A: Chem.* **2004**, *220*, 113–119. [CrossRef]
56. Antonio, M.R.; Barbour, R.L.; Blum, P.R. Interlayer coordination environments of iron, cobalt, and nickel in vanadyl phosphate dihydrate, $VOPO_4 \cdot 2H_2O$, intercalation compounds. *Inorg. Chem.* **1987**, *26*, 1235–1243. [CrossRef]
57. Beneš, L.; Melánová, K.; Zima, V.; Trchová, M.; Čapková, P.; Koudelka, B. Vanadyl phosphate intercalated with dimethyl sulfoxide. *J. Phys. Chem. Solids* **2006**, *67*, 956–960. [CrossRef]

**Disclaimer/Publisher's Note:** The statements, opinions and data contained in all publications are solely those of the individual author(s) and contributor(s) and not of MDPI and/or the editor(s). MDPI and/or the editor(s) disclaim responsibility for any injury to people or property resulting from any ideas, methods, instructions or products referred to in the content.

Article

# Self-Doped Carbon Dots Decorated TiO$_2$ Nanorods: A Novel Synthesis Route for Enhanced Photoelectrochemical Water Splitting

Chau Thi Thanh Thuy [1], Gyuho Shin [1], Lee Jieun [1], Hyung Do Kim [2], Ganesh Koyyada [1,*] and Jae Hong Kim [1,*]

[1] Department of Chemical Engineering, Yeungnam University, 214-1, Daehak-ro 280, Gyeongsan 712-749, Gyeongbuk-do, Korea
[2] Graduate School of Engineering, Department of Polymer Chemistry, Kyoto University, Katsura, Nishikyo-ku, Kyoto 615-8510, Japan
* Correspondence: ganeshkoyyada@gmail.com or ganeshkoyyada@ynu.ac.kr (G.K.); jaehkim@ynu.ac.kr (J.H.K.)

**Abstract:** Herein, we have successfully prepared self-doped carbon dots with nitrogen elements (NCD) in a simple one-pot hydrothermal carbonization method, using L-histidine as a new precursor. The effect of as-prepared carbon dots was studied for photoelectrochemical (PEC) water splitting by decorating NCDs upon TiO$_2$ nanorods systematically by changing the loading time from 2 h to 8 h (TiO$_2$@NCD2h, TiO$_2$@NCD4h, TiO$_2$@NCD6h, and TiO$_2$@NCD8h). The successful decorating of NCDs on TiO$_2$ was confirmed by FE-TEM and Raman spectroscopy. The TiO$_2$@NCD4h has shown a photocurrent density of 2.51 mA.cm$^{-2}$, 3.4 times higher than the pristine TiO$_2$. Moreover, TiO$_2$@NCD4h exhibited 12% higher applied bias photon-to-current efficiency (ABPE) than the pristine TiO$_2$. The detailed IPCE, Mott–Schottky, and impedance (EIS) analyses have revealed the enhanced light harvesting property, free carrier concentration, charge separation, and transportation upon introduction of the NCDs on TiO$_2$. The obtained results clearly portray the key role of NCDs in improving the PEC performance, providing a new insight into the development of highly competent TiO$_2$ and NCDs based photoanodes for PEC water splitting.

**Keywords:** TiO$_2$ photoanode; L-histidine; nitrogen-doped carbon dots; photoelectrochemical; light harvesting

## 1. Introduction

Rapidly spiking global energy demands and pollution caused by the depletion of fossil fuels necessitated the development of natural and renewable sources of energy [1]. Hydrogen is an excellent contender capable of replacing fossil fuels owing to its both eco-friendly and reusable nature. Photoelectrochemical (PEC) water splitting is the most reliable and popular method employed for converting solar light energy into clean and sustainable chemical fuels, such as hydrogen [2,3]. The initial study on photocatalytic water splitting using TiO$_2$ was published way back in 1972 [4]. Since then, different types of semiconductor materials including ZnO, [5] BiVO$_4$, [6] WO$_3$, [7] Fe$_2$O$_3$, [8] SrTiO$_3$, [9] C$_3$N$_4$ [10], and Ta$_3$N [11] were reported as photoelectrodes for PEC. The TiO$_2$ material is considered as the most competent semiconductor for investigating PEC devices due to its characteristics such as advantageous band-edge positions, ease of fabrication, abundance, excellent photo-corrosion resistance, eco-friendliness, and cost effective nature [12]. However, application of TiO$_2$ in PEC has been constrained by comparatively greater band gaps for its rutile (3.0 eV) and anatase (3.2 eV) phases [12], severe bulk charge recombination, and slow OER kinetics [13]. As a result, numerous attempts were made to surpass the limitations, such as use of dopants [14], formation of heterojunctions [15], surface modification [16], introduction of defects [17], and quantum dot sensitization [18].

Recently, carbon dots (CDs) have been gaining enormous attention by virtue of their fascinating characteristics such as low cost, simple synthesis, functionalization, superior chemical inertness, and photobleaching resistance. Most essentially, CDs are a viable alternative for heavy metal-based QDs and organic dye, owing to its low toxicity with environmental friendliness [18–20]. Since the last decade, astonishing progress has been made in the preparation of CDs either in the top-down or bottom-up route [21,22]. However, new inexpensive, large-scale, and green synthetic approaches of CDs still need to be developed. For instance, a study on the CQDs/BiVO$_4$ and CQDs/NiFe-LDH/BiVO$_4$ demonstrated that after the decoration of CDs on their respective semiconductor, negatively shifted onsite potentials and enhanced charge injection rate were observed in PEC water splitting [22–24]. In addition, CDs, such as CQDs/TiO$_2$ [11] CQDs/ZnO [25], CQDs/WO$_3$ [26], CQDs/BiVO$_4$ [1], and CQDs/bFe$_2$O$_3$ [27], etc., can improve the light harvesting nature of photoanode in ultraviolet region and expand the range of visible region.

The CDs decorated TiO$_2$ films have been reported earlier from different origin materials by different methods and utilized as photoanode for PEC [23]. Zhou et al. utilized glucose as precursor and alkali-assisted ultrasonic chemical method to prepare CDs; and spin-coated TiO$_2$ film with CDs solution [15]. Wang et al. employed a hydrothermal method to synthesize CDs from phloroglucinol [24]. Usually, for photo-driven reactions, N-doped carbon dots exhibit improved activity both theoretically and experimentally than CDs, owing to beneficial quantum confinement and were capable of creating defect-rich heterostructures [25,26]. Based on the N-doping source material, light-harvesting ability and energy levels can be modulated [27], while the functionality of NCDs may interpret the interaction with the semi-conducting material [28]. Han et al. described the process of preparation of N-doped CDs (NCDs) anchored to TiO$_2$ photoanode in electrochemical and hydrothermal methods by using graphite rods and ammonia to obtain a nitrogen-doped CDs (NCDs) solution. This report has demonstrated the enhanced PEC efficiency due to an increased interface charge transfer [12,29]. The report by Wang et al. on NCDs@TiO$_2$ showed an enhanced photocatalytic property owing to its extended light responses with narrowed bandgap upon introduction of NCDs [30]. However, due to the complexity of NCDs with regards to energy states and chemical structure, the mechanism of NCDs in boosting PEC performance remains unknown [25,31]. Moreover, synthesis of CDs and preparation of photoanode was proceeded in multiple steps, which again increases the preparation cost of the electrode [32,33]. Therefore, it is of critical importance to prepare at low cost, as well as understand the nanostructure of NCDs, their interfacial interactions with semiconductor materials and further developments of NCDs.

In the present study, we report the synthesis of new NCDs decorated TiO$_2$ film in a simple one-pot hydrothermal method using L-histidine as source material. The effect of NCDs on TiO$_2$ nanorod film for PEC water splitting has been analyzed systematically by changing the NCDs' loading time from 2 h–8 h. The prepared photoanodes are named as TiO$_2$@NCD2h, TiO$_2$@NCD4h, TiO$_2$@NCD6h, and TiO$_2$@NCD8h. NCDs loaded photoanodes showed higher PEC performance than pristine TiO$_2$, suggesting the contribution of NCDs towards enhancing the performance of PEC. The highest efficiency was found for TiO$_2$@NCD4h (2.51 mA.cm$^{-2}$), 3.4 times greater than pristine TiO$_2$ (0.73 mA.cm$^{-2}$). The higher photocurrent for TiO$_2$@NCDs could be ascribed to the improved light harvesting property, decreased rate of recombination, and increased charge carrier density. The detailed characterization of NCDs and NCD loaded TiO$_2$ and PEC water splitting performance analysis were performed and discussed.

## 2. Results and Discussion

### 2.1. Characterization

FE-SEM and HR-TEM analyses were executed in order to assess the successful loading of NCDs on TiO$_2$ and their morphology and the obtained images are illustrated in Figures 1 and 2. The FE-SEM analysis of pristine TiO$_2$ film (Figure 1a) has shown dense nanorod morphology of TiO$_2$ which have perpendicularly grown on FTO glass showing an average length of ~2.8 μm and width of ~150 nm. Moreover, no obvious changes in the size and morphology of TiO$_2$ were observed in Figure 1b, even after dipping for 8 h in NCDs solution. Further, HR-TEM (Figure 1c) analysis confirmed the nanorod morphology of TiO$_2$. Moreover, the observed lattice fringes' distance in Figure 1d was 0.35 nm, which corresponds to the d-spacings of the rutile TiO$_2$ (101) planes, which has well-matched with XRD results [12]. Further, HRTEM image of TiO$_2$@NCD4h (Figure 2) showed that NCDs are uniformly loaded on the TiO$_2$ nanorods and appeared in a sphere and ellipsoidal morphology with particle size ranging from 4 to 10 nm. In addition, 0.21 nm lattice spacing was observed for NCDs particle, associated with the (100) facet of NCDs (Figure S3) [15,18,34]. Moreover, to further investigate the distribution of Ti, O, C, and N elements, the elemental mapping analysis was executed, and the respective results are displayed in Figure 2b. The obtained results have shown even distribution of C and N elements on TiO$_2$ nanorods' surface, suggesting the successful decoration of NCDs on the TiO$_2$.

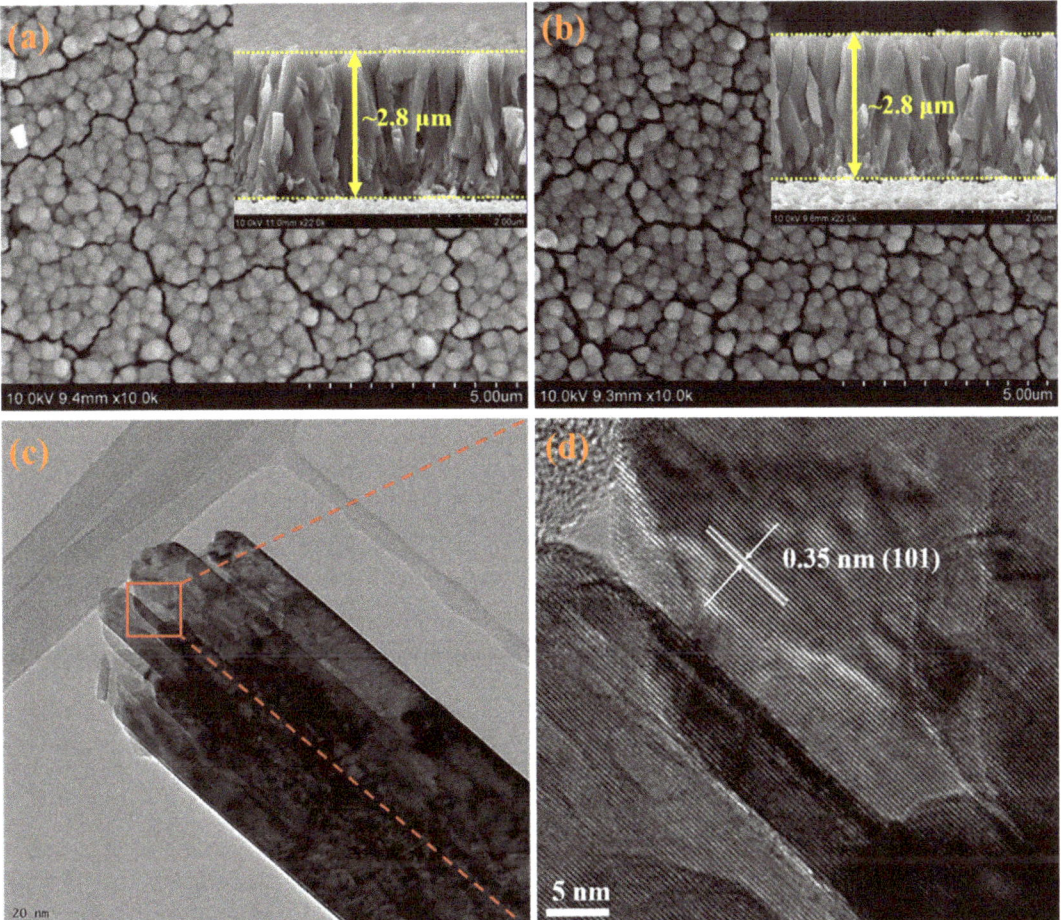

**Figure 1.** (**a**,**b**) Typical SEM of TiO$_2$, TiO$_2$@NCD4h. The corresponding cross-sectional SEM images are shown in the insets. (**c**) FETEM images of TiO$_2$. (**d**) FETEM images of TiO$_2$@NCD4h.

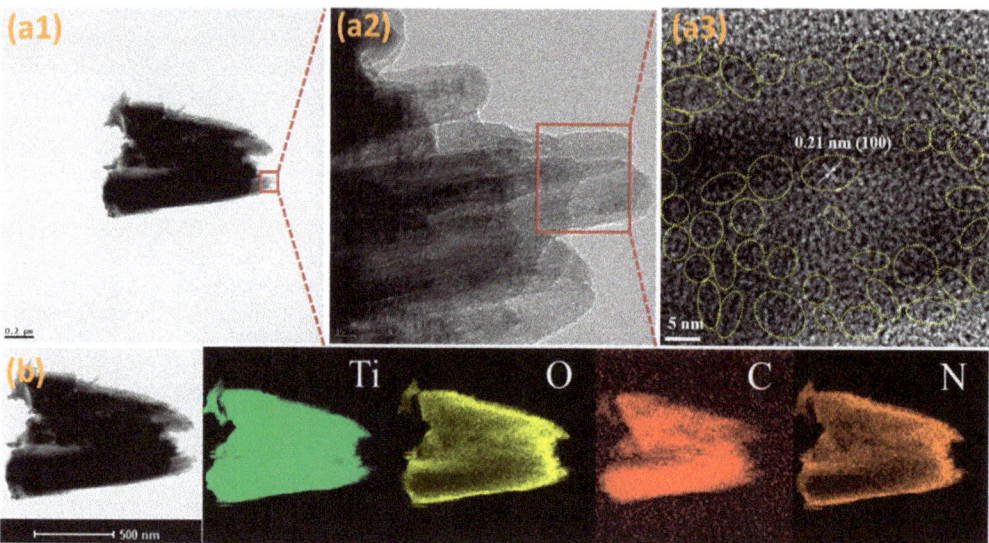

**Figure 2.** (**a1–a3**) FETEM of TiO$_2$@NCD4h. (**b**) HAADF-STEM of TiO$_2$@NCD4h and elemental mapping for Ti, O, C, and N.

The crystalline structure of the as-synthesized TiO$_2$ and the effect of NCDs loading time on TiO$_2$ crystallinity (TiO$_2$@NCD2h, TiO$_2$@NCD4h, TiO$_2$@NCD6h, and TiO$_2$@NCD8h) and the orientation growth were examined using XRD analysis. The obtained XRD peaks are displayed in Figure 3. The diffraction peaks of pristine TiO$_2$ films appeared at 36.10°, 41.27°, 54.39°, 62.86°, and 69.80° and correspond to the (101), (111), (211), (002), and (112) crystal planes of tetragonal rutile structure [15,18,35,36]. Moreover, regardless of the loading time of NCDs, the peak positions are the same, but the (101) plane intensity has increased with NCDs loading time. The results suggest that the TiO$_2$ nanorod crystal structure does not get affected by loading NCDs but size of the crystal and preferred orientation directions sparsely get affected. Furthermore, no noticeable diffraction peak of NCDs was observed for TiO$_2$-NCDs, which could be attributed to the modest load of NCDs, lower than the minimum limitation of XRD detection [15,35].

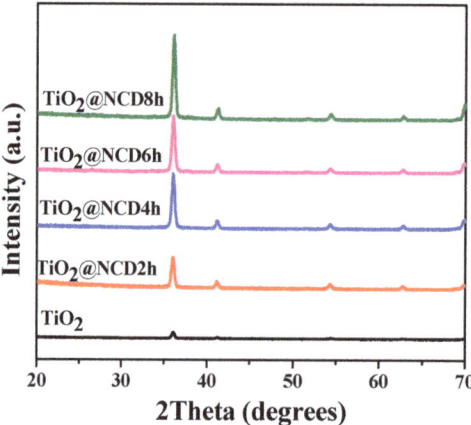

**Figure 3.** XRD patterns of TiO$_2$ and TiO$_2$@NCD2h-8h.

As seen in Figure 4, the Raman peaks of TiO$_2$ located at 615.2, 450.5, and 240 cm$^{-1}$ correspond to (A$_{1g}$), (E$_g$), and multi-photon scattering process, respectively, and represent the TiO$_2$ rutile phase. The peaks which appeared at 1580 and 1333 cm$^{-1}$ can be attributed to the D (disordered $sp^2$) band and G band of NCDs, respectively [37,38]. Thus, the Raman spectrum of TiO$_2$@NCD4h has shown five peaks, indicating successful fabrication of NCDs in TiO$_2$ nanorods. Furthermore, the enhancement of Raman intensity might be contributed by the increased crystallinity, and it is consistent with the XRD results.

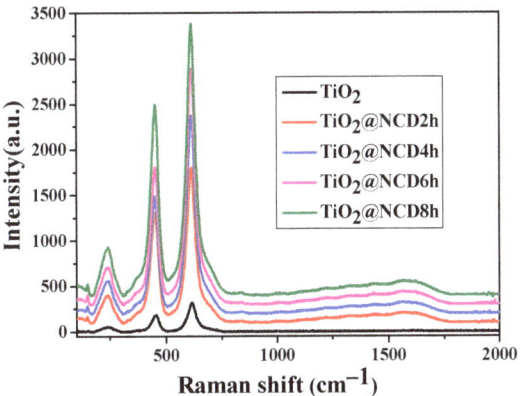

**Figure 4.** Raman spectra of TiO$_2$ and TiO$_2$@NCD4h.

The elemental composition and chemical binding of NCDs decorated on TiO$_2$ catalyst were determined by the XPS analysis (Figure 5). The survey scans illustrated the elements in two structures, e.g., C, N, Ti, and O elements in TiO$_2$@NCDs4h, whereas C, Ti, and O elements were in the pristine TiO$_2$ [12,39] (Figure 5a). An increase in the carbon content and the presence of N1s peak compared with the bare TiO$_2$ evidently confirm that the NCDs were successfully decorated on TiO$_2$. Ti2p spectra (Figure 5b) showed two representative peaks at 464.0 and 458.4 eV (difference: 5.6 eV), which correspond to the spin-orbit coupling for Ti2p1/2 and Ti2p3/2, respectively, and was identical to those for TiO$_2$ [15]. Furthermore, Ti2p peaks of the TiO$_2$@NCDs4h structure have considerably shifted (by 0.2 eV) compared to bare TiO$_2$ (Figure 5b). This is due to the electronegativity of C/Ns, which increased the binding ability of extra-nuclear electrons, hence raising the binding energy. The C 1s spectra (Figure 5c) is fitted with three peaks corresponding to C-C, C-O & C-N, and C=O & C=N bonds at 284.8, 286.0, and 288.3 eV, respectively [40]. The N1s spectra (Figure 5d) shows three peaks which appeared at 396.7 eV, 401.4 eV, and 403.7 eV, ascribed to the pyridine-N, pyrrole-N, and Graphitic-N, respectively, indicating the carbon dot doped with the N element [41,42]. In Figure S4, the peak in the O1 s region of TiO$_2$@NCDs@4h was deconvoluted into three peaks at 532.3, 530.1, and 529.6 eV, which are assigned to O-H, C-O, or O-N, and Ti-O bonds in TiO$_2$@NCDs4h, respectively. In agreement with the above microstructure analysis, XPS results suggest that NCDs were successfully deposited on the TiO$_2$ [40,42,43].

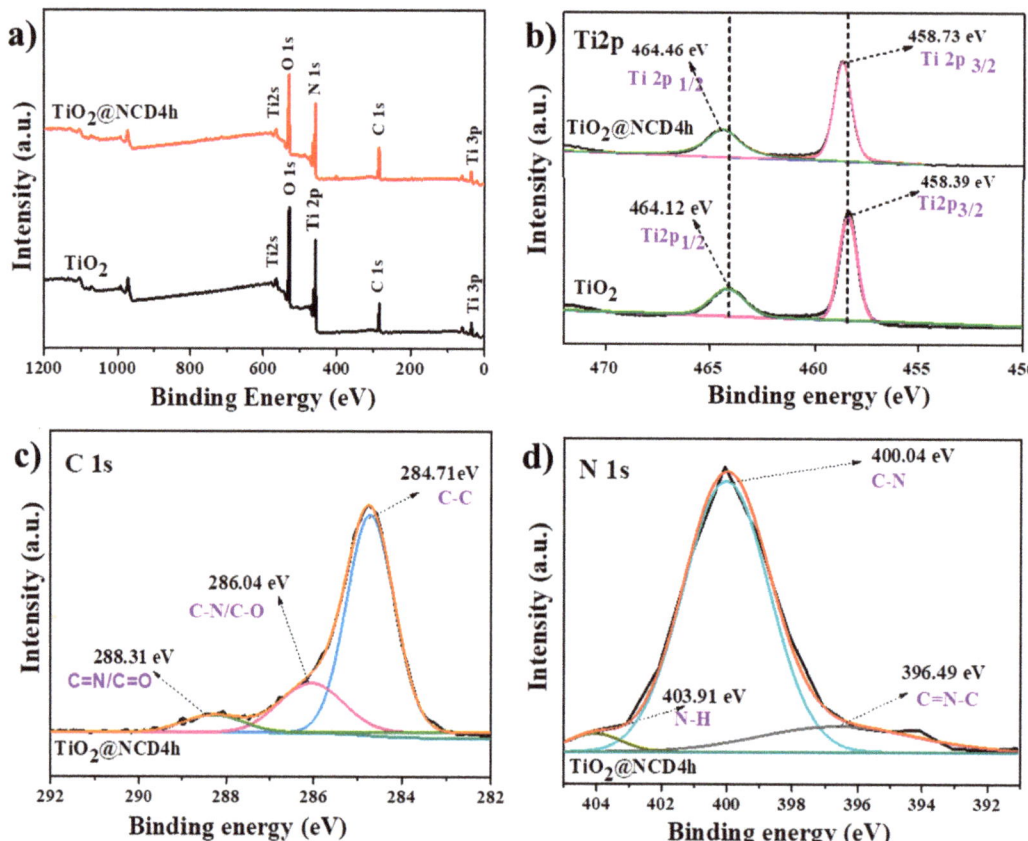

**Figure 5.** XPS spectra of pristine TiO$_2$ and TiO$_2$@NCD4h (**a**) survey scan (**b**) Ti 2p (**c**) C 1s (**d**) N 1s.

The absorption properties of the materials are important parameters to estimate the light harvesting nature and energy levels to be used in solar energy conversions. The optical properties of the prepared NCDs in solution and on TiO$_2$ photoanode with changing time were analyzed systematically. The UV-vis spectra of NCDs in solution and TiO$_2$@NCDs thin films are shown in Figure 6a,b and c by changing the loading time. The absorption band of pristine TiO$_2$ ~400 nm represents the rutile TiO$_2$ band edge [44]. After introduction of the NCDs on TiO$_2$, the light harvesting property was enhanced with the increased amount of NCDs loading. The results suggest the successful decoration of NCDs and their contribution in improving the light harvesting property. Moreover, the Tauc plot Equation (S1) was employed to calculate the bandgap (Eg) of pristine TiO$_2$ and TiO$_2$@NCD photoanodes [36,45,46]. The calculated Eg values of TiO$_2$, TiO$_2$@NCD2h, TiO$_2$@NCD4h, TiO$_2$@NCD6h, and TiO$_2$@NCD8h were 3.12, 3.07, 3.03, 3.04, and 3.05 eV, respectively.

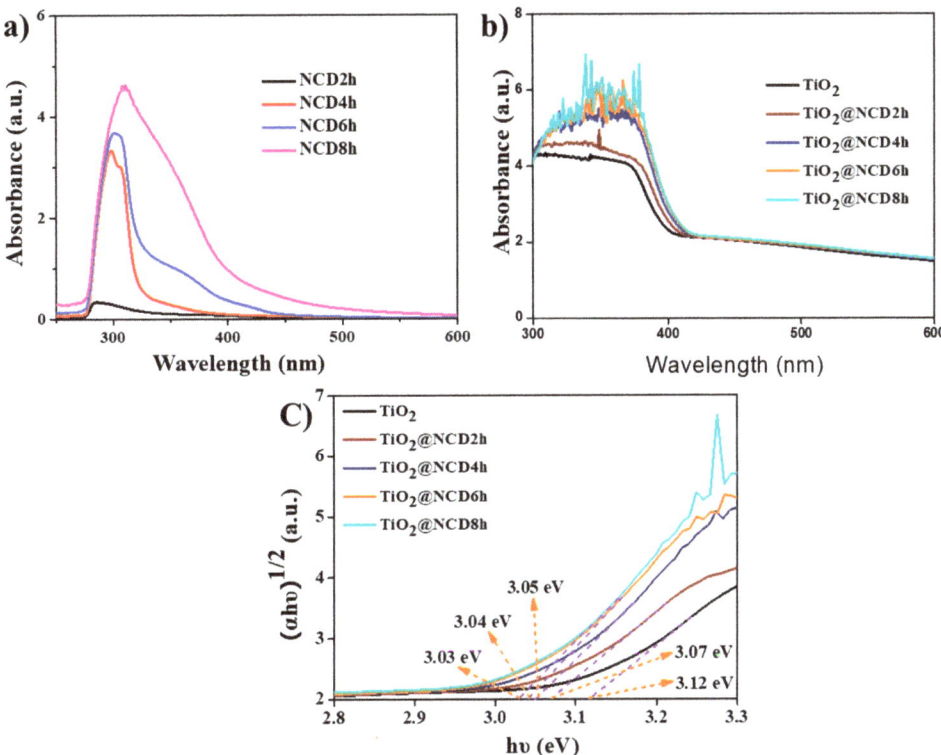

**Figure 6.** Absorption spectrum of (**a**) NCDs solutions. (**b**) Pristine TiO$_2$ and TiO$_2$@NCDs thin films (**c**) bandgap energy of Pristine TiO$_2$ and TiO$_2$@NCDs thin films.

## 2.2. PEC Performance of the Photoanodes

The effect of newly prepared NCDs on photoelectrochemical water oxidation was studied systematically by changing the loading time of NCDs on TiO$_2$ film and comparing it with the pristine TiO$_2$. Figure 7a shows linear sweep voltammetry performance of TiO$_2$, TiO$_2$@2hNCD, TiO$_2$@4hNCDs, TiO$_2$@6h NCDs, and TiO$_2$@8hNCDs, and their photocurrent data at 1.23 V are illustrated in Table S2. The pristine TiO$_2$ film displayed a 0.73 mA.cm$^{-2}$ photocurrent at 1.89 V vs. RHE. After loading NCDs upon the TiO$_2$ film, an enhanced photocurrent was observed compared the pristine TiO$_2$, which instigated to perform optimization studies to improve the NCDs loading and thereby achieve optimum PEC performance using TiO$_2$ with NCDs. In order to optimize the TiO$_2$@NCDs photoanodes, the decorated NCDs on TiO$_2$ was controlled by monitoring the hydrothermal reaction. In Figure 7a, the photocurrents of four NCDs decorated TiO$_2$ (TiO$_2$@NCDs)-based photoanodes showed higher photocurrent than pristine TiO$_2$, indicating the contribution of NCDs in enhancing PEC performance of TiO$_2$. Significantly, the photoanode corresponding to TiO$_2$@NCD2h has displayed an improved photocurrent density of 2.33 mA.cm$^{-2}$ at 1.89 V vs. RHE, while pristine TiO$_2$ photoanode has shown 0.73 mA.cm$^{-2}$ at 1.89 V vs. RHE. Further, by increasing the loading time from 2 h to 4 h (TiO$_2$@NCD4h), the photocurrent density has also increased to 2.51 mA.cm$^{-2}$ at 1.89 V vs. RHE, which was 3.4 times greater than the pristine TiO$_2$. Moreover, TiO$_2$@NCD4h photoanode possesses both enhanced photocurrent density and smaller onset potential than pristine TiO$_2$. The higher photoresponse of NCD decorated TiO$_2$ might be due to the addition of NCD which could effectively promote the separation of photogenerated electron-hole, and promote the capture of water molecules and intermediates in the process of water decomposition by electrons and holes at the interfaces [15,47,48]. However, further increasing the NCDs loading by increasing loading time to 6 h (TiO$_2$@NCD6h) and 8 h (TiO$_2$@NCD8h) showed declined photocurrent density of 1.99 and 1.85 mA.cm$^{-2}$ at 1.89 V vs. RHE, respectively. In addition, onsite potentials also increased compared to the TiO$_2$@NCD4h-based films, possibly due to the variation in conductivity by decorated NCDs. This phenomenon will be further discussed in electrochemical impedance spectroscopy (EIS) section [29].

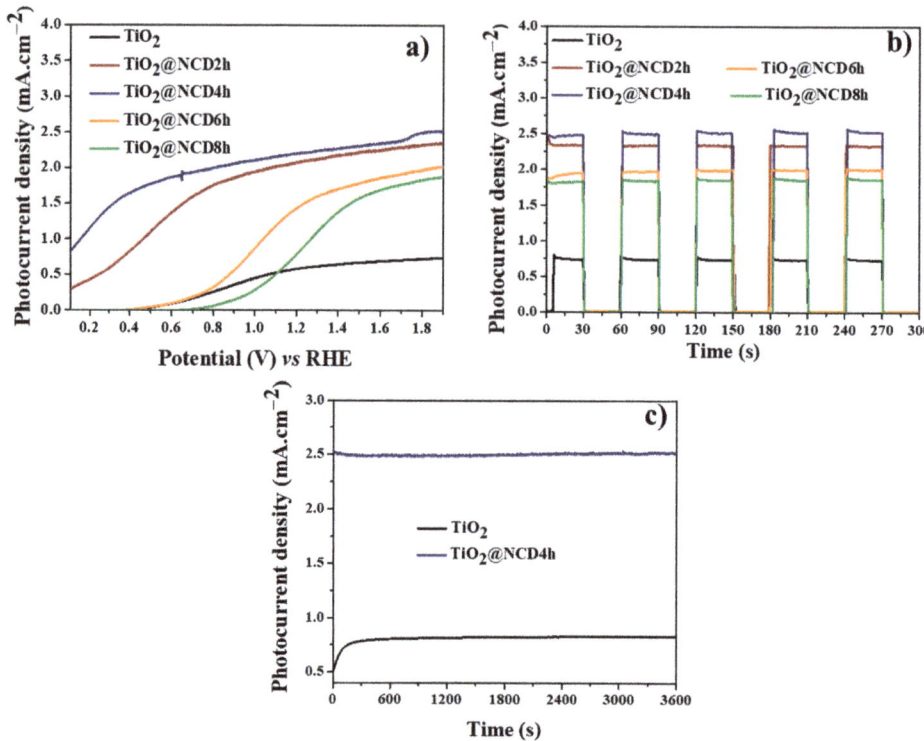

**Figure 7.** (a) Photocurrent density vs. applied potential curves; (b) Transient photocurrent density curves at 1.89 V vs. RHE of the as-prepared photoelectrodes; (c) Stability test of pristine TiO$_2$ and TiO$_2$@NCD4h at 1.89 V vs. RHE.

Chronoamperometric analysis was performed under chopped illumination for all the prepared electrodes at 1.89 V vs. RHE for 30 s to better understand photo response with time and stability. As depicted in Figure 7b, the photocurrent has rapidly increased immediately after illumination and sharply fell to zero upon stopping the illumination. It confirms that the prepared photoanodes have a well-reproducible photocurrent. Meanwhile, after decorating NCDs on TiO$_2$, the response speed was boosted compared with the pristine TiO$_2$, indicating that the presence of NCDs can significantly reduce the charge recombination in the intersection of electrolyte and photoanode surface. The highest photocurrent density was detected for the TiO$_2$@NCD4h, which is consistent with the observed LSV results. In order to understand the durability of the prepared electrode, 1 h of continuous illuminated chronoamperometric analysis was performed with a high performing photoanode (TiO$_2$@NCD4h) by comparing with the pristine TiO$_2$ based photo anode at same experimental condition. As displayed in Figure 7c, after continuous illumination for 1 h, TiO$_2$@NCD4h has retained 99% of its initial activity, which supports the excellent stability of NCDs decorated TiO$_2$ photo anode. The current densities have matched well with the

LSV data. Both TiO$_2$ and TiO$_2$@NCD4h showed excellent stability after 1 h of continuous irradiation without any photocurrent decay.

The incident photon-to-current conversion efficiency (IPCE) was evaluated to analyze the contribution of various photons in obtaining solar photocurrent. The IPCE has been deduced by using the formula (1):

$$\text{IPCE (\%)} = \frac{1240 J(\lambda)}{\lambda P(\lambda)} \times 100 \ (\%) \tag{1}$$

where, $P(\lambda)$, $\lambda$, and $J(\lambda)$ are the intensity of a specific wavelength, wavelength of incident light, and photocurrent density at specific wavelength, respectively. Figure 8a shows enhanced IPCE after decorating NCDs on TiO$_2$ with highest IPCE of 29.76% at ~390 nm for TiO$_2$@NCD4h, which was ~3 times greater than the pristine TiO$_2$ (9.76%). The IPCE trend was consistent with the obtained photocurrent density and the enhanced IPCE region is in good agreement with the optical absorption properties. The improved IPCE after the introduction of NCDs to the TiO$_2$ reveals the contribution of NCDs in obtaining enhanced photocurrent density.

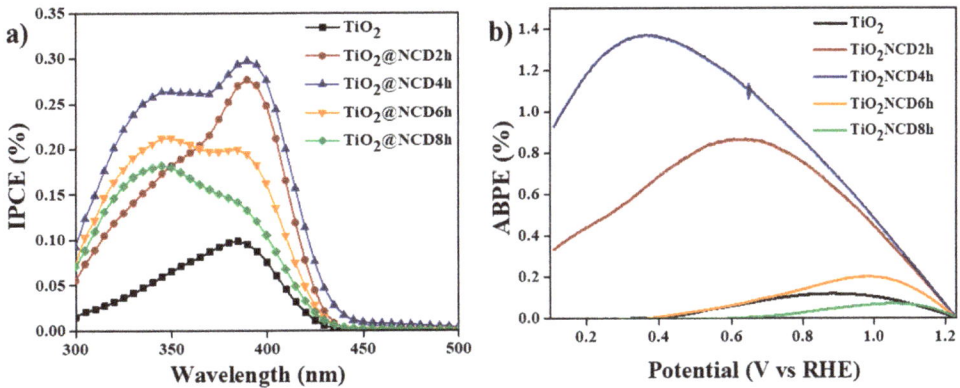

**Figure 8.** (a). Incident photon-to-current conversion efficiency (IPCE) curves; (b) Applied bias ABPE of TiO$_2$, TiO$_2$@NCD2h, TiO$_2$@NCD4h, TiO$_2$@NCD6h, and TiO$_2$@NCD8h measured at 1.23V vs. RHE.

Besides, the applied bias photon-to-current efficiency (ABPE) has been calculated by using the Equation (2):

$$\text{ABPE (\%)} = \frac{J(1.23 - V)}{P} \times 100 \ (\%) \tag{2}$$

where $P$, $J$, and $V$ are the power density of incident light (100 mW cm$^{-2}$), photocurrent density (mA cm$^{-2}$), and the applied bias (V vs. RHE), respectively. As seen in Figure 8b, the pristine TiO$_2$ reached maximum 0.11% ABPE at 0.88 V vs. RHE, while TiO$_2$@NCD4h reached 1.37% photo conversion efficiency at 0.36 V vs. RHE, 12 times greater than the pristine TiO$_2$ ABPE, suggesting an effective electron-hole pairs separation after the introduction of NCDs [12].

To further comprehend the interfacial charge transfer kinetics at the intersection of photoanode and electrolyte, EIS was employed under illumination and the respective Nyquist plots are displayed in Figure 9a. The decreased radius order of semi-circle was TiO$_2$ > TiO$_2$@NCD8h > TiO$_2$@NCD6h > TiO$_2$@NCD2h > TiO$_2$@NCD4h. The smallest arc of the TiO$_2$@NCD4h compared with its counter parts demonstrates the improved interfacial charge transfer kinetics due to the introduction of NCDs [18]. Furthermore, using Zview software program, the EIS curves have been fitted with an analogous circuit model given in Figure 9a inset, where CPE, Rct, and Rs indicate the constant phase element, charge transfer resistance, and series resistance at the electrolyte/electrode interface, respectively. The observed Rct values of TiO$_2$ and NCDs decorated TiO$_2$ films (2 h–8 h) were 376.0 Ω, 273.6 Ω, 248.3 Ω, 277.2 Ω, and 297.4 Ω, respectively. The lowest $R_{ct}$ value of TiO$_2$@NCD4h further demonstrates the advantage of NCDs decorated TiO$_2$ nanorods in enhancing the charge separation and transfer kinetics.

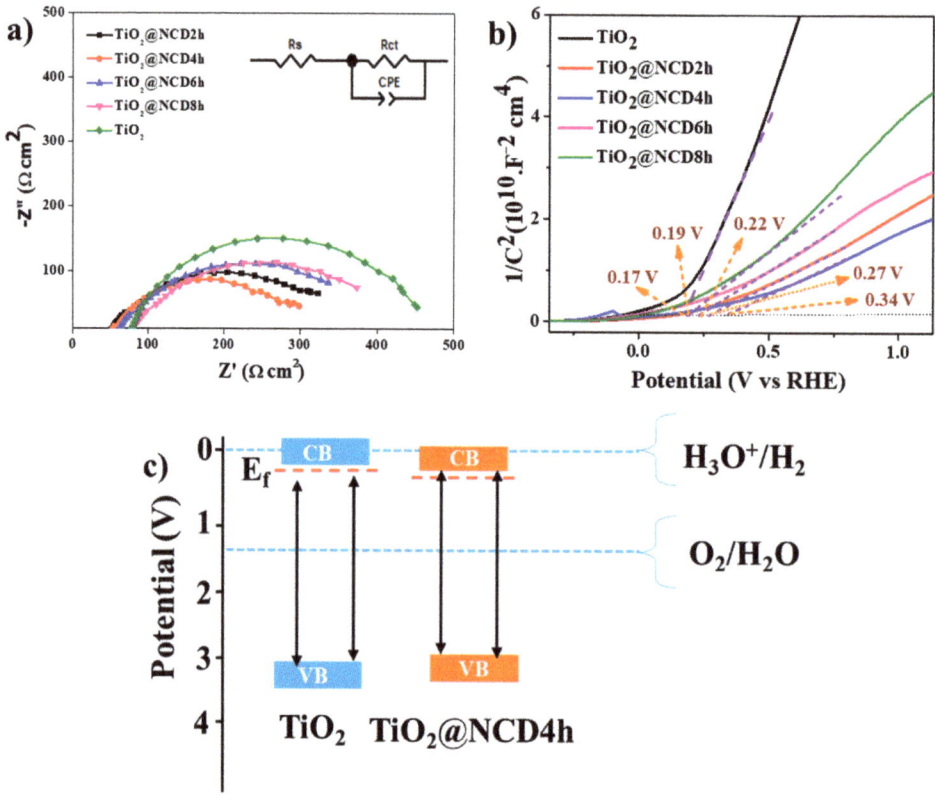

**Figure 9.** (**a**) EIS spectra; (**b**) 0 Mott–Schottky plots of TiO$_2$, TiO$_2$@NCD2h, TiO$_2$@NCD4h, TiO$_2$@NCD6h, and TiO$_2$@NCD8h. (**c**) Schematic energy levels of TiO$_2$, TiO$_2$@NCD4h.

Mott–Schottky analyses have been executed to estimate the energy band position of pristine TiO$_2$ and TiO$_2$@NCDs, and the corresponding curves are displayed in Figure 9b and the data are depicted in Table S1. The positive slope of both curves indicates n-type semiconductor of TiO$_2$ [18]. The flat band (V$_{FB}$) potential can be calculated following the Equation (S1). The obtained V$_{FB}$ values of TiO$_2$ and TiO$_2$@NCDs (2 h, 4 h, 6 h, and 8 h) were 0.17, 0.27, 0.34, 0.22, and 0.19 V vs. NHE, respectively, which could be accomplished by the X-axis intercept. Moreover, NCDs decorated TiO$_2$ films have shown decreased V$_{FB}$ than pristine TiO$_2$, suggesting an increased band bending of the photoanode, favorable to enhance the charge transfer between the photoanode interfaces and electrolyte. Eventually, enhanced PEC was observed for NCDs decorated TiO$_2$ films. As per the available literature [49,50], the bottom of the conduction band (CB) was −0.1 V lower than the V$_{FB}$ of an n-type semiconductor [50]. Therefore, the CB of TiO$_2$ and TiO$_2$@NCDs were estimated to be positioned at less than 0.1 V of their V$_{FB}$ [33]. Based on the V$_{FB}$, the CB edge of TiO$_2$ and TiO$_2$@NCDs (2 h, 4 h, 6 h, and 8 h) were determined to be at 0.07, 0.17, 0.24, 0.12, and 0.09 V, respectively. Particularly, doping of NCDs promotes a downward shift in energy levels towards higher potentials and enhances the carrier density in TiO$_2$ [51].

## 3. Experimental

### 3.1. Materials

All chemicals were used directly without purifying any further. Hydrochloric acid (35%) was obtained from OCI Company Ltd., titanium butoxide (TBOT, 98%) was purchased from Sigma-Aldrich, L-histidine (98%) was procured from Alfa Aesar, and sodium sulfate anhydrous (99%) was obtained from Duksan. For all the experiments, Milli-Q water (MΩ 18) was used.

### 3.2. Preparation of Rutile $TiO_2$ Film ($TiO_2$):

The FTO coated glasses (1.5 mm × 2.5 mm, 8 Ω/cm$^2$) were cleaned ultrasonically using detergent, milli-Q water, ethanol, and acetone for 1 h, respectively. $TiO_2$ film was synthesized by following a reported hydrothermal method with certain modifications [18]. Under continuous stirring, 0.33 mL titanium (IV) butoxide was added dropwise to 20 mL equal volumes of HCl (35%) and milli-Q water mixed solution until it turned translucent. The solution was then moved to a 50 mL autoclave lined with Teflon, and the FTO glass was placed against the walls of Teflon vessel, conducted side down, for 12 h and heated to 150 °C. The $TiO_2$ layer was completely cleaned with milli-Q water and ethanol after cooling to RT, before being sintered in air at 450 °C for 1 h.

### 3.3. Preparation of $TiO_2$@NCDs:

NCDs have been prepared using a hydrothermal approach (Scheme 1). First, 0.2 g of L-histidine was included in 20 mL mixture of milli-Q water and HCl in the ratio of 19:1. The solution was shifted to 50 mL autoclave, and two $TiO_2$ films on FTO glasses were inserted in the Teflon vessel with the $TiO_2$ side facing down. Then, the hydrothermal reaction was performed at 180 °C for 2, 4, 6, 8, and 10 h. The samples were denoted as $TiO_2$@NCD2h, $TiO_2$@NCD4h, $TiO_2$@NCD6h, $TiO_2$@NCD8h, respectively (Figures S1 and S2). The $TiO_2$@NCD films were extensively washed with milli-Q $H_2O$ and ethanol upon cooling to RT. Then, copper wires and as prepared photoanodes were adhered using silver paint. The samples were air dried for 3 h. Finally, the samples were encased by nonconductive epoxy with the illuminated area of 1 cm$^2$ and left to rest in air for at least 3 h.

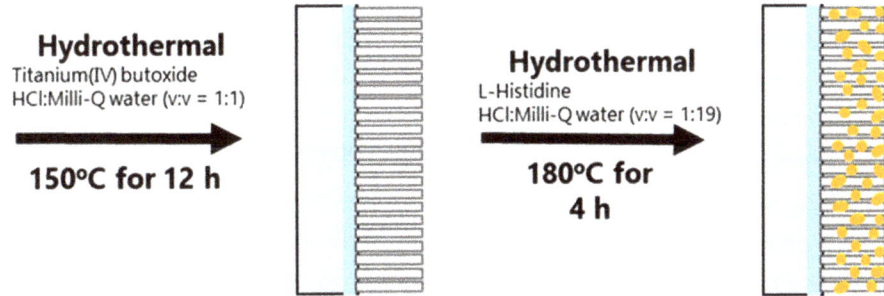

**Scheme 1.** Schematic illustration of $TiO_2$@NCDs.

## 4. Conclusions

In conclusion, the new NCDs were successfully prepared in a simple one-pot hydrothermal synthesis method using L-histidine as an initial precursor and the as-prepared NCDs were decorated on the $TiO_2$ nanorod-based photoanode. The as-prepared NCDs and NCD decorated $TiO_2$ nanorods were well characterized using FE-SEM, HR-TEM, EDS elemental mapping, which revealed the nanorod morphology of $TiO_2$ and uniform distribution of CDs on $TiO_2$ surface while, XPS and Raman analyses have confirmed the successful self nitrogen element doping, preparation and decorating of NCDs on $TiO_2$ nanorod. The effect of NCDs decorated $TiO_2$ was tested for photoelectrochemical water splitting analysis systematically by changing the loading time of NCDs from 2 h to 8 h. The highest efficiency was observed for the TiO$_2$@NCD4h-based photoanode (2.51 mA.cm$^{-2}$), which was a 3.4 times higher photocurrent density than the pristine TiO$_2$-based photoanode. It might be attributed to the increased light harvesting property with charge separation and transportation. The observed IPCE of TiO$_2$@NCD4h has shown 3 times higher quantum yield (29.76%) than pristine TiO$_2$ (9.76). In addition, the calculated ABPE was 12% higher for TiO$_2$@NCD4h than the pristine TiO$_2$, which revealed the enhanced light harvesting property of photoanodes upon loading the NCDs. Moreover, the reduced charge transfer resistance and higher charge carrier density, as observed from EIS and Mott–Schottky analyses, respectively, further support the advantage of newly prepared NCDs in enhancing the PEC performance by promoting effective charge separation and transportation. This study may open up new insights into the rational design and synthesis of highly efficient photoanodes for PEC water splitting.

**Supplementary Materials:** The following supporting information can be downloaded at: https://www.mdpi.com/article/10.3390/catal12101281/s1; Figure S1: Imagies of NCDs' solution under natural and UV light. Figure S2. NCDs solution based on the reaction time. Figure S3. TEM of NCD. Figure S4. XPS O 1s spectra of TiO$_2$@NCDs4h. Table S1. EIS Data of TiO$_2$ and NCD decorated TiO$_2$ photoanodes Table S2. The photocurrent densities of the TiO$_2$ and TiO$_2$@NCDs photoanodes.

**Author Contributions:** Conceptualization, G.K., C.T.T.T.; Experiments, C.T.T.T., G.S., L.J.; data curation, C.T.T.T., G.K.; formal analysis, H.D.K.; funding acquisition, J.H.K.; supervision, J.H.K. and G.K. writing—review and editing, C.T.T.T., G.K. and J.H.K. All authors have read and agreed to the published version of the manuscript.

**Funding:** This work was supported by the Korea Institute of Energy Technology Evaluation and Planning (KETEP) and the Ministry of Trade, Industry & Energy (MOTIE) of the Republic of Korea (20214000000720). This work was supported by "Human Resources Program in Energy Technology" of the Korea Institute of Energy Technology Evaluation and Planning (KETEP), granted financial resource from the Ministry of Trade, Industry & Energy, Republic of Korea. (No. 20204010600100).

**Data Availability Statement:** The data presented in this study are available upon request from the corresponding author.

**Conflicts of Interest:** The authors declare no conflict of interest.

## References

1. Li, Q.; Guo, B.; Yu, J.; Ran, J.; Zhang, B.; Yan, H.; Gong, J.R. Highly efficient visible-light-driven photocatalytic hydrogen production of CdS-cluster-decorated graphene nanosheets. *J. Am. Chem. Soc.* **2011**, *133*, 10878–10884. [CrossRef] [PubMed]
2. Li, H.; Yu, H.; Sun, L.; Zhai, J.; Han, X. A self-assembled 3D Pt/TiO$_2$ architecture for high-performance photocatalytic hydrogen production. *Nanoscale* **2015**, *7*, 1610–1615. [CrossRef] [PubMed]
3. Alshorifi, F.T.; Ali, S.L.; Salama, R.S. Promotional Synergistic Effect of Cs–Au NPs on the Performance of Cs–Au/MgFe$_2$O$_4$ Catalysts in Catalysis 3, 4-Dihydropyrimidin-2 (1H)-Ones and Degradation of RhB Dye. *J. Inorg. Organomet. Polym. Mater.* **2022**, *32*, 3765–3776. [CrossRef]
4. Fujishima, A.; Honda, K. Electrochemical photolysis of water at a semiconductor electrode. *Nature* **1972**, *238*, 37–38. [CrossRef] [PubMed]
5. Zhang, W.-D.; Jiang, L.-C.; Ye, J.-S. Photoelectrochemical study on charge transfer properties of ZnO nanowires promoted by carbon nanotubes. *J. Phys. Chem. C* **2009**, *113*, 16247–16253. [CrossRef]
6. Kim, T.W.; Ping, Y.; Galli, G.A.; Choi, K.-S. Simultaneous enhancements in photon absorption and charge transport of bismuth vanadate photoanodes for solar water splitting. *Nat. Commun.* **2015**, *6*, 8769. [CrossRef]
7. Hou, Y.; Zuo, F.; Dagg, A.P.; Liu, J.; Feng, P. Branched WO$_3$ nanosheet array with layered C$_3$N$_4$ heterojunctions and CoOx nanoparticles as a flexible photoanode for efficient photoelectrochemical water oxidation. *Adv. Mater.* **2014**, *26*, 5043–5049. [CrossRef]
8. Liu, S.; Zheng, L.; Yu, P.; Han, S.; Fang, X. Novel composites of α-Fe$_2$O$_3$ tetrakaidecahedron and graphene oxide as an effective photoelectrode with enhanced photocurrent performances. *Adv. Funct. Mater.* **2016**, *26*, 3331–3339. [CrossRef]
9. Liang, Z.; Hou, H.; Song, K.; Zhang, K.; Fang, Z.; Gao, F.; Wang, L.; Chen, D.; Yang, W.; Zeng, H. Boosting the photoelectrochemical activities of all-inorganic perovskite SrTiO$_3$ nanofibers by engineering homo/hetero junctions. *J. Mater. Chem. A* **2018**, *6*, 17530–17539. [CrossRef]
10. Ye, L.J.; Wang, D.; Chen, S.J. Fabrication and Enhanced Photoelectrochemical Performance of MoS$_2$/S-Doped g-C$_3$N$_4$ Heterojunction Film. *ACS Appl. Mater. Interfaces* **2016**, *8*, 5280–5289. [CrossRef]

11. Liu, G.; Fu, P.; Zhou, L.; Yan, P.; Ding, C.; Shi, J.; Li, C. Efficient Hole Extraction from a Hole-Storage-Layer-Stabilized Tantalum Nitride Photoanode for Solar Water Splitting. *Chemistry* **2015**, *21*, 9624–9628. [CrossRef] [PubMed]
12. Liang, Z.; Hou, H.; Fang, Z.; Gao, F.; Wang, L.; Chen, D.; Yang, W. Hydrogenated $TiO_2$ Nanorod Arrays Decorated with Carbon Quantum Dots toward Efficient Photoelectrochemical Water Splitting. *ACS Appl. Mater. Interfaces* **2019**, *11*, 19167–19175. [CrossRef] [PubMed]
13. Zhou, T.; Chen, S.; Wang, J.; Zhang, Y.; Li, J.; Bai, J.; Zhou, B. Dramatically enhanced solar-driven water splitting of $BiVO_4$ photoanode via strengthening hole transfer and light harvesting by co-modification of CQDs and ultrathin β-FeOOH layers. *Chem. Eng. J.* **2021**, *403*, 126350. [CrossRef]
14. Wang, D.-H.; Jia, L.; Wu, X.-L.; Lu, L.-Q.; Xu, A.-W. One-step hydrothermal synthesis of N-doped $TiO_2$/C nanocomposites with high visible light photocatalytic activity. *Nanoscale* **2012**, *4*, 576–584. [CrossRef]
15. Zhou, T.S.; Chen, S.; Li, L.S.; Wang, J.C.; Zhang, Y.; Li, J.H.; Bai, J.; Xia, L.G.; Xu, Q.J.; Rahim, M.; et al. Carbon quantum dots modified anatase/rutile $TiO_2$ photoanode with dramatically enhanced photoelectrochemical performance. *Appl. Catal. B-Environ.* **2020**, *269*, 118776. [CrossRef]
16. Wen, P.; Su, F.J.; Li, H.; Sun, Y.H.; Liang, Z.Q.; Liang, W.K.; Zhang, J.C.; Qin, W.; Geyer, S.M.; Qiu, Y.J.; et al. A $Ni_2P$ nanocrystal cocatalyst enhanced $TiO_2$ photoanode towards highly efficient photoelectrochemical water splitting. *Chem. Eng. J.* **2020**, *385*, 123878. [CrossRef]
17. Cheng, X.; Dong, G.; Zhang, Y.; Feng, C.; Bi, Y. Dual-bonding interactions between $MnO_2$ cocatalyst and $TiO_2$ photoanodes for efficient solar water splitting. *Appl. Catal. B: Environ.* **2020**, *267*, 118723. [CrossRef]
18. Zhou, T.; Li, L.; Li, J.; Wang, J.; Bai, J.; Xia, L.; Xu, Q.; Zhou, B. Electrochemically reduced $TiO_2$ photoanode coupled with 426 oxygen vacancy-rich carbon quantum dots for synergistically improving photoelectrochemical performance. *Chem. Eng. J.* **2021**, *425*, 131770. [CrossRef]
19. Wang, H.-J.; Yu, T.-T.; Chen, H.-L.; Nan, W.-B.; Xie, L.-Q.; Zhang, Q.-Q. A self-quenching-resistant carbon dots powder with tunable solid-state fluorescence and their applications in light-emitting diodes and fingerprints detection. *Dye. Pigment.* **2018**, *159*, 245–251. [CrossRef]
20. Tangy, A.; Kumar, V.B.; Pulidindi, I.N.; Kinel-Tahan, Y.; Yehoshua, Y.; Gedanken, A. In-situ transesterification of Chlorella vulgaris using carbon-dot functionalized strontium oxide as a heterogeneous catalyst under microwave irradiation. *Energy Fuels* **2016**, *30*, 10602–10610. [CrossRef]
21. Feng, Z.; Adolfsson, K.H.; Xu, Y.; Fang, H.; Hakkarainen, M.; Wu, M. Carbon dot/polymer nanocomposites: From green synthesis to energy, environmental and biomedical applications. *Sustain. Mater. Technol.* **2021**, *29*, e00304. [CrossRef]
22. Beutier, C.; Serghei, A.; Cassagnau, P.; Heuillet, P.; Cantaloube, B.; Selles, N.; Morfin, I.; Sudre, G.; David, L.J.P. In situ coupled mechanical/electrical/WAXS/SAXS investigations on ethylene propylene diene monomer resin/carbon black nanocomposites. *Polymer* **2022**, *254*, 125077. [CrossRef]
23. Sendão, R.M.S.; Esteves da Silva, J.C.G.; Pinto da Silva, L. Photocatalytic removal of pharmaceutical water pollutants by $TiO_2$—Carbon dots nanocomposites: A review. *Chemosphere* **2022**, *301*, 134731. [CrossRef]
24. Wang, X.; Wang, M.; Liu, G.; Zhang, Y.; Han, G.; Vomiero, A.; Zhao, H. Colloidal carbon quantum dots as light absorber for 444 efficient and stable ecofriendly photoelectrochemical hydrogen generation. *Nano Energy* **2021**, *86*, 106122. [CrossRef]
25. Luo, H.; Dimitrov, S.; Daboczi, M.; Kim, J.-S.; Guo, Q.; Fang, Y.; Stoeckel, M.-A.; Samorì, P.; Fenwick, O.; Jorge Sobrido, A.B.; et al. Nitrogen-Doped Carbon Dots/$TiO_2$ Nanoparticle Composites for Photoelectrochemical Water Oxidation. *ACS Appl. Nano Mater.* **2020**, *3*, 3371–3381. [CrossRef]
26. Hola, K.; Sudolská, M.; Kalytchuk, S.; Nachtigallová, D.; Rogach, A.L.; Otyepka, M.; Zboril, R. Graphitic nitrogen triggers red fluorescence in carbon dots. *ACS Nano* **2017**, *11*, 12402–12410. [CrossRef] [PubMed]
27. Hu, R.; Li, L.; Jin, W.J. Controlling speciation of nitrogen in nitrogen-doped carbon dots by ferric ion catalysis for enhancing fluorescence. *Carbon* **2017**, *111*, 133–141. [CrossRef]
28. Xie, S.; Su, H.; Wei, W.; Li, M.; Tong, Y.; Mao, Z. Remarkable photoelectrochemical performance of carbon dots sensitized $TiO_2$ under visible light irradiation. *J. Mater. Chem. A* **2014**, *2*, 16365–16368. [CrossRef]
29. Han, Y.; Wu, J.; Li, Y.; Gu, X.; He, T.; Zhao, Y.; Huang, H.; Liu, Y.; Kang, Z. Carbon dots enhance the interface electron transfer and photoelectrochemical kinetics in $TiO_2$ photoanode. *Appl. Catal. B: Environ.* **2022**, *304*, 120983. [CrossRef]
30. Wang, Q.; Cai, J.; Biesold-McGee, G.V.; Huang, J.; Ng, Y.H.; Sun, H.; Wang, J.; Lai, Y.; Lin, Z. Silk fibroin-derived nitrogen-doped carbon quantum dots anchored on $TiO_2$ nanotube arrays for heterogeneous photocatalytic degradation and water splitting. *Nano Energy* **2020**, *78*, 105313. [CrossRef]
31. Tian, J.; Leng, Y.; Zhao, Z.; Xia, Y.; Sang, Y.; Hao, P.; Zhan, J.; Li, M.; Liu, H. Carbon quantum dots/hydrogenated $TiO_2$ nanobelt heterostructures and their broad spectrum photocatalytic properties under UV, visible, and near-infrared irradiation. *Nano Energy* **2015**, *11*, 419–427. [CrossRef]
32. Ning, X.; Huang, J.; Li, L.; Gu, Y.; Jia, S.; Qiu, R.; Li, S.; Kim, B.H. Homostructured rutile $TiO_2$ nanotree arrays thin film electrodes with nitrogen doping for enhanced photoelectrochemical performance. *J. Mater. Sci. Mater. Electron.* **2019**, *30*, 16030–16040. [CrossRef]
33. Altass, H.M.; Morad, M.; Khder, A.E.-R.S.; Mannaa, M.A.; Jassas, R.S.; Alsimaree, A.A.; Ahmed, S.A.; Salama, R.S. Enhanced catalytic activity for CO oxidation by highly active Pd nanoparticles supported on reduced graphene oxide/copper metal organic framework. *J. Taiwan Inst. Chem. Eng.* **2021**, *128*, 194–208. [CrossRef]

34. Liu, Y.; Wang, J.; Wu, J.; Zhao, Y.; Huang, H.; Liu, Y.; Kang, Z. Critical roles of $H_2O$ and $O_2$ in $H_2O_2$ photoproduction over biomass derived metal-free catalyst. *Appl. Catal. B: Environ.* **2022**, *319*, 121944. [CrossRef]
35. Masuda, Y.; Kato, K. Synthesis and phase transformation of $TiO_2$ nano-crystals in aqueous solutions. *J. Ceram. Soc. Jpn.* **2009**, *117*, 373–376. [CrossRef]
36. Hu, A.; Zhang, X.; Luong, D.; Oakes, K.D.; Servos, M.R.; Liang, R.; Kurdi, S.; Peng, P.; Zhou, Y. Adsorption and Photocatalytic Degradation Kinetics of Pharmaceuticals by $TiO_2$ Nanowires During Water Treatment. *Waste Biomass Valorization* **2012**, *3*, 443–449. [CrossRef]
37. Hanaor, D.A.; Sorrell, C.C. Review of the anatase to rutile phase transformation. *J. Mater. Sci.* **2011**, *46*, 855–874. [CrossRef]
38. Ma, H.L.; Yang, J.Y.; Dai, Y.; Zhang, Y.B.; Lu, B.; Ma, G.H. Raman study of phase transformation of $TiO_2$ rutile single crystal irradiated by infrared femtosecond laser. *Appl. Surf. Sci.* **2007**, *253*, 7497–7500. [CrossRef]
39. Jiang, D.; Xu, Y.; Hou, B.; Wu, D.; Sun, Y. Synthesis of visible light-activated $TiO_2$ photocatalyst via surface organic modification. *J. Solid State Chem.* **2007**, *180*, 1787–1791. [CrossRef]
40. Wei, N.; Liu, Y.; Feng, M.; Li, Z.; Chen, S.; Zheng, Y.; Wang, D. Controllable $TiO_2$ core-shell phase heterojunction for efficient photoelectrochemical water splitting under solar light. *Appl. Catal. B: Environ.* **2019**, *244*, 519–528. [CrossRef]
41. Yang, J.; Bai, H.; Tan, X.; Lian, J. IR and XPS investigation of visible-light photocatalysis—Nitrogen–carbon-doped $TiO_2$ film. *Appl. Surf. Sci.* **2006**, *253*, 1988–1994. [CrossRef]
42. Wu, D.; Zhang, W.; Feng, Y.; Ma, J. Necklace-like carbon nanofibers encapsulating $V_3S_4$ microspheres for ultrafast and stable potassium-ion storage. *J. Mater. Chem. A* **2020**, *8*, 2618–2626. [CrossRef]
43. Song, J.; Zheng, M.; Yuan, X.; Li, Q.; Wang, F.; Ma, L.; You, Y.; Liu, S.; Liu, P.; Jiang, D.; et al. Electrochemically induced $Ti^{3+}$ self-doping of $TiO_2$ nanotube arrays for improved photoelectrochemical water splitting. *J. Mater. Sci.* **2017**, *52*, 6976–6986. [CrossRef]
44. Zhuang, H.; Zhang, S.; Lin, M.; Lin, L.; Cai, Z.; Xu, W. Controlling interface properties for enhanced photocatalytic performance: A case-study of $CuO/TiO_2$ nanobelts. *Mater. Adv.* **2020**, *1*, 767–773. [CrossRef]
45. Koyyada, G.; Goud, B.S.; Devarayapalli, K.C.; Shim, J.; Vattikuti, S.P.; Kim, J.H. $BiFeO_3/Fe_2O_3$ electrode for photoelectrochemical water oxidation and photocatalytic dye degradation: A single step synthetic approach. *Chemosphere* **2022**, *303*, 135071. [CrossRef]
46. Mohamad, M.; Ul Haq, B.; Ahmed, R.; Shaari, A.; Ali, N.; Hussain, R. A density functional study of structural, electronic and optical properties of titanium dioxide: Characterization of rutile, anatase and brookite polymorphs. *Mater. Sci. Semicond. Process.* **2015**, *31*, 405–414. [CrossRef]
47. Wang, T.; Long, X.; Wei, S.; Wang, P.; Wang, C.; Jin, J.; Hu, G. Boosting Hole Transfer in the Fluorine-Doped Hematite Photoanode by Depositing Ultrathin Amorphous FeOOH/CoOOH Cocatalysts. *ACS Appl. Mater. Interfaces* **2020**, *12*, 49705–49712. [CrossRef]
48. Fan, X.; Gao, B.; Wang, T.; Huang, X.; Gong, H.; Xue, H.; Guo, H.; Song, L.; Xia, W.; He, J. Layered double hydroxide modified $WO_3$ nanorod arrays for enhanced photoelectrochemical water splitting. *Appl. Catal. A: Gen.* **2016**, *528*, 52–58. [CrossRef]
49. Goud, B.S.; Koyyada, G.; Jung, J.H.; Reddy, G.R.; Shim, J.; Nam, N.D.; Vattikuti, S.P. Surface oxygen vacancy facilitated Z-scheme $MoS_2/Bi_2O_3$ heterojunction for enhanced visible-light driven photocatalysis-pollutant degradation and hydrogen production. *Int. J. Hydrogen Energy* **2020**, *45*, 18961–18975. [CrossRef]
50. Chen, W.-Q.; Li, L.-Y.; Li, L.; Qiu, W.-H.; Tang, L.; Xu, L.; Xu, K.-J.; Wu, M.-H. MoS2/ZIF-8 hybrid materials for environmental catalysis: Solar-driven antibiotic-degradation engineering. *Engineering* **2019**, *5*, 755–767. [CrossRef]
51. Kalanur, S.S. Structural, Optical, Band Edge and Enhanced Photoelectrochemical Water Splitting Properties of Tin-Doped $WO_3$. *Catalysts* **2019**, *9*, 456. [CrossRef]

*Article*

# Eco-Friendly Synthesis of Functionalized Carbon Nanodots from Cashew Nut Skin Waste for Bioimaging

Somasundaram Chandra Kishore [1,†], Suguna Perumal [2,†], Raji Atchudan [3,*,†], Thomas Nesakumar Jebakumar Immanuel Edison [4,†], Ashok K. Sundramoorthy [5,†], Muthulakshmi Alagan [6], Sambasivam Sangaraju [7] and Yong Rok Lee [3,*]

[1] Department of Biomedical Engineering, Saveetha School of Engineering, Saveetha Institute of Medical and Technical Sciences, Saveetha Nagar, Chennai 602105, Tamil Nadu, India
[2] Department of Chemistry, Sejong University, Seoul 143747, Republic of Korea
[3] School of Chemical Engineering, Yeungnam University, Gyeongsan 38541, Republic of Korea
[4] Department of Chemistry, Sethu Institute of Technology, Virudhunagar District, Kariapatti 626115, Tamil Nadu, India; jebakumar84@gmail.com
[5] Department of Prosthodontics, Saveetha Institute of Medical and Technical Sciences, Saveetha Dental College and Hospitals, Poonamallee High Road, Velappanchavadi, Chennai 600077, Tamil Nadu, India
[6] Department of Civil and Environmental Engineering, National Institute of Technical Teachers Training and Research, Chennai 600113, Tamil Nadu, India
[7] National Water and Energy Center, United Arab Emirates University, Al Ain P.O. Box 15551, United Arab Emirates
* Correspondence: atchudanr@yu.ac.kr (R.A.); yrlee@yu.ac.kr (Y.R.L.)
† These authors contributed equally to this work.

**Citation:** Kishore, S.C.; Perumal, S.; Atchudan, R.; Edison, T.N.J.I.; Sundramoorthy, A.K.; Alagan, M.; Sangaraju, S.; Lee, Y.R. Eco-Friendly Synthesis of Functionalized Carbon Nanodots from Cashew Nut Skin Waste for Bioimaging. *Catalysts* **2023**, *13*, 547. https://doi.org/10.3390/catal13030547

Academic Editors: Indra Neel Pulidindi, Archana Deokar and Aharon Gedanken

Received: 26 December 2022
Revised: 8 March 2023
Accepted: 8 March 2023
Published: 9 March 2023

**Copyright:** © 2023 by the authors. Licensee MDPI, Basel, Switzerland. This article is an open access article distributed under the terms and conditions of the Creative Commons Attribution (CC BY) license (https://creativecommons.org/licenses/by/4.0/).

**Abstract:** In this study, *Anacardium occidentale* (*A. occidentale*) nut skin waste (cashew nut skin waste) was used as a raw material to synthesize functionalized carbon nanodots (F-CNDs). *A. occidentale* biomass-derived F-CNDs were synthesized at a low temperature (200 °C) using a facile, economical hydrothermal method and subjected to XRD, FESEM, TEM, HRTEM, XPS, Raman Spectroscopy, ATR-FTIR, and Ultraviolet-visible (UV–vis) absorption and fluorescence spectroscopy to determine their structures, chemical compositions, and optical properties. The analysis revealed that dispersed, hydrophilic F-CNDs had a mean diameter of 2.5 nm. XPS and ATR-FTIR showed F-CNDs had a crystalline core and an amorphous surface decorated with $-NH_2$, $-COOH$, and $C=O$. In addition, F-CNDs had a quantum yield of 15.5% and exhibited fluorescence with maximum emission at 406 nm when excited at 340 nm. Human colon cancer (HCT-116) cell assays showed that F-CNDs readily penetrated into the cells, had outstanding biocompatibility, high photostability, and minimal toxicity. An MTT assay showed that the viability of HCT-116 cells incubated for 24 h in the presence of F-CNDs (200 µg mL$^{-1}$) exceeded 95%. Furthermore, when stimulated by filters of three different wavelengths (405, 488, and 555 nm) under a laser scanning confocal microscope, HCT-116 cells containing F-CNDs emitted blue, red, and green, respectively, which suggests F-CNDs might be useful in the biomedical field. Thus, we describe the production of a fluorescent nanoprobe from cashew nut waste potentially suitable for bioimaging applications.

**Keywords:** cashew nut skin; carbon nanodot; human colon cancer cell; cell viability; bioimaging

## 1. Introduction

Carbon nanodots (CNDs) [1], carbonized polymer dots [2], carbon quantum dots [3], graphene quantum dots [4], and other nanoscale carbon particles with dimensions of ~≤10 nm are all regarded as carbon dots (CDs) and are considered a new class of fluorescent carbon-based nanomaterials. Xu et al. accidentally discovered CDs in 2004 while purifying carbon nanotubes [5]. Ever since, a wide range of CDs with various chemical and optical properties have been produced using a number of different techniques. The characteristic features of CDs, which include tunable fluorescence emission [6], aqueous dispersibility [7], chemical inertness [8], biocompatibility [9], and ease of functionalization [10], make them powerful alternative semiconducting nanomaterials. Various biomedical utilities, such as nanoplatforms for biosensors [11], bioimaging [12], drug delivery vehicles [13], and gene transfer [14], are made possible by the ability of CDs to coexist with biological tissues without causing adverse effects. Because of their high fluorescence quantum yields, CDs are used as fluorescent probes in biological samples [15], and can be easily functionalized with biomolecules such as peptides or antibodies. In addition, they have a low photobleaching characteristic nature [16]. CDs are particularly useful for in vivo bioimaging and can be functionalized for targeted drug delivery. Worldwide, one in every six deaths is caused by cancer, which is the second most frequent cause of death. Uncontrolled cell growth is a main characteristic of cancer. For the effective treatment of cancer, early diagnosis is essential. Early cancer detection can help choose the best course of treatment and increase patient survival. Important information about a disease's course and a patient's response to therapy is provided by a diagnosis, which aids in modifying the patient's treatment plan while they are undergoing it [17]. The intriguing physicochemical and optical characteristics of CDs have great potential in the diagnosis and treatment of cancer. Furthermore, CDs can increase the efficacy and delivery of molecules because they are readily absorbed by cells, but more research is required to determine their safety and efficacy for in vivo applications.

Regardless of the synthetic process used to produce nanomaterials, the production of CDs can be categorized as top-down, bottom-up, or physical or chemical. In general, physical techniques involving arc discharge [18], laser ablation [19], and electrochemical etching [20] are hazardous to the environment and difficult to manage. Hydrothermal (HT) [21], ultrasonic [22], microwave-assisted [23], thermal decomposition [24], and electrochemical [25] processes are examples of chemical methods. The HT approach is usually used to produce CDs because it uses mild chemicals and is inexpensive. This approach has been widely employed to prepare a variety of carbon compounds because HT synthesis has negligible toxicological impact. Furthermore, HT conditions can cause reagent solubility, enhance chemical and physical reactions, and enable carbonaceous structures to develop. Conventionally, developing materials with high carbon contents, such as carbon nanotubes, mesoporous carbon, graphene, and graphitic carbon compounds, requires high temperatures (300–800 °C), whereas those produced by dehydration and polymerization are produced at lower temperatures (<300 °C), and often possess various surface functional groups after carbonization. In general, CNDs can be functionalized and doped with heteroatoms to enhance their fluorescence characteristics and quantum yields. Particularly, the HT method has become more popular for the synthesis of functionalized CNDs (F-CNDs) because it is a one-step procedure without additional oxidation and passivation, has gentle reaction parameters, and requires inexpensive equipment.

CDs are noted for their photophysical characteristics, particularly their fluorescence properties [26], which, like structure, morphology, and composition, are sensitive to the precursors and preparation techniques used [27]. In general, CDs are composed of crystalline carbon cores and decorated with carboxylic acid, alcohol, and amine functional groups [28]. Several biosources, such as lemon juice [29], leaf extract [30], grape juice [31], honey [32], hair [33], carrot juice [34], garlic [35], egg [36], betel leaf [37], and food waste [38] are used as CD precursors.

It is generally known that the transitions between intrinsic states cannot fully account for CD optical properties. The emission of many CDs, however, appears to be primarily influenced by surface-related extrinsic contributions, such as emissions from surface defects and surface charge traps. A proper passivation procedure is essential to produce highly fluorescent CDs, and solvents and pH significantly impact CD fluorescence. Research goals in the engineering area include tuning the photophysical and electrochemical properties of CDs by altering ground and excited state properties [39] and modifying the form, chirality, composition, size, and surface chemistry of CDs [40]. In the present study, we sought to develop non-toxic, <10 nm sized CDs compatible with aqueous environments using *Anacardium occidentale* (*A. occidentale*) nut skin waste as a precursor for the synthesis of F-CNDs.

Cashew is the popular name for *A. occidentale* (AO), a member of the *Anacardiaceae* family, and it is commonly grown in tropical areas of India, Brazil, and Africa. An essential by-product generated during the processing of cashews is the testa (skin) of the cashew kernel. The resulting testa is a potential candidate for commercial exploitation given that cashew kernels are consumed worldwide on an annual basis in excess of 1,000,000 tonnes. It is said to be an excellent source of hydrolyzable tannins. The cashew nut is a significant cash crop worldwide. India produces and exports the most cashew kernels worldwide, making up nearly 50% of all exports. A brown skin, known as testa, completely envelops cashew nuts, and this skin is one of the best sources of hydrolyzable tannins such as catechin, epicatechin, and epigallocatechin [41,42]. The seed testa has the greatest proportion of phenolic compounds that serve as a barrier of protection for the cotyledon in seeds. Furthermore, it also contains high levels of three phenolic acids, viz. syringic, gallic, and p-coumaric acids [43], which confer significant antioxidant activity [44]. In order to understand the possible mechanisms behind the formation of F-CNDs, it is presumed that the testa of cashew nuts consists of hydrolyzable tannins, phenolic acids, and various other molecules. These constituents undergo the process of dehydration, polymerization, and carbonization to form F-CNDs. We investigated AO biomass-derived F-CNDs synthesized using the HT approach at a lower temperature of 200 °C. The best quality F-CNDs with significant fluorescence properties were subjected to cellular imaging of human colon cancer cells.

## 2. Results and Discussion

FESEM images of F-CNDs at different magnifications are provided in Figure 1a–c. F-CNDs formed a thin layer over the surface of the sample holder. EDX revealed the elements present on the surface of F-CNDs (Figure 1d–g). Elements were identified by color, e.g., green, red, and yellow indicated carbon (C), oxygen (O), and nitrogen (N), respectively. O and N were distributed evenly over carbon substrates. EDX peaks shown in Figure 1h confirmed the presence of carbon, nitrogen, oxygen, silicon, and platinum. Silicon and platinum were attributed to sample preparation. For FESEM analysis, F-CNDs were spin-coated on silicon wafers and sputtered with platinum.

HRTEM was used to determine F-CND morphology and sizes. The morphological features of F-CNDs are well demonstrated by the micrographs in Figure 2a–c. F-CNDs were observed as spherical, well-dispersed dark dots with a few aggregations. In the high magnification, it is clear that F-CNDs were composed of graphitic layers with an interlayer spacing of 0.21 nm (inset in Figure 2c). The particle size distribution of F-CNDs is shown as a histogram in Figure 2d, which was derived via Gaussian particle-size-distribution fitting and by measuring the sizes of 100 randomly selected particles in HRTEM images (Figure 2a. F-CND sizes ranged from 1.5 to 4 nm with a mean particle size of ~2.5 nm).

X-ray powder diffraction was used to determine the crystal phases in F-CNDs. Figure 3a shows that the XRD spectrum of F-CNDs contained a broad peak at $2\theta = 23°$, corresponding to the (0 0 2) carbon lattice [45]. The shoulder peak at $2\theta = 43°$ was ascribed to the (1 0 0) plane, and the corresponding d-spacing value was 0.21 nm, which agreed well with TEM results. The absence of a sharp peak, corresponding to the formation of an amorphous layer on F-CNDs, suggested the presence of surface functional groups. F-CNDs were also subjected to Raman spectroscopy to determine the purity and degree of graphitization of samples. The Raman spectrum of F-CNDs is shown in Figure 3b. Two prominent peaks corresponding to carbon D and G bands were observed at 1360 and 1585 cm$^{-1}$, respectively [46]. These bands correspond to the disorder (vibration of sp$^3$ carbon atom) and graphitic nature (vibration of sp$^2$ carbon atom) of carbon materials and had an intensity ratio ($I_D/I_G$) of 0.63 [47,48], which confirmed a graphitic nature and a few surface defects [48,49]. The deconvoluted Raman spectrum shown in Figure 3c was used to assess the degree of graphitization in F-CNDs. Areas of the D and G bands ($A_D$ and $A_G$, respectively) were used to calculate the areal D to G ratio ($A_D/A_G$), which was 0.65. This value indicates the formation of graphitized F-CNDs with minimal surface disorder or few defects. Surface disorder could be due to functional groups or edge effects. An

ATR-FTIR (attenuated total reflectance-Fourier transform infrared) spectrum of F-CNDs provided information about surface functional groups (Figure 3d). The hydrophilic nature of the F-CNDs was confirmed by the presence of N–H and O–H stretching vibrations at 3500–3100 cm$^{-1}$ [50,51]. Peaks between 2870 and 2962 cm$^{-1}$ were assigned to the C–H asymmetric and symmetric stretch [52]. The presence of carboxyl/carbonyl groups was confirmed by C=O and C=C stretching vibration peaks at 1670 and 1575 cm$^{-1}$, respectively [53]. The peaks between 1021 and 1120 cm$^{-1}$ indicated the presence of the C–O–C group, and peaks at 1445, 1260, and 1397 cm$^{-1}$ were ascribed to –C–N, C–OH and bending vibrations of N–H and O–H, respectively [54]. Out-of-plane stretching vibrations of C–H were confirmed by an absorption band at 665 cm$^{-1}$ and were attributed to the carboxylic groups on F-CNDs [55]. These findings show that F-CNDs were composed of C, N, and O and decorated with –COOH, –OH, and –C–N groups.

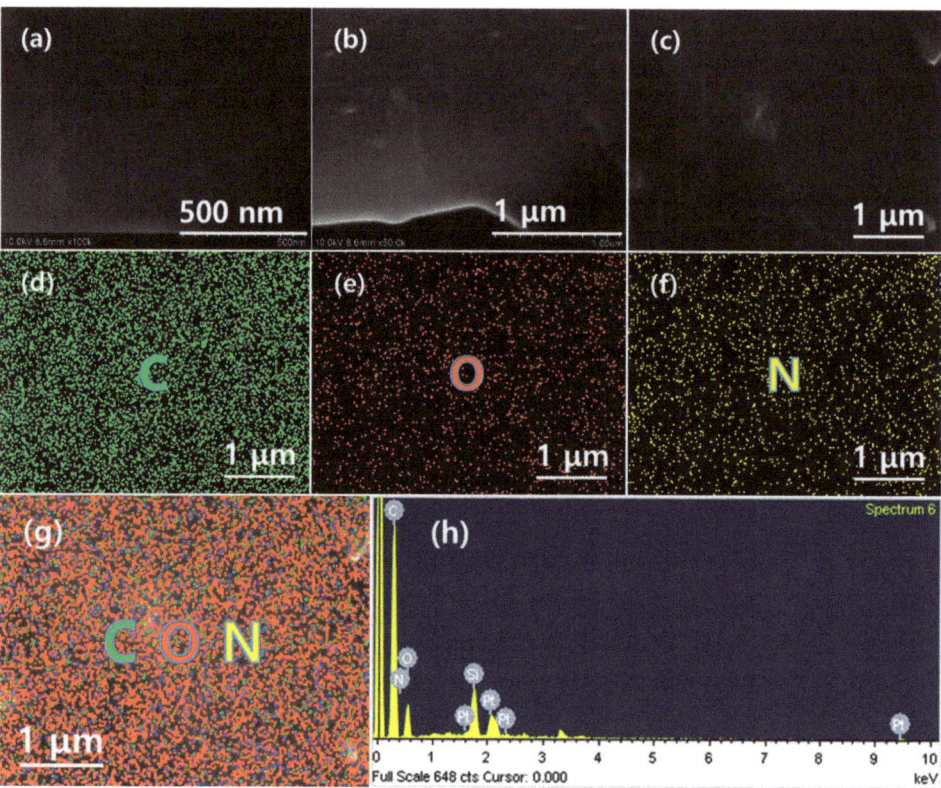

**Figure 1.** (**a–c**) FE-SEM images of functionalized carbon nanodots (F-CNDs; (**d–g**) EDX elemental mapping images of F-CNDs (**d**) carbon, (**e**) oxygen, (**f**) nitrogen, and (**g**) overlapping image showing all elements; (**h**) EDX spectrum of F-CNDs.

X-ray photoelectron spectroscopy (XPS) was used to determine the elemental composition, type of bonding, and nature of functional groups. An XPS spectrum of F-CNDs is provided in Figure 4a. The peaks observed at the binding energies (BEs) of 285, 400, and 532 eV indicated the presence of C 1 s, N 1 s, and O 1 s, respectively. Interestingly, the atomic ratio of carbon to other elements was 3:1, and the atomic weight percentages of carbon, nitrogen, and oxygen were 75, 4, and 21%, respectively. Furthermore, the high-resolution XPS spectrum of C 1 s (Figure 4b) was deconvoluted into five distinct peaks. The binding energy (BE) of the peak at 284.5 eV corresponded to the C=C/C–C bond of the sp$^2$ and sp$^3$ graphitic structure of F-CNDs [56,57]. The binding energy peak at 285.1 eV corresponded to the pyridinic C–N–C bonds of F-CNDs. The presence of C–OH/C–O–C was confirmed by the peak at 286.1 eV, corresponding to hydroxyl bound to carbon [58]. The peak at 287.0 eV corresponded to C=N/C=O bonds representing pyrrolic nitrogen and carbonyl groups (–C=O) [58], whereas the presence of carboxyl groups (O=C–OH) was confirmed by the peak at 288.5 eV [57]. Figure 4c depicts the XPS spectrum of N 1 s, which exhibited three deconvoluted peaks signifying the presence of pyridinic nitrogen (C–N–C), pyrrolic nitrogen (C–N–H), and graphitic nitrogen (C$_3$–N bonds) with Bes of 399.2, 400.2, and 401.7 eV, respectively [59,60], and showing that F-CND carbon had been doped with nitrogen. Notably, fluorescence results from the ability of excited nitrogen-doped carbon to emit light. The chemical type and concentration of nitrogen, carbon structure, and the conditions used for material synthesis can all affect the mechanism of nitrogen-doped carbon fluorescence. However, in most cases, movements of nitrogen electrons to lower energy levels are responsible. The XPS spectrum of O 1 s (Figure 4d) had two deconvoluted peaks at BE 531.5 and 533.1 eV corresponding to C=O/C–OH and C–O–C/O–C=OH, respectively [61]. These findings imply that the surfaces of F-CNDs had –OH, –C–N, and –COOH groups, which provide hydrophilicity and dispersibility in water. Furthermore, ATR-FTIR results were in line with XPS results.

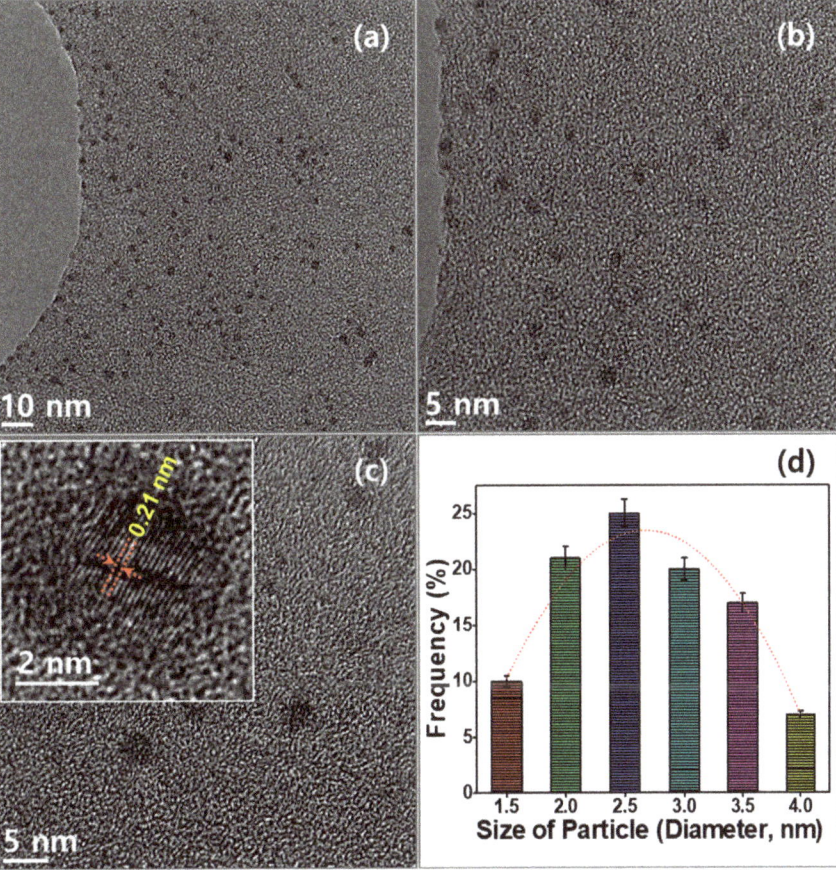

**Figure 2.** (**a**–**c**) HRTEM images of synthesized F-CNDs at different magnifications and (**d**) a particle size distribution histogram.

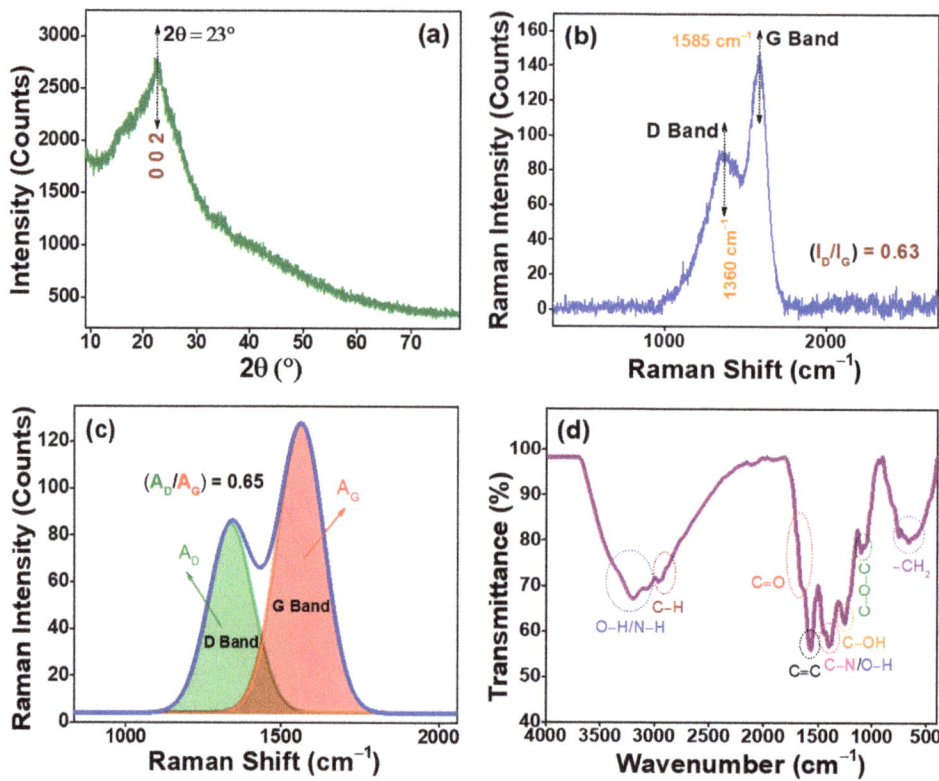

**Figure 3.** (a) Powder XRD pattern, (b) Raman spectrum, (c) deconvoluted Raman spectrum, and (d) ATR-FTIR spectrum of synthesized F-CNDs.

The optical properties of F-CNDs were evaluated using Ultraviolet-visible (UV–vis) absorption and fluorescence spectroscopy. The UV–vis absorption spectrum of F-CNDs (Figure 5a) exhibited two prominent peaks at 217 and 275 corresponding to $\pi$–$\pi$* transitions of C–C/C=C and C=C, respectively. In addition, a shoulder was observed at 323, corresponding to the n–$\pi$* transition of C=O or C=N [62]. The inset in Figure 5a demonstrates the dispersion of F-CNDs in water and the difference between exposure to daylight or 365 nm UV light. F-CNDs were dispersed thoroughly in aqueous solvents, and UV exposure resulted in a color change from pale yellow to cyan. This phenomenon was ascribed to the different functional groups on F-CNDs.

F-CNDs exhibited maximum fluorescence at 406 nm when excited at 340 nm (Figure 5b); that is, a Stokes shift of 66 nm occurred. The magnitude of a Stokes shift can significantly impact the practical use of fluorescence. For instance, a significant Stokes shift can improve biological imaging by lowering background noise and increasing the signal-to-noise ratio. However, in some situations, such as in fluorescence resonance energy transfer, a slight spectral overlap between excitation and emission spectra is required to enable energy transfer between fluorescent molecules. The effects of fluorescence excitation wavelengths in the range of 330–420 nm on the emission spectrum of F-CNDs are shown in Figure 5c. Interestingly, the intensity of the emission spectrum increased upon increasing the excitation wavelength from 330 to 340 nm but reduced upon increasing it from 340 to 420 nm, and maximum emission intensity was observed at 340 nm. A normalized excitation-dependent emission spectrum (Figure S1) implies a redshift in the 395 to 495 nm wavelength range. The shift primarily results from electron transfer from the conjugated surface functional groups narrowing the energy gap. Presumably, if an emitting molecule or fluorophore is in a different environment than the absorbing molecule, a redshift in emission could also occur. In addition, some types of fluorescence, such as two-photon fluorescence, in which two photons of lower energy are simultaneously absorbed, can also cause a redshift. The photostability of F-CNDs was studied by continuously irradiating them with 365 nm UV light at a power of 4 W for 0–120 min (Figure 5d). The intensities of the emission spectra obtained were unchanged without any decay in emission, which confirmed the photostability of F-CNDs. Furthermore, prolonged UV exposure for 120 min caused no color change or precipitate formation (inset of Figure 5d). In addition, the quantum yield of F-CNDs was calculated to be 15.5%. These characteristics of F-CNDs might be due to a wide range of particle sizes, interactions caused by quantum confinement, and the presence of different functional groups.

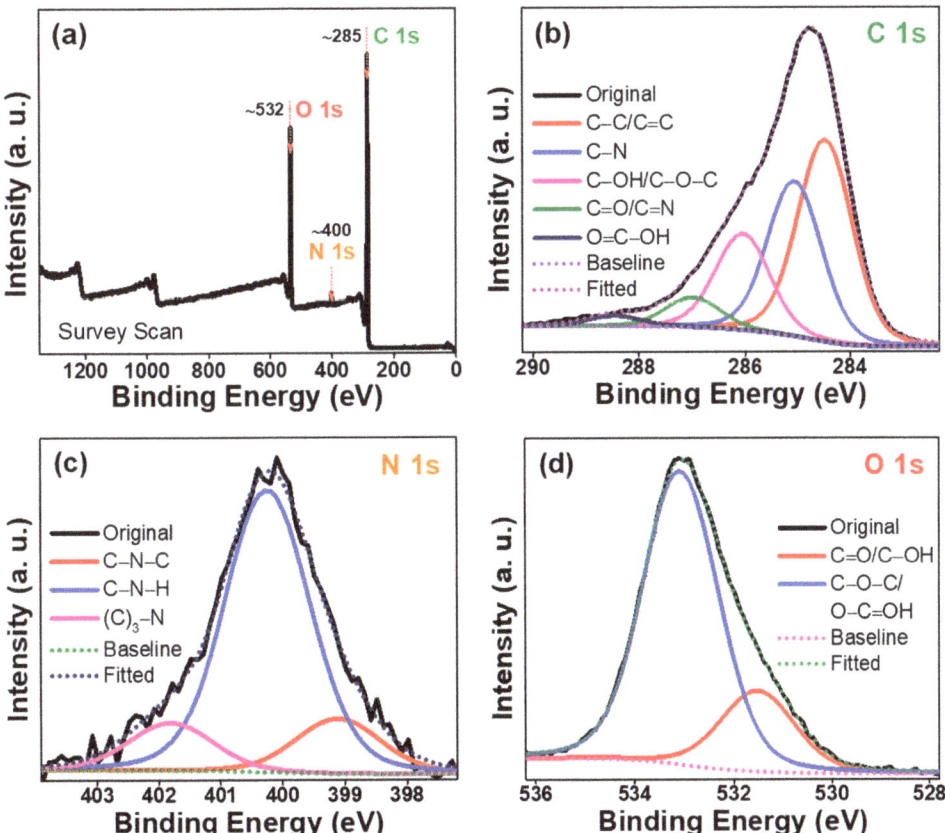

**Figure 4.** (a) XPS-survey spectrum and high-resolution XPS spectra of (b) C 1 s, (c) N 1 s, and (d) O 1 s of synthesized F-CNDs.

In the carbon core of CDs, $sp^2$-conjugated frameworks are typically accompanied by a number of imperfect $sp^2$ domains. These areas will generate or induce surface energy traps that can serve as exciton capture sites, leading to fluorescence associated with the surface defect state. Therefore, surface flaws are responsible for visible light multicolor emissions from CDs. The band gap primarily controls the emission wavelength and is influenced by a wide range of variables, including CD surface chemistry, synthesis techniques, and edge configuration. Due to the epoxy, carboxyl, and hydroxyl groups present in the $sp^2$

clusters, which encompass an extensive spectrum of size distribution, various band gap energies exhibit a variety of emission spectra. Two main types of mechanisms underlie luminescence, namely, quantum confinement in nanometric structures and those involving radiative relaxation of excited states attained by different functional groups within CDs [63]. Furthermore, pyrolytic processing and partial thermal decomposition of precursors cause the formation of intermediate organic fluorophores [64]. Based on our results, we suggest the emission properties of F-CDs are probably due to radiative transitions within or between functional groups on the surfaces of F-CNDs.

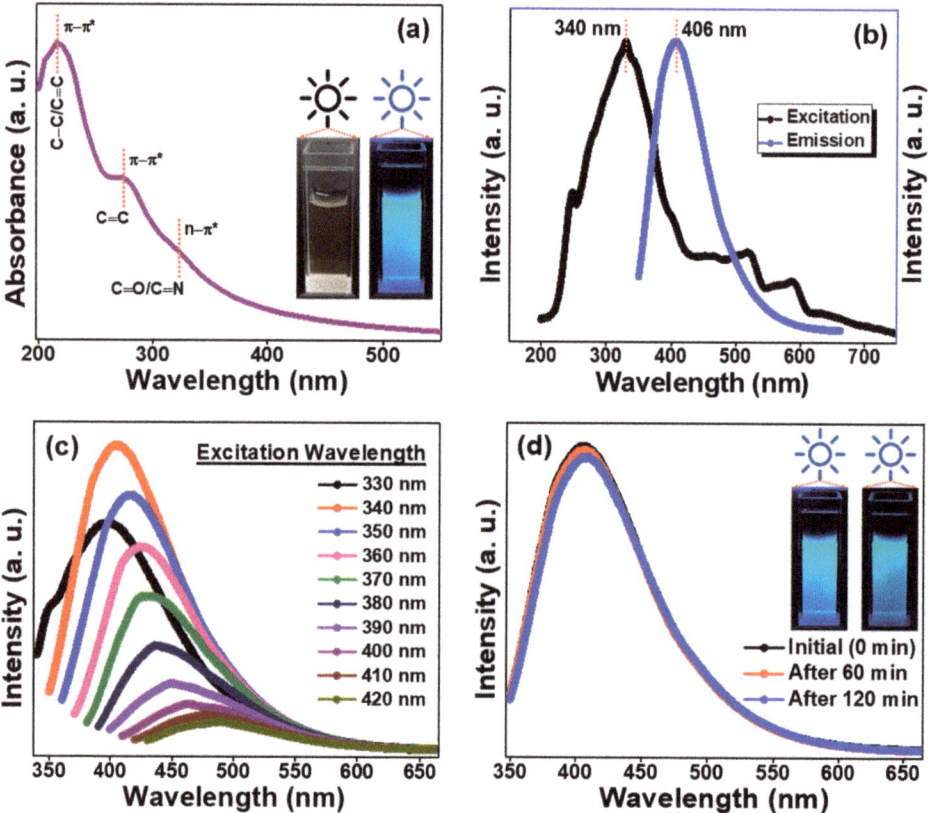

**Figure 5.** (**a**) UV-vis absorption spectrum (inset: photographic images of synthesized F-CNDs in aqueous solution under daylight (**left**) and 365 nm UV light (**right**)); (**b**) fluorescence excitation and emission spectra, and (**c**) fluorescence excitation-dependent emission spectra of synthesized F-CNDs. (**d**) Fluorescence emission spectra of synthesized F-CNDs before and after continuous irradiation with 365 nm UV light (inset: photographic images of synthesized F-CNDs in aqueous suspension under 365 nm UV light before (0 min) and after (120 min) continuous irradiation with 365 nm UV).

F-CNDs emit controllable fluorescence, have appropriate quantum yields, high water dispersibility, low cytotoxicity, and excellent biocompatibility, and do not exhibit photobleaching. The produced F-CNDs were used for cellular imaging without modification. MTT cell viability test results for HCT-116 cells (a human colon cancer cell line) at F-CND concentrations of 0 to 200 µg mL$^{-1}$ are shown in Figure S2. The bar chart provides a comparison between the viabilities of F-CND treated and untreated cells (controls) and shows a slight decrease (from 100 to 97%) in cell viabilities with increasing concentration of F-CNDs from 0 to 200 µg mL$^{-1}$. This observation indicated good compatibility and low cytotoxicity of F-CNDs with a human colon cancer cell line, and cytotoxicity does not lead to cell death even at higher concentrations of 200 μg mL$^{-1}$, which is an essential property required for F-CNDs to make them suitable for bioimaging of cells. To comprehend the dynamics, one must first understand how F-CNDs become internalized within cells, tissues, or cellular cytoskeleton components. Actin filaments, Microtubules, and intermediate filaments are intracellular components that actively collaborate with cancer cells. Confocal microscopy was used to investigate the bioimaging characteristics of F-CNDs in human colon cancer cells. Figure 6 contains confocal microscopy photographs of HCT-116 cells, treated or not with F-CNDs, taken using different wavelength filters, viz. 405 (blue), 488 (green), and 555 nm (red) after exposure to bright field illumination for 12 or 24 h. No emission was observed from untreated HCT-116 cells, whereas fluorescence was observed from human colon cancer cells treated with F-CNDs when 405, 488, or 555 nm filters were used, which produced blue, green, and red emissions, respectively. The overlapping image was multi-colored (Figure 6), indicating excitation wavelength-dependent emission characteristics. Upon increasing the exposure time from 12 h to 24 h, enhancement in the intensity of fluorescence is well observed from the image. It has been well established that F-CNDs are easily internalized and uniformly distributed in human colon cancer cells. Therefore, these results show that F-CNDs are candidate fluorescent nanoprobes for imaging human colon cancer cells.

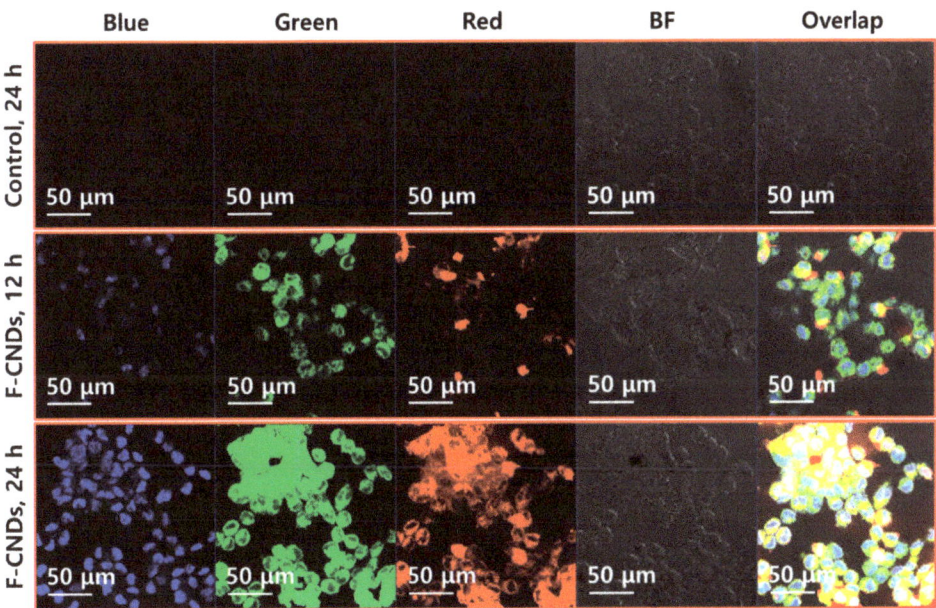

**Figure 6.** Confocal microscopy fluorescence images of human colon cancer cells treated with or without F-CNDs and the synthesized F-CNDs treated for 12 and 24 h with the concentration of 100 μg mL$^{-1}$ using different excitation filters 405, 488, and 555 nm (blue, green, and red, respectively) as well as bright-field illumination.

## 3. Materials and Methods

### 3.1. Materials

Cashew nut skin waste was collected from Tamil Nadu, India. Aqueous ammonia (NH$_4$OH, 25%) was purchased from Sigma-Aldrich, Republic of Korea. Phosphate buffered saline (PBS), N-(2-hydroxyethyl)piperazine-N'-(2-ethane sulfonic acid) (HEPES), p-formaldehyde, quinine sulfate, and dimethyl sulfoxide (DMSO) were purchased from Sigma-Aldrich, Republic of Korea. 3-(4,5-dimethylthiazol-2-yl)-2,5-diphenyltetrazolium bromide (MTT) was purchased from Generay Biotech, Shanghai, China. HCT-116 human colon cancer cells were purchased from ATCC, CCL-247, Manassas, VA, USA. All the chemicals were used as purchased and distilled water was used throughout this study.

*3.2. The Synthesis of Functionalized Carbon Nanodots*

F-CNDs were synthesized using washed, dried, and ground cashew nut skins. The whitish-brown powder obtained was added to 50 mL of water with 1 mL of 25% ammonium hydroxide solution and placed in an autoclave at 200 °C for 24 h. Large carbon particles were eliminated via filtration, and the filtrate was passed through a mixed cellulose ester membrane filter with a pore size of 0.22 µm, frozen in liquid nitrogen, and dried at below −80 °C in a freeze dryer. The F-CNDs obtained were used in subsequent experiments. Scheme 1 shows the synthesis procedure of F-CNDs from cashew nut skin waste using the hydrothermal-carbonization.

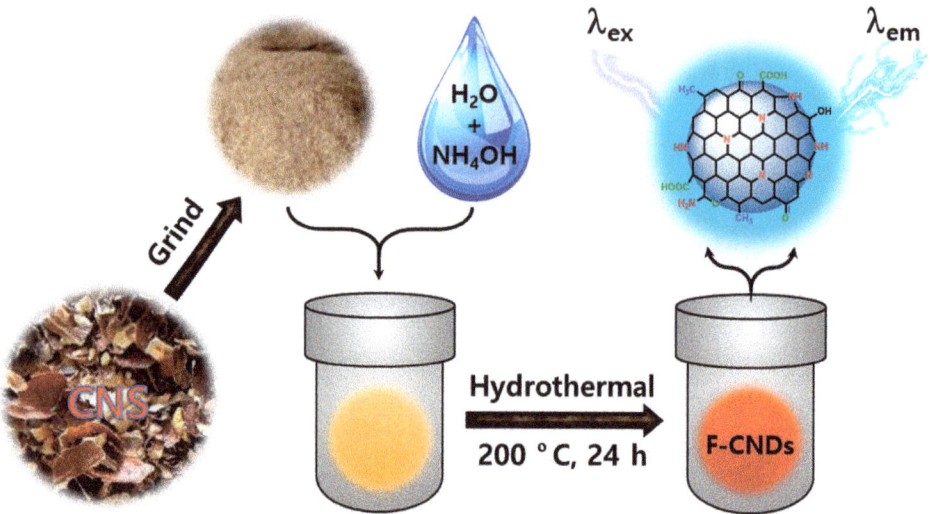

**Scheme 1.** Hydrothermal synthesis of functionalized carbon nanodots from cashew nut skin waste.

## 4. Conclusions

Using a single-step method, cashew nut skin waste was used to synthesize F-CNDs using a simple hydrothermal route at a very low temperature without any further modifications that were quite economical. The formations of F-CNDs were considered to be due to the dehydration, polymerization, and carbonization of hydrolyzable tannins, and phenolic acids present in the testa of cashew nuts. F-CNDs exhibited a graphitic structure at the core with few surface defects as determined by XRD and Raman Spectroscopy. F-CNDs had a mean particle size of 2.5 nm and were composed of carbon, nitrogen, and oxygen decorated with functional groups (C=O, –OH, –NH$_2$, and –COOH), as determined by XPS and ATR-FTIR, which conferred F-CNDs with significant hydrophilicity and dispersibility. F-CNDs had excellent fluorescent properties and exhibited maximum emission at 406 nm when excited at 340 nm due to radiative transitions within or between functional groups present on the surfaces of F-CNDs. F-CNDs were photostable, had a quantum yield of 15.5%, and at concentrations of 0–200 µg mL$^{-1}$ returned MTT viability greater than 95% for HCT-116 cells. F-CNDs thus proved to have remarkable biocompatibility and low cytotoxicity with the cancer cell line. Confocal microscopy of human colon cancer cells treated with or without F-CNDs revealed blue, green, and red emissions when exposed to 405, 488, and 555 nm light, respectively, in addition to a bright field. After increasing the time of exposure from 12 h to 24 h, significant enhancement in the intensity of fluorescence was observed.

Our results show nano-sized, cashew-nut-skin-derived F-CNDs have a graphitized core structure and are surface functionalized by organic moieties. They are suitable nanoprobes for bioimaging, drug delivery, and cell labeling. In the near future, a material that is safe for the delivery of anticancer drugs could be developed using the successful integration of F-CNDs with anticancer drugs.

**Supplementary Materials:** The following are available online at https://www.mdpi.com/article/10.3390/catal13030547/s1. Instrumentation methods, quantum yield measurements, photobleaching measurements, cell culture, cell viability assay, and microscopy results. Figure S1. Fluorescence excitation-dependent emission normalized-spectra of synthesized F-CNDs; Figure S2. Cell viability MTT assay results. The bar chart provides a comparison of the viabilities of F-CND treated cells and untreated controls.

**Author Contributions:** Investigation and writing of the original draft, S.C.K.; visualization, reviewing the original draft, and editing, S.P.; conceptualization, data curation, formal analysis, investigation, and writing the original draft, R.A.; formal analysis and investigation T.N.J.I.E.; investigation and visualization, A.K.S.; investigation and validation, M.A.; formal analysis and visualization, S.S.; project administration and supervision, Y.R.L. The authors contributed equally to this work. All authors have read and agreed to the published version of the manuscript.

**Funding:** This research received no external funding.

**Data Availability Statement:** Not applicable.

**Conflicts of Interest:** The authors declare no conflict of interest.

# References

1. Wang, B.; Lu, S. The light of carbon dots: From mechanism to applications. *Matter* **2022**, *5*, 110–149. [CrossRef]
2. Yao, X.; Lewis, R.E.; Haynes, C.L. Synthesis Processes, Photoluminescence Mechanism, and the Toxicity of Amorphous or Polymeric Carbon Dots. *Acc. Chem. Res.* **2022**, *55*, 3312–3321. [CrossRef]
3. Yuan, T.; Meng, T.; Shi, Y.; Song, X.; Xie, W.; Li, Y.; Li, X.; Zhang, Y.; Fan, L. Toward phosphorescent and delayed fluorescent carbon quantum dots for next-generation electroluminescent displays. *J. Mater. Chem. C* **2022**, *10*, 2333–2348. [CrossRef]
4. Ghaffarkhah, A.; Hosseini, E.; Kamkar, M.; Sehat, A.A.; Dordanihaghighi, S.; Allahbakhsh, A.; van der Kuur, C.; Arjmand, M. Synthesis, applications, and prospects of graphene quantum dots: A comprehensive review. *Small* **2022**, *18*, 2102683. [CrossRef] [PubMed]
5. Xu, X.; Ray, R.; Gu, Y.; Ploehn, H.J.; Gearheart, L.; Raker, K.; Scrivens, W.A. Electrophoretic analysis and purification of fluorescent single-walled carbon nanotube fragments. *J. Am. Chem. Soc.* **2004**, *126*, 12736–12737. [CrossRef]
6. Li, J.; Gong, X. The Emerging Development of Multicolor Carbon Dots. *Small* **2022**, *18*, 2205099. [CrossRef]
7. Falara, P.P.; Zourou, A.; Kordatos, K.V. Recent advances in Carbon Dots/2-D hybrid materials. *Carbon* **2022**, *195*, 219–245. [CrossRef]
8. Hebbar, A.; Selvaraj, R.; Vinayagam, R.; Varadavenkatesan, T.; Kumar, P.S.; Duc, P.A.; Rangasamy, G. A critical review on the environmental applications of carbon dots. *Chemosphere* **2022**, *313*, 137308. [CrossRef]
9. Biswal, M.R.; Bhatia, S. Carbon Dot Nanoparticles: Exploring the Potential Use for Gene Delivery in Ophthalmic Diseases. *Nanomaterials* **2021**, *11*, 935. [CrossRef]
10. Guo, R.; Li, L.; Wang, B.; Xiang, Y.; Zou, G.; Zhu, Y.; Hou, H.; Ji, X. Functionalized carbon dots for advanced batteries. *Energy Storage Mater.* **2021**, *37*, 8–39. [CrossRef]
11. Wang, F.-T.; Wang, L.-N.; Xu, J.; Huang, K.-J.; Wu, X. Synthesis and modification of carbon dots for advanced biosensing application. *Analyst* **2021**, *146*, 4418–4435. [CrossRef]
12. Kaur, P.; Verma, G. Converting fruit waste into carbon dots for bioimaging applications. *Mater. Today Sustain.* **2022**, *18*, 100137. [CrossRef]
13. Tang, J.; Kong, B.; Wu, H.; Xu, M.; Wang, Y.; Wang, Y.; Zhao, D.; Zheng, G. Carbon nanodots featuring efficient FRET for real-time monitoring of drug delivery and two-photon imaging. *Adv. Mater.* **2013**, *25*, 6569–6574. [CrossRef]
14. Kashkoulinejad-Kouhi, T.; Sawalha, S.; Safarian, S.; Arnaiz, B. A carbon-based nanocarrier for efficient gene delivery. *Ther. Deliv.* **2021**, *12*, 311–323. [CrossRef]
15. Shi, L.; Yang, J.H.; Zeng, H.B.; Chen, Y.M.; Yang, S.C.; Wu, C.; Zeng, H.; Yoshihito, O.; Zhang, Q. Carbon dots with high fluorescence quantum yield: The fluorescence originates from organic fluorophores. *Nanoscale* **2016**, *8*, 14374–14378. [CrossRef] [PubMed]
16. Xiong, Y.; Schneider, J.; Reckmeier, C.J.; Huang, H.; Kasák, P.; Rogach, A.L. Carbonization conditions influence the emission characteristics and the stability against photobleaching of nitrogen doped carbon dots. *Nanoscale* **2017**, *9*, 11730–11738. [CrossRef] [PubMed]

17. Pourmadadi, M.; Rahmani, E.; Rajabzadeh-Khosroshahi, M.; Samadi, A.; Behzadmehr, R.; Rahdar, A.; Ferreira, L.F.R. Properties and application of carbon quantum dots (CQDs) in biosensors for disease detection: A comprehensive review. *J. Drug Deliv. Sci. Technol.* **2023**, *80*, 104156. [CrossRef]
18. Nagarajan, D.; Gangadharan, D.; Venkatanarasimhan, S. Synthetic strategies toward developing carbon dots via top-down approach. In *Carbon Dots in Analytical Chemistry*; Elsevier: Amsterdam, The Netherlands, 2023; pp. 1–13.
19. Hu, S.-L.; Niu, K.-Y.; Sun, J.; Yang, J.; Zhao, N.-Q.; Du, X.-W. One-step synthesis of fluorescent carbon nanoparticles by laser irradiation. *J. Mater. Chem.* **2009**, *19*, 484–488. [CrossRef]
20. Bao, L.; Zhang, Z.L.; Tian, Z.Q.; Zhang, L.; Liu, C.; Lin, Y.; Qi, B.; Pang, D.W. Electrochemical tuning of luminescent carbon nanodots: From preparation to luminescence mechanism. *Adv. Mater.* **2011**, *23*, 5801–5806. [CrossRef] [PubMed]
21. Atchudan, R.; Kishore, S.C.; Gangadaran, P.; Edison, T.N.J.I.; Perumal, S.; Rajendran, R.L.; Alagan, M.; Al-Rashed, S.; Ahn, B.-C.; Lee, Y.R. Tunable fluorescent carbon dots from biowaste as fluorescent ink and imaging human normal and cancer cells. *Environ. Res.* **2022**, *204*, 112365. [CrossRef] [PubMed]
22. Manoharan, P.; Dhanabalan, S.C.; Alagan, M.; Muthuvijayan, S.; Ponraj, J.S.; Somasundaram, C.K. Facile synthesis and characterisation of green luminescent carbon nanodots prepared from tender coconut water using the acid-assisted ultrasonic route. *Micro Nano Lett.* **2020**, *15*, 920–924. [CrossRef]
23. Jiang, J.; He, Y.; Li, S.; Cui, H. Amino acids as the source for producing carbon nanodots: Microwave assisted one-step synthesis, intrinsic photoluminescence property and intense chemiluminescence enhancement. *Chem. Commun.* **2012**, *48*, 9634–9636. [CrossRef] [PubMed]
24. Ortega-Liebana, M.; Chung, N.; Limpens, R.; Gomez, L.; Hueso, J.; Santamaria, J.; Gregorkiewicz, T. Uniform luminescent carbon nanodots prepared by rapid pyrolysis of organic precursors confined within nanoporous templating structures. *Carbon* **2017**, *117*, 437–446. [CrossRef]
25. Ming, H.; Ma, Z.; Liu, Y.; Pan, K.; Yu, H.; Wang, F.; Kang, Z. Large scale electrochemical synthesis of high quality carbon nanodots and their photocatalytic property. *Dalton Trans.* **2012**, *41*, 9526–9531. [CrossRef]
26. Atchudan, R.; Kishore, S.C.; Edison, T.N.J.I.; Perumal, S.; Vinodh, R.; Sundramoorthy, A.K.; Babu, R.S.; Alagan, M.; Lee, Y.R. Highly fluorescent carbon dots as a potential fluorescence probe for selective sensing of ferric ions in aqueous solution. *Chemosensors* **2021**, *9*, 301. [CrossRef]
27. Lou, Q.; Yang, X.; Liu, K.; Ding, Z.; Qin, J.; Li, Y.; Lv, C.; Shang, Y.; Zhang, Y.; Zhang, Z. Pressure-induced photoluminescence enhancement and ambient retention in confined carbon dots. *Nano Res.* **2022**, *15*, 2545–2551. [CrossRef]
28. Ji, Z.; Sheardy, A.; Zeng, Z.; Zhang, W.; Chevva, H.; Allado, K.; Yin, Z.; Wei, J. Tuning the functional groups on carbon nanodots and antioxidant studies. *Molecules* **2019**, *24*, 152. [CrossRef]
29. Ding, H.; Zhou, X.; Qin, B.; Zhou, Z.; Zhao, Y. Highly fluorescent near-infrared emitting carbon dots derived from lemon juice and its bioimaging application. *J. Lumin.* **2019**, *211*, 298–304. [CrossRef]
30. Duarah, R.; Karak, N. Facile and ultrafast green approach to synthesize biobased luminescent reduced carbon nanodot: An efficient photocatalyst. *ACS Sustain. Chem. Eng.* **2017**, *5*, 9454–9466. [CrossRef]
31. Wang, S.; Huo, X.; Zhao, H.; Dong, Y.; Cheng, Q.; Li, Y. One-pot green synthesis of N, S co-doped biomass carbon dots from natural grapefruit juice for selective sensing of Cr (VI). *Chem. Phys. Impact* **2022**, *5*, 100112. [CrossRef]
32. Mandani, S.; Dey, D.; Sharma, B.; Sarma, T.K. Natural occurrence of fluorescent carbon dots in honey. *Carbon* **2017**, *119*, 569–572. [CrossRef]
33. Sun, D.; Ban, R.; Zhang, P.-H.; Wu, G.-H.; Zhang, J.-R.; Zhu, J.-J. Hair fiber as a precursor for synthesizing of sulfur-and nitrogen-co-doped carbon dots with tunable luminescence properties. *Carbon* **2013**, *64*, 424–434. [CrossRef]
34. Liu, Y.; Liu, Y.; Park, M.; Park, S.-J.; Zhang, Y.; Akanda, M.R.; Park, B.-Y.; Kim, H.Y. Green synthesis of fluorescent carbon dots from carrot juice for in vitro cellular imaging. *Carbon Lett.* **2017**, *21*, 61–67. [CrossRef]
35. Zhao, S.; Lan, M.; Zhu, X.; Xue, H.; Ng, T.-W.; Meng, X.; Lee, C.-S.; Wang, P.; Zhang, W. Green synthesis of bifunctional fluorescent carbon dots from garlic for cellular imaging and free radical scavenging. *ACS Appl. Mater. Interfaces* **2015**, *7*, 17054–17060. [CrossRef] [PubMed]
36. Pramanik, S.; Chatterjee, S.; Kumar, G.S.; Devi, P.S. Egg-shell derived carbon dots for base pair selective DNA binding and recognition. *Phys. Chem. Chem. Phys.* **2018**, *20*, 20476–20488. [CrossRef] [PubMed]
37. Atchudan, R.; Gangadaran, P.; Edison, T.N.J.I.; Perumal, S.; Sundramoorthy, A.K.; Vinodh, R.; Rajendran, R.L.; Ahn, B.-C.; Lee, Y.R. Betel leaf derived multicolor emitting carbon dots as a fluorescent probe for imaging mouse normal fibroblast and human thyroid cancer cells. *Phys. E: Low-Dimens. Syst. Nanostructures* **2022**, *136*, 115010. [CrossRef]
38. Park, S.Y.; Lee, H.U.; Park, E.S.; Lee, S.C.; Lee, J.-W.; Jeong, S.W.; Kim, C.H.; Lee, Y.-C.; Huh, Y.S.; Lee, J. Photoluminescent green carbon nanodots from food-waste-derived sources: Large-scale synthesis, properties, and biomedical applications. *ACS Appl. Mater. Interfaces* **2014**, *6*, 3365–3370. [CrossRef]
39. Lou, Q.; Ni, Q.; Niu, C.; Wei, J.; Zhang, Z.; Shen, W.; Shen, C.; Qin, C.; Zheng, G.; Liu, K. Carbon nanodots with nearly unity fluorescent efficiency realized via localized excitons. *Adv. Sci.* **2022**, *9*, 2203622. [CrossRef]
40. Song, Z.; Shang, Y.; Lou, Q.; Zhu, J.; Hu, J.; Xu, W.; Li, C.; Chen, X.; Liu, K.; Shan, C.X. A Molecular Engineering Strategy for Achieving Blue Phosphorescent Carbon Dots with Outstanding Efficiency Above 50%. *Adv. Mater.* **2022**, *35*, 2207970. [CrossRef]
41. Paramashivappa, R.; Kumar, P.P.; Vithayathil, P.; Rao, A.S. Novel method for isolation of major phenolic constituents from cashew (Anacardium occidentale L.) nut shell liquid. *J. Agric. Food Chem.* **2001**, *49*, 2548–2551. [CrossRef]

42. Chandrasekara, N.; Shahidi, F. Antioxidative potential of cashew phenolics in food and biological model systems as affected by roasting. *Food Chem.* **2011**, *129*, 1388–1396. [CrossRef]
43. Mathew, A.; Parpia, H. Polyphenols of cashew kernel testa. *J. Food Sci.* **1970**, *35*, 140–143. [CrossRef]
44. Edison, T.N.J.I.; Atchudan, R.; Sethuraman, M.G.; Lee, Y.R. Reductive-degradation of carcinogenic azo dyes using Anacardium occidentale testa derived silver nanoparticles. *J. Photochem. Photobiol. B Biol.* **2016**, *162*, 604–610. [CrossRef]
45. Jayaweera, S.; Yin, K.; Ng, W.J. Nitrogen-doped durian shell derived carbon dots for inner filter effect mediated sensing of tetracycline and fluorescent ink. *J. Fluoresc.* **2019**, *29*, 221–229. [CrossRef]
46. Temerov, F.; Belyaev, A.; Ankudze, B.; Pakkanen, T.T. Preparation and photoluminescence properties of graphene quantum dots by decomposition of graphene-encapsulated metal nanoparticles derived from Kraft lignin and transition metal salts. *J. Lumin.* **2019**, *206*, 403–411. [CrossRef]
47. Dager, A.; Baliyan, A.; Kurosu, S.; Maekawa, T.; Tachibana, M. Ultrafast synthesis of carbon quantum dots from fenugreek seeds using microwave plasma enhanced decomposition: Application of C-QDs to grow fluorescent protein crystals. *Sci. Rep.* **2020**, *10*, 12333. [CrossRef] [PubMed]
48. Hussain, S.; Shah, K.A.; Islam, S. Investigation of effects produced by chemical functionalization in single-walled and multi-walled carbon nanotubes using Raman spectroscopy. *Mater. Sci.-Pol.* **2013**, *31*, 276–280. [CrossRef]
49. Atchudan, R.; Muthuchamy, N.; Edison, T.N.J.I.; Perumal, S.; Vinodh, R.; Park, K.H.; Lee, Y.R. An ultrasensitive photoelectrochemical biosensor for glucose based on bio-derived nitrogen-doped carbon sheets wrapped titanium dioxide nanoparticles. *Biosens. Bioelectron.* **2019**, *126*, 160–169. [CrossRef]
50. Tammina, S.K.; Yang, D.; Koppala, S.; Cheng, C.; Yang, Y. Highly photoluminescent N, P doped carbon quantum dots as a fluorescent sensor for the detection of dopamine and temperature. *J. Photochem. Photobiol. B Biol.* **2019**, *194*, 61–70. [CrossRef]
51. Atchudan, R.; Edison, T.N.J.I.; Chakradhar, D.; Perumal, S.; Shim, J.-J.; Lee, Y.R. Facile green synthesis of nitrogen-doped carbon dots using Chionanthus retusus fruit extract and investigation of their suitability for metal ion sensing and biological applications. *Sens. Actuators B: Chem.* **2017**, *246*, 497–509. [CrossRef]
52. Souza da Costa, R.; Ferreira da Cunha, W.; Simenremis Pereira, N.; Marti Ceschin, A. An Alternative Route to Obtain Carbon Quantum Dots from Photoluminescent Materials in Peat. *Materials* **2018**, *11*, 1492. [CrossRef]
53. Campos, B.; Abellán, C.; Zougagh, M.; Jimenez-Jimenez, J.; Rodríguez-Castellón, E.; da Silva, J.E.; Ríos, A.; Algarra, M. Fluorescent chemosensor for pyridine based on N-doped carbon dots. *J. Colloid Interface Sci.* **2015**, *458*, 209–216. [CrossRef] [PubMed]
54. Wu, G.; Feng, M.; Zhan, H. Generation of nitrogen-doped photoluminescent carbonaceous nanodots via the hydrothermal treatment of fish scales for the detection of hypochlorite. *RSC Adv.* **2015**, *5*, 44636–44641. [CrossRef]
55. Gong, J.; An, X.; Yan, X. A novel rapid and green synthesis of highly luminescent carbon dots with good biocompatibility for cell imaging. *New J. Chem.* **2014**, *38*, 1376–1379. [CrossRef]
56. Wang, M.; Wan, Y.; Zhang, K.; Fu, Q.; Wang, L.; Zeng, J.; Xia, Z.; Gao, D. Green synthesis of carbon dots using the flowers of Osmanthus fragrans (*Thunb.*) Lour. as precursors: Application in Fe3+ and ascorbic acid determination and cell imaging. *Anal. Bioanal. Chem.* **2019**, *411*, 2715–2727. [CrossRef]
57. Atchudan, R.; Edison, T.N.J.I.; Perumal, S.; Vinodh, R.; Sundramoorthy, A.K.; Babu, R.S.; Lee, Y.R. Morus nigra-derived hydrophilic carbon dots for the highly selective and sensitive detection of ferric ion in aqueous media and human colon cancer cell imaging. *Colloids Surf. A Physicochem. Eng. Asp.* **2022**, *635*, 128073. [CrossRef]
58. Zhang, R.; Chen, W. Nitrogen-doped carbon quantum dots: Facile synthesis and application as a "turn-off" fluorescent probe for detection of Hg2+ ions. *Biosens. Bioelectron.* **2014**, *55*, 83–90. [CrossRef] [PubMed]
59. Hu, Q.; Gong, X.; Liu, L.; Choi, M.M. Characterization and analytical separation of fluorescent carbon nanodots. *J. Nanomater.* **2017**, *2017*, 1804178. [CrossRef]
60. Atchudan, R.; Edison, T.N.J.I.; Perumal, S.; Muthuchamy, N.; Lee, Y.R. Hydrophilic nitrogen-doped carbon dots from biowaste using dwarf banana peel for environmental and biological applications. *Fuel* **2020**, *275*, 117821. [CrossRef]
61. Siddique, A.B.; Pramanick, A.K.; Chatterjee, S.; Ray, M. Amorphous carbon dots and their remarkable ability to detect 2, 4, 6-trinitrophenol. *Sci. Rep.* **2018**, *8*, 9770. [CrossRef]
62. Wang, W.; Chen, J.; Wang, D.; Shen, Y.; Yang, L.; Zhang, T.; Ge, J. Facile synthesis of biomass waste-derived fluorescent N, S, P co-doped carbon dots for detection of Fe 3+ ions in solutions and living cells. *Anal. Methods* **2021**, *13*, 789–795. [CrossRef] [PubMed]
63. Stan, L.; Volf, I.; Stan, C.S.; Albu, C.; Coroaba, A.; Ursu, L.E.; Popa, M. Intense Blue Photo Emissive Carbon Dots Prepared through Pyrolytic Processing of Ligno-Cellulosic Wastes. *Nanomaterials* **2023**, *13*, 131. [CrossRef] [PubMed]
64. Fu, M.; Ehrat, F.; Wang, Y.; Milowska, K.Z.; Reckmeier, C.; Rogach, A.L.; Stolarczyk, J.K.; Urban, A.S.; Feldmann, J. Carbon dots: A unique fluorescent cocktail of polycyclic aromatic hydrocarbons. *Nano Lett.* **2015**, *15*, 6030–6035. [CrossRef] [PubMed]

**Disclaimer/Publisher's Note:** The statements, opinions and data contained in all publications are solely those of the individual author(s) and contributor(s) and not of MDPI and/or the editor(s). MDPI and/or the editor(s) disclaim responsibility for any injury to people or property resulting from any ideas, methods, instructions or products referred to in the content.

*Review*

# Green Carbon Dots: Applications in Development of Electrochemical Sensors, Assessment of Toxicity as Well as Anticancer Properties

Madushmita Hatimuria [1], Plabana Phukan [1], Soumabha Bag [1], Jyotirmoy Ghosh [2], Krishna Gavvala [3], Ashok Pabbathi [1,*] and Joydeep Das [4,*]

[1] Department of Industrial Chemistry, School of Physical Sciences, Mizoram University, Aizawl 796004, Mizoram, India
[2] Department of Chemistry, Banwarilal Bhalotia College, Asansol 713303, West Bengal, India
[3] Department of Chemistry, Indian Institute of Technology-Hyderabad, Hyderabad 502285, Telangana, India
[4] Department of Chemistry, School of Physical Sciences, Mizoram University, Aizawl 796004, Mizoram, India
\* Correspondence: pabbathi@mzu.edu.in (A.P.); joydeep@mzu.edu.in (J.D.)

**Abstract:** Carbon dots are one of the most promising nanomaterials which exhibit a wide range of applications in the field of bioimaging, sensing and biomedicine due to their ultra-small size, high photostability, tunable fluorescence, electrical properties, etc. However, green carbon dots synthesized from several natural and renewable sources show some additional advantages, such as favorable biocompatibility, wide sources, low cost of production and ecofriendly nature. In this review, we will provide an update on the latest research of green carbon dots regarding their applications in cancer therapy and in the development of electrochemical sensors. Besides, the toxicity assessment of carbon dots as well as the challenges and future direction of research on their anticancer and sensing applications will be discussed.

**Keywords:** green carbon dots; toxicity; electrochemical sensor; anticancer agent; biomedicine

## 1. Introduction

Carbon-based fluorescent nanoparticles, or carbon dots (CDs), are classified into three main categories, namely carbon nanodots, carbon quantum dots and graphene quantum dots [1]. These three different materials have several applications in various fields [2–4]. In 2004, while purifying single-walled carbon nanotubes (SWCNTs), Xu and colleagues discovered carbon quantum dots (CQDs). This group of carbon allotropes has a particle size of less than 10 nm [5]. Since its discovery, CQDs have attracted the attention of scientists due to their unique properties, such as favorable biocompatibility, low toxicity, large surface-to-volume ratio, stable photoluminescence and excellent hydrophilicity, along with tunable optical and electrical properties [6,7]. CQDs have been used in a wide range of applications, which includes carbon fixation, gas storage, cell biology, cancer imaging, drug administration, etc. [8]. Since its discovery, many synthetic routes for the preparation of CQDs have been established, including green means. Although the green approach to prepare CQDs is advantageous, the as-prepared CQDs offers limited practical applications due to its low fluorescence intensity and single emission wavelength. These shortcomings of green CQDs have been improved later upon incorporating special strategies that enhance the intensity and multicolor emission of CQDs [9].

Recently, green carbon dots (GCDs) have gained immense attention, owing to the availability of plenty of natural resources, excellent photophysical properties of the as-prepared GQDs and several other advantages over the CQDs (Table 1). The natural precursors for GCDs include several plant parts, fruits, amino acids, etc. [10–12]. GCDs are more biocompatible, which makes them highly suitable for a wide variety of biological

applications, including drug delivery, antibacterial and anticancer agents [13–15]. GCDs possess large surface areas due to their small size, which facilitates their applications in sensor development [15,16]. One of the advantages of GCDs is their tunable luminescence intensity and surface area by size control. In addition, the reaction conditions, precursors and surface modifications can also play crucial roles for tuning the properties of GCDs.

**Table 1.** Advantages of GCDs over CQDs.

| Applications and Advantages | Green Carbon Dots (GCDs) | Carbon Quantum Dots (CQDs) |
|---|---|---|
| Precursors availability | High | Low |
| Preparation cost | Low | High |
| Aqueous solubility | Generally high | Generally low |
| Biocompatibility/therapeutic applications | High | Low |
| Requirement of additional surface passivation/Doping | Not/Less required | Highly required |
| Biodegradability | Generally high | Generally low |

GCDs have tremendous potential applications in biomedicine and bioimaging [13]. For advancements in biological applications, it is important to test the biocompatibility of GCDs. Several in vitro studies have been carried out to test the cytotoxicity of GCDs and found that GCDs have shown low toxicity on different cell types [17–19]. However, studies related to GCDs toxicity evaluation in vivo are limited [20,21]. For example, Atchudan et al. [22] evaluated the toxicity of GCDs on *C. elegans* and found low toxicity. GCDs are also reported to be potential anticancer agents [19,23] and are used for bioimaging [21] applications. Although most of the GCDs are highly biocompatible and exhibit low cytotoxicity toward cancer cells, they can be used as an effective drug delivery carrier for various cancerous cells due to their small size, which enable them to be absorbed by the cancer cells easily. Both hydrophilic and hydrophobic anticancer drugs can be loaded on the GCD surface via electrostatic, covalent, hydrophobic or pi-pi stacking interaction, and the drug-loaded GCDs show sustained release pattern of the drug under mildly acidic conditions [24]. Besides, GCDs can also be used for the delivery of hydrophobic photosensitizer molecules and other near infrared active substances into cancer cells for photothermal or photodynamic therapy [25].

Fluorescent GCDs can be employed in the sensing of metal ions and organic compounds [15]. Very recently, GCDs are emerging as an excellent probe in electrochemical sensing applications due to their excellent electronic properties, excellent conductivity, low cost of production, and facile surface modification [15]. The electrical conductivity and stability of GCDs can be tuned by changing precursors or synthesis conditions to obtain desired GCDs for electrochemical sensing applications. GCDs offer high active areas and many hydrophilic functional groups, which are important for electrochemical sensing [4,16]. Because of these attractive properties, carbon dots are widely used in sensing applications as modifiers for the fabrication of electrode materials which can improve the rate of electron transferring process [26–29].

However, there are not many reviews focusing on the toxicity and electrochemical sensing applications of GCDs. Herein, first, we briefly discussed a few of the recent studies regarding the green synthesis and optical properties of GCDs. Thereafter, we summarized the recent literature related to the toxicity of GCDs, anticancer activity and electrochemical sensing applications. We further pointed out the challenges and opportunities for GCDs in toxicity assessment, anticancer and electrochemical applications.

## 2. Green Synthesis Methods and Optical Properties of GCDs

The green synthetic method of carbon dots, also referred as GCD in this review, is carried out by using renewable precursor and solvents that are nontoxic and environmentally benign [14]. Either top-down or bottom-up approaches are mostly adopted for the synthesis of GCDs; however, each method offers unique advantages and challenges. We have reviewed recent bottom-up synthetic methods, such as, microwave pyrolysis, hydrothermal (solvothermal process) and electrochemical etching techniques here. GCDs synthetic routes follow three common steps: (i) high temperature pyrolysis of the carbon sources, which results in (ii) carbonization and nucleation, followed by (iii) surface passivation using stable surfactants. The microwave pyrolysis method is selected for the rapid carbon dot synthesis. On the other hand, the hydrothermal (or solvothermal) method produces isotropic (typically below 10 nm diameter) carbon dots which also offers strong emission properties with high quantum yield (QY). The electrochemical etching-based method to prepare carbon dots is simple and convenient to operate; therefore, it can easily be carried out in the laboratory condition. By using these methods, the sizes of GCDs and its luminescent (PL) performance are easily tailored [8,16,30–32].

Each of the green precursor needed to meet certain requirements to be used in these techniques for the synthesis of desired carbon dots. For instance, the carbohydrates with C:H:O as 1:2:1 and which are easy to dehydrate were selected as carbon dot precursor for hydrothermal synthesis [31]. Therefore, carbohydrates sourced from leaves, roots or flowers, fruits and seeds from plants or plant biomass, such as bark, shells, kernels, peels, etc., were used as green carbon dot source [14].

Huo et al. reported a hydrothermal method for the preparation of GCDs from grapefruits. They prepared three types of GCDs: undoped green carbon dots (UGCDs), UGCDs-peel and nitrogen-doped green carbon dots (N-GCDs) [9]. The three types of GCDs showed size-dependent absorption and emission properties plausibly due to their difference in surface energy states. As the size increases in the UGCDs-peel, it exhibited a red-shifted PL emission compared to others [9]. Size dependency was also reported by Ahmadian Fard Fini et al. in a separate hydrothermal method [33] as shown in the scanning electron microscopy (SEM) image in Figure 1 [33].

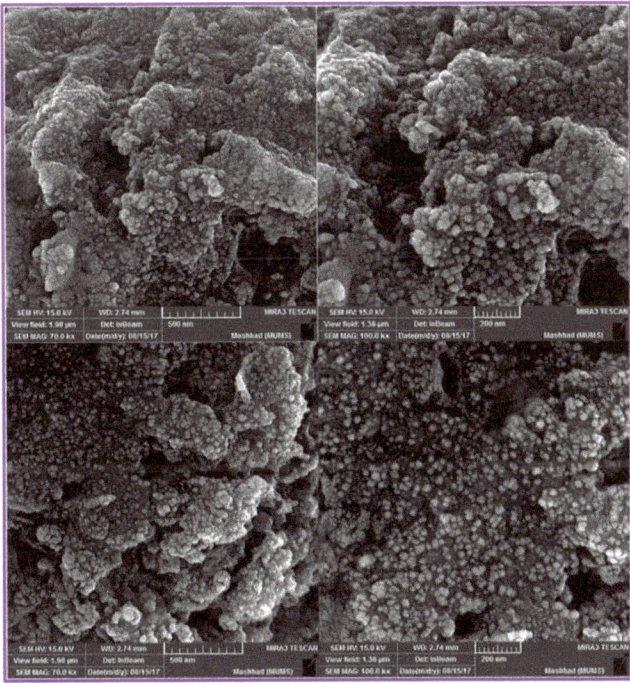

**Figure 1.** SEM images of carbon dots with different sizes. Reproduced with permission from ref. [33].

Zhu et al. decomposed alkali lignin (AL) to GCD through hydrothermal method. As part of the process, the AL were hydrolyzed in presence of mild organic acids (e.g., 4-aminobenzoic acid, benzenesulphonic acid, 4-aminobenzenesulphonic acid, 2,4-diaminobenzenesulphonic acid) before the hydrothermal treatment. These GCDs showed stability in a wide pH range (3–11). They also exhibited yellowish green fluorescence below pH 7 due to the protonation, and green due to deprotonation in basic medium [7]. The emission property also changed with respect to the sizes of the GCDs as shown in Figure 2.

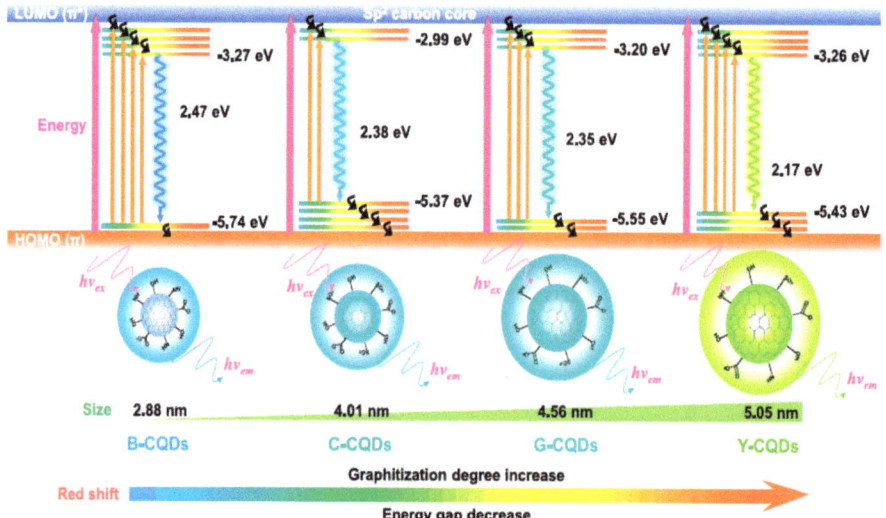

**Figure 2.** Classification of GCD based on their size and emission properties. While smaller GCDs emit blue color, the red shifting of emission occurs as the GCD size increases. Reproduced with permission from Zhu et al. [7].

Zheng et al. used 2,7-dihydroxynaphthalene as the carbon source for the synthesis of GCDs with high product yield through an effective one-step solvothermal method. With a particle size close to 3.31 nm (Table 2), these GCDs offer high quantum yield (QY) [34]. From the results of their study, Zheng et al. found that the red-green-blue (RGB) spectral composition was 93.86% for the GCDs [34].

**Table 2.** GCDs quantum yield (QY) values under different reaction conditions [34].

| V(EDA) (mL) | QY% | GCDs Size | Reaction Duration (h) | Reaction Temperature (°C) |
|---|---|---|---|---|
| 0 | 5.29 | 3.31 nm | 12 | 180 |
| 2 | 41.07 | 3.31 nm | 12 | 180 |
| 4 | 62.98 | 3.31 nm | 12 | 180 |
| 8 | 36.22 | 3.31 nm | 12 | 180 |
| 4 | 15.07 | 3.31 nm | 12 | 160 |
| 4 | 62.98 | 3.31 nm | 12 | 180 |
| 4 | 43.05 | 3.31 nm | 12 | 200 |
| 4 | 25.42 | 3.31 nm | 10 | 180 |
| 4 | 62.98 | 3.31 nm | 12 | 180 |
| 4 | 24.02 | 3.31 nm | 14 | 180 |

Hoan et al. reported hydrothermally prepared GCDs from lemon juice (at 120 to 280 °C for 12 h) which were amorphous in nature. Dynamic light scattering measurement confirmed the ~50 nm size of the GCDs. The zeta potential value of the GCDs was ~9.48 mV, confirming that they are relatively stable. Under UV irradiation, the GCD dispersion emits green color. The fluorescence intensity is only influenced by the hydrothermal reaction times (3, 6, 9 and 12 h) as emission increases with the increase in reaction time. Size distribution, band gaps and the excitation-emission wavelength of the as-prepared GCDs are impacted by the reaction temperature as shown in Table 3. Temperature-dependent aggregation and the absence of functional groups contribute toward the properties of the GCDs [31]. Mathew et al. prepared GCDs from *Simarouba glauca* (SG) leaves through hydrothermal method, where the SG leaves are of carbon source. The prepared GCDs showed selective electrochemical sensing properties toward doxycycline [35].

**Table 3.** Quantum yield values of hydrothermally prepared GCDs under different temperature conditions [31].

| $V_{(Lemon\ Juice)}$ (mL) | Reaction Temperature (°C) | Reaction Duration (h) | QY% |
|---|---|---|---|
| 40 | 150 | 12 | 14.8 |
| 40 | 200 | 12 | 16.87 |
| 40 | 240 | 12 | 21.37 |
| 40 | 280 | 12 | 24.89 |

In another synthesis of GCDs, Asghar et al. reported the preparation of 2–7 nm GCDs from honey through microwave digestion. As-prepared GCDs showed D (presence of $sp^3$ defects) and G (in-plane stretching vibration of $sp^2$ carbon atoms) band in Raman spectrum [36]. Zhao et al. studied blue-emitting GCDs (prepared through hydrothermal method) where they used biomass water hyacinth as a carbon source [37].

Visheratina et al. used *D*- and *L*-cysteine for the preparation of chiral GCDs by hydrothermal carbonization. At the identical hydrothermal condition (150 °C for 4 and 20 h), *L*-GCDs size ranges from 4.4 ± 0.5 to 5.3 ± 0.3 nm, which are all larger in diameter than *D*-GCDs [38].

Yen et al. selected a simple three-electrode (graphite-coated rod, Ag/AgCl reference electrode and platinum (Pt) wire as the counter electrode) electrochemical method to make high-quality GCDs in pure water electrolyte (i.e., in the absence of acids and bases). With the use of this facile electrochemical fabrication, smaller than 5 nm GCD-water suspension could be prepared in a single step, without separate workup, such as filtering, dialysis, centrifugation, column chromatography, or gel-electrophoresis [39]. Summary of the above-mentioned reports are shown in Table 4.

Hydrothermal technique is therefore the most widely used method for the preparation of GCDs because of its easiness, low cost, high scalability and eco-friendly nature. However, in order to bring GCDs into commercial applications, more study should be conducted on the search for high-quality natural precursors for the synthesis of GCDs with improved chemical and photo-stability, which will allow GCDs to become an excellent alternative to existing quantum dots or other dyes for bioimaging applications.

Table 4. Size and optical properties of the hydrothermally (solvothermally) prepared GCDs are summarized.

| Author | Treatment | GCDs Source | Size (nm) | Excitation Wavelength (nm) | Emission Wavelength (nm) | Ref. |
|---|---|---|---|---|---|---|
| Huo et al. | Hydrothermal | Natural grapefruit | 4.74–8.20 | 320–360 | 411–420 | [9] |
| Zhu et al. | Hydrolysis followed by hydrothermal | Alkali lignin | 2.88–5.05 | 450 | 520 | [7] |
| Zheng et al. | Solvothermal | 2,7-dihydroxynaphthalene | 3.31 | 460 | 513 | [34] |
| Hoan et al. | Hydrothermal | Lemon juice | 3–5 | 410–480 | 500–550 | [31] |
| Mathew et al. | Hydrothermal | Simarouba glauca leaves | 2.64 | 365 | 445 | [35] |
| Asghar et al. | Microwave | Honey | 2–7 | - | - | [36] |
| Visheratina et al. | Hydrothermal | L-cysteine  D-cysteine | 4.4 and 5.3  4.4 and 5.3 | 350 | ~430 | [38] |
| Yen et al. | Electrochemical | Graphite-coated rod | 0.5–4 | 365 | 500 | [39] |

### 3. Electrochemical Sensing Ability of GCDs

In the case of GCDs, the smaller the particle size, the more the surface-to-area ration increases, which in turn enhances the contact area in the electrode. This observation makes GCD a good candidate for electrochemical sensing. To study the electrochemical behavior of GCDs, Borna et al. modified the glassy carbon electrode (GCE) with CQDs and used cyclic voltammetry (CV) as well as linear sweep voltammetry (LSV) to detect the anticancer drug, letrozole. In CV, the GCE modified with GCDs exhibit the highest intensity of the anodic and cathodic peak current, plausibly due to the small size of GCDs. The enhanced surface area increased the contact area of the modified electrode with the analyte, and boosted the current intensity. These findings also demonstrate the function of GCDs as an electrocatalyst. In the presence of letrozole, a significant current increase was seen in the modified electrode with electrochemically synthesized GCDs at a current intensity of 100 mA. As a result, the GCE modified with GCDs was chosen as the best electrode for use in the construction of sensors and drug analysis (letrozole analysis) as shown in Table 5 [6].

Table 5. Detection limit of different sensors [6].

| Sensor | Material | Detection Limit (M) | Sensitivity (A/M) | GCDs Synthesized Method | Required Current for GCDs Synthesis | GCDs Size Range |
|---|---|---|---|---|---|---|
| GCE/GCDs | Letrozole | $1.85 \times 10^{-5}$ | 0.111 | Electrochemical | 100 mA | 1–10 nm |
| GCE/GCDs | Clomifene | $70 \times 10^{-5}$ | 0.041 | Electrochemical | 100 mA | 1–10 nm |
| GCE/GCDs | Letrozole | $4.23 \times 10^{-5}$ | 0.076 | Electrochemical | 200 mA | 1–10 nm |
| GCE/GCDs | Clomifene | $85 \times 10^{-5}$ | 0.033 | Electrochemical | 200 mA | 1–10 nm |
| GCE/GCDs | Letrozole | $5.15 \times 10^{-5}$ | 0.067 | Electrochemical | 300 mA | 1–10 nm |
| GCE/GCDs | Clomifene | $90 \times 10^{-5}$ | 0.028 | Electrochemical | 300 mA | 1–10 nm |
| GCE/GCDs | Letrozole | $4.27 \times 10^{-5}$ | 0.069 | Hydrothermal | - | 1–10 nm |
| GCE/GCDs | Clomifene | $87 \times 10^{-5}$ | 0.031 | Hydrothermal | - | 1–10 nm |

Zhou et al. reported an increase in oxidation efficiency of ascorbic acid (AA) using $NH_2$-GCDs-modified GCE. The $NH_2$-GCDs could increase the conductivity of the electrode surface, and the positively charged amine groups allow them to interact with the dienol hydroxyl groups in AA through electrostatic contact, thereby detecting AA with a high degree of specificity. They found that, compared to many other previously reported quantum dot materials, the $NH_2$-GCDs had a superior effect on detecting AA in both electrochemistry and fluorescence approaches. It is due to this reason that the $NH_2$-GCDs is able to boost the rate of electron transfer via direct connection with dienol hydroxyl groups in AA. Additionally, $NH_2$-GCDs can also be joined to AA by hydrogen bonds [40].

Ran et al. used WP6-N-GCD (water soluble pillar [6] arenes nitrogen-doped green carbon dot) nanocomposite as an electrode material to build a sensitive electrochemical sensing platform for trinitrotoluene (TNT) detection. The electrochemical analysis demonstrated that WP6-N-GCDs outperformed the β-CD-N-GCDs in terms of supramolecular recognition and enrichment capabilities, and displayed a higher peak current toward TNT. The WP6-N-GCD-modified electrode showed a linear response ranging from 0.001 μM to 1 μM and form 1 μM to 20 μM with a LOD of 0.95 nM due to the synergistic effects of WP6 and N-GCDs. These results indicate that WP-N-GCD composites are ideal materials for electrochemical sensing platforms as shown in Table 6 [41].

**Table 6.** Comparison of some electrode materials for electrochemical sensing of TNT [41].

| Materials | Linear Range (μM) | LOD (nM) |
|---|---|---|
| Ionic liquid-graphene | 0.13–6.6 | 17.6 |
| Boron-doped diamond | 0.088–1.76 | 44 |
| Ordered mesoporous carbon | - | 0.88 |
| Nitrogen-doped graphene | 0.53–8.8 | 129.9 |
| Deposited graphene | 0.0044–0.88 | 0.88 |
| N-rich carbon nanodots | 5–30 | 1 |
| PtPd-rGONRs | 0.044–13.2 | 3.5 |
| Vanadium dioxide | 0.44–4.4 | 4.4 |
| N-doped graphene nanodots | 0.0044–1.76 | 0.88 |
| WP6-N-GCDs | 0.001–1; 1–20 | 0.95 |

GCDs have also been used to determine hydrogen peroxide ($H_2O_2$), a common industrial oxidant electrochemically. Hassanvand et al. [16] demonstrated the use of GCE modified with GCDs as an amperometric $H_2O_2$ sensor. When compared to octahedral $Cu_2O$, GCDs/octahedral $Cu_2O$ exhibited more favorable electrocatalysis behavior for the glucose oxidation and $H_2O_2$ reduction reactions, as shown in Table 7 [16].

**Table 7.** $H_2O_2$ and glucose detection by GCDs-based electrochemical sensors [16].

| Modified Electrode | Target Compounds | Detection Limit | GCD Size | Electrochemical Method |
|---|---|---|---|---|
| GCDs/GCE | $H_2O_2$ | $3 \times 10^{-9}$ M | - | Amperometry |
| GCDs/$Cu_2O$/NF/GCE | $H_2O_2$ | $2.8 \times 10^{-6}$ M | 10 nm | Amperometry |
| CuO/GCDs/CHNS/GCE | $H_2O_2$ | $2.4 \times 10^{-9}$ M | ~4–6 nm | Amperometry |
| GCDs/$Cu_2O$/GCE | Glucose | $6 \times 10^{-6}$ M | - | Amperometry |
| GCDs/AuNPs-GOx/Au | Glucose | $17 \times 10^{-6}$ M | - | Amperometry |
| GCDs/Au-NPs-Gox/GDAE | Glucose | $13.6 \times 10^{-6}$ M | - | Amperometry |

Accurate clinical dopamine (DA) diagnosis can help to address several neurological disorders. Modifying GCE with N-doped green carbon dots (NGCDs) can be used for DA detection with a wide linear range and a low detection limit, as shown in Table 8 [16].

**Table 8.** DA detection using different GCDs-based sensors [16].

| Modified Electrode | Method | Target Compound | GCDs Average Size | Detection Limit |
|---|---|---|---|---|
| GCDs/MoS$_2$/Mo foil | CV | DA | - | 0.0090 μM |
| NF/NGCDs/GCE | DPV | DA | 7.4 nm | 1.0 nM |
| GCDs/GCE | LSV | DA | 3.3 nm | 2.7 μM |
| β-CD/GCDs/GCE | DPV | DA | 7.6 nm | 0.14 μM |

Wang et al. have shown that when GCDs are mixed separately with layered-double-hydroxides (LDHs), metal sulfides, and metal phosphides, etc., they can be used as electrocatalysts for oxygen reduction reaction (ORR), oxygen evolution reaction (OER), hydrogen evolution reaction (HER), CO$_2$ reduction reaction (CO$_2$RR), etc. OER activity of CoP/GCDs composite is better with 400 mV overpotential in alkaline electrolytes than pure CoP. The CoP/GCDs composite's electrical catalytic performance is increased due to the presence of functional groups, its small size, good conductivity, and fast electron transfer of GCDs. If specific surface area is high, the presence of electrochemical active sites will be high, and contact area with the electrolyte will be large, which results in increasing HER performance [30].

Lin et al. [15] described the potential use of GCDs as electrochemical sensors as it can be synthesized through different methods, have rich functional groups, and can make composites with other materials easily. When GCE was modified with GCDs, it showed an increase in response current and potential compared to bare GCE. It also showed a greater electrochemical reaction due to the presence of many functional groups in GCDs/GCE [15]. When Mathew et al. compared the bare GCE and the developed ternary sensor, they found that the anodic peak current becomes threefold in the case of the developed ternary sensor. This shows that the developed ternary sensor was able to detect small levels of doxycycline accurately. They conducted the electrochemical sensing in phosphate buffer solution, where the potential range was 0 to 1 V for cyclic voltametric measurement, with a scan rate of 50 mV/s. The modified GCE was remarkably stable and had great repeatability and reusability [35].

From the above discussion, it is clear that GCDs are widely used for the sensitive and selective detection of a large number of biomolecules, such as ascorbic acids, hydrogen peroxide, drug molecules, as well as explosives (trinitrotoluene). However, only limited studies are reported till date. Therefore, more research should be conducted to explore the potential applications of GCDs.

## 4. Toxicity Assessment and Anticancer Properties of Green Carbon Dots (GCDs)

GCDs synthesized from various green sources have many potential biological applications. Here, we will discuss about the biocompatibility of these synthesized GCDs studied by several researchers with normal and cancerous cells.

Ensafi et al. prepared fluorescent carbon dots from a naturally sourced saffron. Cytotoxicity of saffron carbon dots was tested on olfactory mucosa cells and bone marrow cells after incubating for 24 h, and low toxicity for all the different concentrations (Table 9) of carbon dots was observed [42]. Zhang et al. prepared green carbon dots using schizonepetae herba carbonisata (SHC). Cytotoxicity of the prepared carbon dots was evaluated in RAW 264.7 cells in various concentrations (Table 9). Cell viability was negligible, but with increasing concentration from 840–10,000 μg/mL, cell viability started to decrease [10].

Table 9. Toxicity of GCDs for normal cells.

| Sl No. | Material | Source of Green Carbon Dot Synthesis | | Average Size (nm) | Toxicity Assay in Cell Line | Concentration | Remark | Reference |
|---|---|---|---|---|---|---|---|---|
| 1 | Fluorescent carbon dots | Saffron | | 6.0 | Olfactory mucosa cells and bone marrow cells | 0.005–1.5 mg/mL | Low toxicity (more than 70% cell viability remains) | [42] |
| 2 | Carbon dots | Schizonepetae herba carbonisata (SHC) | | 0.8–4.0 | RAW 264.7 cells | 39.06–10,000 µg/mL | Negligible cytotoxicity up to 840 µg/mL concentration | [10] |
| 3 | Carbon dots | Phellodendri cortex (PC) | | 1.2–4.8 | RAW 264.7 cells | 0.01–10,000 µg/mL | Negligible cytotoxicity up to 1000 µg/mL concentration | [17] |
| 4 | Enantiomeric carbon dots | L-lysine | | 4.0 | SH-SY5Y cells | 0.2 and 0.4 mg/mL | L-lysine carbon dots showed negligible cytotoxicity | [12] |
|   |   | D-lysine | |   |   |   |   |   |
| 5 | Carbon dots | Gum tragacanth (GT) | | 70–90 | Human umbilical vein endothelial cells (HUVEC cell line) | 0–50 µg/mL | Low cytotoxicity (more than 80% cell viability remains) | [18] |
|   |   | Gum tragacanth (GT) and chitosan | |   |   |   |   |   |
| 6 | Re-based carbon dots | Ginsenoside Re, citric acid and EDA | | 4.6 | Human renal epithelial cells 293T, HL-7702 (L-02), MCF-10A and NSFbs | 0–1.0 mg/mL | Low toxicity (after 24 h of incubation, cell viability was more than 90%) | [23] |
| 7 | Fluorescent carbon dots | Cinnamon | | 3.4 | Human kidney cells (HK-2) | 0.1–2.0 mg/mL | Low toxicity (more than 80% cell viability remains) | [19] |
|   |   | Red chili | | 3.1 |   | 0.1–2.0 mg/mL |   |   |
|   |   | Turmeric | | 4.3 |   | 0.1–2.0 mg/mL |   |   |
|   |   | Black pepper | | 3.5 |   | 0.1–4.0 mg/mL |   |   |
| 8 | Carbon dots | Kiwi | | 4.4 | Epithelial human kidney cells (HK-2) | 0.25–5.0 mg/mL | Low toxicity (up to 1 mg/mL concentration cell viability more than 60%) | [11] |
|   |   | Avocado | | 4.4 |   |   |   |   |
|   |   | Pear | | 4.1 |   |   |   |   |
| 9 | Fluorescent NP-carbon dots | Wet algal biomass | | 4.7 | HEK-293 (normal human embryonic kidney cell line) | 5–75 µg/mL | Negligible cytotoxicity | [43] |
| 10 | Fluorescent carbon dots | Cyanobacteria powder | | 2.5 | PC12 cells | 0–1000 µg/mL | Low cytotoxicity | [44] |
| 11 | Carbon dots | Banana peel waste | | 5 | Nematode | 0–200 µg/mL | Negligible cytotoxicity | [45] |
| 12 | Carbon dots | *Fusobacterium nucleatum* cells | | 4.1 | BEAS-2B (Lung normal epithelial cells) | 12.5–200 µg/mL | Low cytotoxicity (more than 80% cell viability remains) | [46] |
| 13 | Fluorescent carbon dots | Fresh mint leaves | | 6.5 | Primary H8 cells | 0–200 µg/mL | Negligible cytotoxicity | [47] |
| 14 | Carbon dots (CDs) | E-CD | Citric acid and EDA | 10 | EA. hy926 cells | 0.1–3.2 mg/mL | Low cytotoxicity | [48] |
|   |   | U-CD | Urea and citric acid | 5 |   |   |   |   |

From the natural source of Phellodendri cortex (PC), Liu and his coworkers prepared green carbon dots. Cytotoxicity of these green carbon dots was evaluated in RAW 264.7 cells with eight different concentrations (Table 9). After 24 h of incubation, cell viability was negligible by up to 1000 µg/mL of concentration. However, when the concentration increased to more than that amount, cytotoxicity also started to increase [17]. Malishev et al. prepared two types of green carbon dots using amino acids (Table 9) as a natural source. Cytotoxicity of these amino acid-based carbon dots was evaluated in SH-SY5Y cells. For the cytotoxicity assay, Aβ42 was induced in both types of the carbon dots. Cytotoxicity was measured for Aβ42 alone and Aβ42-induced carbon dots. Without carbon dots, in the presence of Aβ42, cell viability decreased to 25%. In the case of pre-incubated Aβ42 with D-lysine carbon dots at the above concentrations (Table 9), the results did not show much difference with Aβ42 alone. However, in the case of pre-incubated Aβ42 with L-lysine carbon dots at both the concentrations (Table 9), the results have shown significantly low cytotoxicity [12].

Two types of green carbon dots (GT carbon dots and GT/Chitosan carbon dots) were prepared by Moradi and his coworkers using natural and ecofriendly sources, gum tragacanth and chitosan. Cytotoxicity of these prepared carbon dots was tested in human umbilical vein endothelial cells (HUVEC cell line) at different concentrations (Table 9) after 24 h of incubation, and low cytotoxicity and high biocompatibility were observed [18].

Carbon dots prepared from natural sources (Table 9) were used by Yao and his coworkers to evaluate the cytotoxicity assay in human renal epithelial cells 293T, HL-7702(L-02), MCF-10A and NSFbs. After 24 h of incubation of the cell lines with various concentrations (Table 9) of prepared carbon dots, they observed low cytotoxicity for all the concentrations [23].

Vasimalai et al. prepared four different types of carbon dots from four different spices, i.e., cinnamon, red chili, turmeric and black pepper. Cytotoxicity of these prepared carbon dots were evaluated in different concentrations (Table 9) on human kidney cells (HK-2) after 24 h of incubation. For all the carbon dots, cell viability remained at more than 80% at 0–2.0 mg/mL concentration, but in the case of black pepper carbon dots at higher concentrations (3.0 and 4.0 mg/mL), results showed high toxicity for the cell line [19]. Cytotoxicity of these four carbon dots in different concentrations on HK-2 and LN229 cells is shown in the Figure 3.

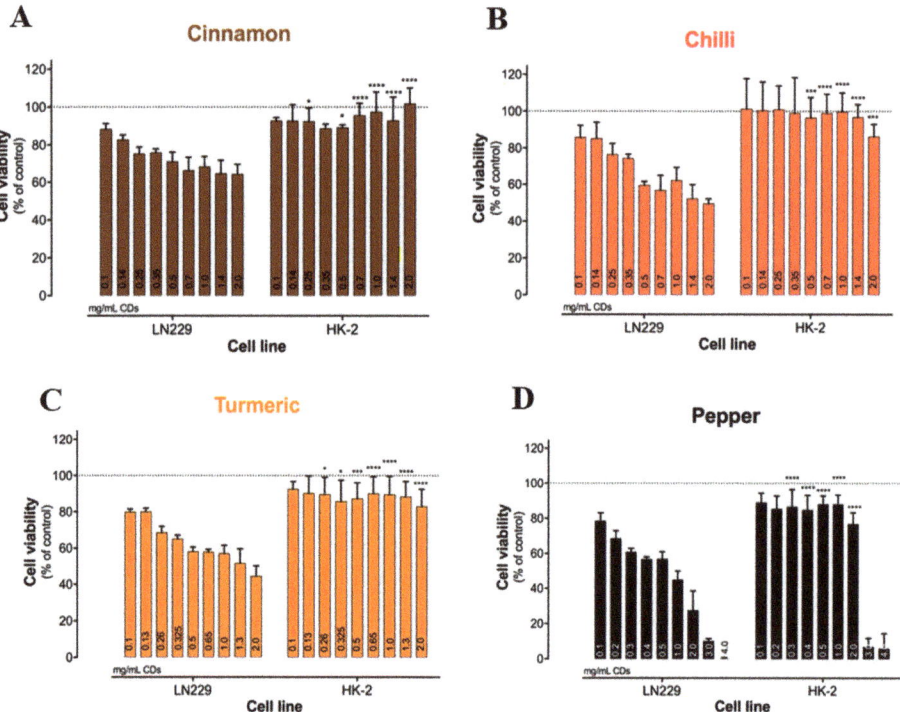

**Figure 3.** Cell viability evaluation in two different types of cells (cancerous LN229 cells and normal HK-2 cells) after a 24 h incubation with increasing concentrations of: (**A**) cinnamon, (**B**) red chili, (**C**) turmeric and (**D**) black pepper carbon dots. Represented results are mean ± SD of at least three independent experiments. * $p < 0.05$, *** $p < 0.001$ and **** $p < 0.0001$ are for statistical significance of differences between the effect of the same concentration of spice-based carbon dots in other cell line (LN229). Adapted from ref. [19].

Fruit-based carbon dots were prepared by Dias and his team from kiwi, avocado and pear. For the cytotoxicity evaluation at different concentrations (Table 9) of the prepared carbon dots, epithelial human kidney cell (HK-2) line was used. Cells were incubated for two different time periods (48 and 72 h) with different concentrations of carbon dots. After

72 h of incubation, cell viability was more than that measured at 48 h. When concentration is more than 1 mg/mL, cell viability started to decrease. Overall, pear-based carbon dots showed more cell viability and kiwi-based carbon dots showed less cell viability [11]. Wet algal biomass source was used by Singh and his coworker to prepare green carbon dots. HEK-293 (normal human embryonic kidney) cell line was used for the cytotoxicity evaluation of these carbon dots. After 24 h of incubation in different concentrations of carbon dots (Table 9), more than 80% of cell viability was observed, even in the highest concentration (Table 9) [43]. Wang et al. evaluated the cytotoxicity of cyanobacteria-based carbon dots in PC12 cells in different concentrations (Table 9), and 94.4% of cell viability was observed, up to 100 µg/mL concentration. Even at a high concentration (500 µg/mL), cell viability was observed to be more than 80%, which concludes low cytotoxicity and biocompatibility [44].

From the banana peel waste, Atchudan and his team prepared green carbon dots, and for the cytotoxicity evaluation in different concentrations (Table 9), live-cell line (nematode) was used. From their observations, overall, only 5% cell viability was decreased even at the highest concentration, thereby indicating negligible cytotoxicity with excellent biocompatibility [45]. Liu et al. prepared carbon dots from *Fusobacterium nucleatum* cells, and a cytotoxicity test was evaluated in BEAS-2B (lung normal epithelial cells) cell line. After incubation for 48 h at different carbon dots concentrations (Table 9), low cytotoxicity was observed for all concentrations [46]. From fresh mint leaves, Raveendran and Kizhakayil prepared green carbon dots and performed a cytotoxicity assay with different concentrations (Table 9) in primary H8 cells. For all the concentrations, no significant change was observed in cell viability [47].

Cytotoxicity of curcumin-loaded green carbon dots (Curc-GCDs) and prepared GCDs were studied in different concentrations (Table 9) by Arvapalli et al. [48] in EA. hy926 cells. They prepared two different types of carbon dots (Table 9) and then loaded the carbon dots with curcumin. Their observations showed very low cytotoxicity for both Curc-GCDs and GCDs at EA. hy926 cells, even at the highest concentration (Table 9) after 24 h of incubation [48].

Most of the GCDs mentioned in the list had shown excellent biocompatibility in the normal cells. Therefore, all the green sources used in Table 9 has great potential toward GCDs preparation. However, most of the studies used low concentrations for the cytotoxicity evaluation. Thus, future studies should focus on toxicity evaluation using a wide range of concentrations.

Most of the prepared GCDs mentioned in Table 10 showed low toxicity and excellent biocompatibility in all the cancerous cells. Chatzimitakos et al. [49] prepared GCDs from two different green sources (Table 10) and studied the cytotoxicity of prepared GCDs in various cancerous cells (Table 10). Results showed that after 48 h of incubation, cell viability for citrus sinensis CD and citrus limon CD is more than 90% and 95%, respectively [49]. GCDs prepared from green sources (water chestnut and onion) were used by Hu and his team to evaluate the cytotoxicity in the cancerous cell (Table 10) and observed low cytotoxicity in all the concentrations (Table 10) [50].

Among all the discussed literature in Table 10, only limited GCDs showed cytotoxicity toward some selected cancer cells and are therefore useful for cancer therapy. Yao et al. prepared carbon dots from natural sources (Ginsenoside Re, citric acid and EDA) to evaluate their toxicity in cancer cell lines, i.e., HepG2, MCF-7 and A375 cells and normal cells (Table 9). After 24 h of incubation with various concentrations (Table 10) of prepared carbon dots, they observed that with increasing concentrations, cell viability decreased to 50%, 40% and 30% for HepG2, MCF-7 and A375 cells, respectively. A375 cells showed the highest inhibition than other cells. Further LDH-release assay and ROS generation was checked in A375 cells. From the LDH-release assay, they observed a concentration-dependent LDH release in the medium. When compared to the normal cell, ROS generation was comparatively high in cancer cells, thereby causing oxidative damage and apoptosis in the cancer cells. Therefore, these carbon dots can be considered as potential anticancer agents [23].

Table 10. Toxicity assay of GCDs for cancerous cells.

| Sl No. | Material | Source of Green Carbon Dot Synthesis | Average Size (nm) | Toxicity Assay in Cell Line | Concentration | Remark | Reference |
|---|---|---|---|---|---|---|---|
| 1 | Luminescent carbon dots | Citrus sinensis | 6.5 | HeLa, A549, MDA-MB-231 and HEK-293 cells | 400 µg/mL | Extremely low cytotoxicity | [49] |
| | | Citrus limon peels | 4.5 | | | | |
| 2 | S and N co-doped carbon dots | Water chestnut and onion | 3.5 | Human bladder cancer T24 cells | 0–300 µg/mL | Low cytotoxicity (after 24 h of incubation, cell viability remained at more than 80% for all the concentrations) | [50] |
| 3 | Nitrogen- and sulfur-co-doped carbon dots | Ginkgo leaves juice | 2.2 | HeLa cells | N/A | Low cytotoxicity | [51] |
| 4 | NIR-light emission carbon dots | Fresh spinach | 3–11 | A549 cells | 0–200 µg/mL | Negligible toxicity (after 24 h of co-incubation, cell viability showed above 94.2% at all the concentrations) | [52] |
| 5 | Multicolor luminescent carbon dots | ATP | 3.8 | HeLa cells | 0–500 µg/mL | Negligible toxicity (very less change was observed between 24 h and 48 h incubation) | [53] |
| 6 | N-doped carbon dots | Sucrose and urea | 1.6 | HeLa cells | 0–1.0 mg/mL | Negligible cytotoxicity (cell viability was more than 98.5% after 24 h of incubation, even at a high concentration, i.e., 1.0 mg/mL) | [54] |
| 7 | Hydrophilic carbon dots | Glucose powder | 2.6 | HeLa cells | 0–1.0 mg/mL | Negligible cytotoxicity (cell viability was more than 98% after 24 h of incubation, even at a high concentration, i.e., 1.0 mg/mL) | [55] |
| 8 | Carbon dots | Date kernels | 2.5 | Human MG-63 cells | 200.0 µg/mL | Low cytotoxicity (after 48 h of incubation, cell viability remains more than 85%) | [56] |
| 9 | Carbon dots | Quince fruit (*Cydonia oblonga*) powder | 4.9 | HT-29 cells | 5–1000 µg/mL | Low toxicity | [57] |
| 10 | Re-based carbon dots | Ginsenoside Re, citric acid and EDA | 4.6 | MCF-7, A375 HepG2 cells | 0–1.0 mg/mL | High cytotoxicity * While carbon dots were prepared separately from Ginsenoside; citric acid and EDA, cytotoxicity was relatively low. | [23] |
| 11 | Fluorescent carbon dots | Cinnamon | 3.4 | Human glioblastoma cells (LN-229 cancer cell line) | 0.1–2.0 mg/mL | High toxicity | [19] |
| | | Red chili | 3.1 | | 0.1–2.0 mg/mL | | |
| | | Turmeric | 4.3 | | 0.1–2.0 mg/mL | | |
| | | Black pepper | 3.5 | | 0.1–4.0 mg/mL | | |
| 12 | Carbon dots | Kiwi | 4.4 | Epithelial human colorectal adenocarcinoma cells (Caco-2) | 0.25–5.0 mg/mL | Low toxicity (below 1.5 mg/mL concentration, cell viability was more than 80%, but cell death can induce in higher concentrations) | [11] |
| | | Avocado | 4.4 | | | | |
| | | Pear | 4.1 | | | | |
| 13 | Fluorescent carbon dots | *Prunus cerasifera* fruits juice | 3–5 | HepG2 cells | 0–500 µg/mL | Low toxicity (after 24 h of incubation below 500 µg/mL concentration, cell viability was more than 90%) | [58] |
| 14 | Carbon dots | Celery leaves | 2.1 | HepG2 cells | 0.01–0.022 g/mL | Low toxicity (cell viability was more than 85% for all the concentration after 24 h of incubation) | [59] |
| 15 | Carbon dots | Lychee waste | 3.1 | A375 (Skin melanoma) cells | 0.0–1.2 mg/mL | Low cytotoxicity (after 48 h of incubation, cell viability was more than 89% for the highest concentration, i.e., 1.2 mg/mL) | [60] |

Table 10. Cont.

| Sl No. | Material | Source of Green Carbon Dot Synthesis | Average Size (nm) | Toxicity Assay in Cell Line | Concentration | Remark | Reference |
|---|---|---|---|---|---|---|---|
| 16 | Fluorescent-N-doped carbon dots | Lemon juice and ethylenediamine | 3.0 | Human breast adenocarcinoma (MCF7) cells | 0.312–2.0 mg/mL | Low cytotoxicity (after 24 h of incubation cell viability for 2.0 mg/mL, the highest concentration was more than 88%) | [61] |
| 17 | Fluorescent carbon dots | Fresh lamb | At 200 °C = 2.8<br>At 300 °C = 1.9<br>At 350 °C = 1.7 | HepG2 cells | 2.0 mg/mL | Low cytotoxicity (after 4 h of incubation cell, viability was more than 90% at this particular concentration) | [62] |
| 18 | Carbon dots | *Osmanthus fragrans* flowers | 2.2 | A549 cells | 25–1000 µg/mL | Negligible toxicity (after 24 h of incubation, cell viability showed above 90% at all concentrations) | [63] |
| 19 | Carbon dots | Dried wheat straw | 2.1 | HeLa cells | 0–0.8 mg/mL | Negligible cytotoxicity (cell viability remains more than 90% at all concentrations) | [64] |
| 20 | Carbon dots | Gelatin and papain | 3.8 | A549 cells | 0–300 µg/mL | Negligible cytotoxicity (very less difference between 12 h and 24 h incubation for all concentrations, after 24 h incubation, cell viability remained above 91%) | [65] |
| 21 | Carbon dots | Glucose | 3.0 | HeLa, HepG2 and HEK-293 cells | 0–300 mg/L | Negligible cytotoxicity (no change in cell viability for HeLa and HepG2 cells, but in the case of HEK-293 cell with increasing concentration, cell viability also increases) | [66] |
| 22 | N, B co-doped bright fluorescent carbon dots | *Solanum betaceum* (S. betaceum) fruit extract | 5.0 | HeLa cells | 10–50 µg/mL | Low cytotoxicity (at minimum and maximum concentration, i.e., 10 µg/mL and 50 µg/mL, cell viability were 100% and 70%, respectively) | [67] |
| 23 | Carbon dots | Gelatin | 5.0 | MCF-7 cell line | 20–120 µg/mL | Low cytotoxicity (cell viability was more than 80% even for the highest concentration after 24 h of incubation) | [68] |
| 24 | Zwitterionic carbon dots | Citric acid and L-histidine | 8.5 | A549 cells | 0.01–1.5 mg/mL | Low cytotoxicity (after 24 h of incubation, cell viability was more than 90% even at a high concentration) | [69] |
| 25 | Carbon dots | Corn stalk shell | 1.2–3.2 | A549 cells | 0–100 mg/L | Low cytotoxicity (after incubation for 24 h, cell viability remained more than 90% for all concentrations. Again, after 48 h of incubation, cell viability remained more than 75% for 100 mg/L concentration) | [70] |
| 26 | Carbon dots | *Fusobacterium nucleatum* cells | 4.1 | HeLa cells | 12.5–200 µg/mL | Low cytotoxicity (more than 80% cell viability remains) | [46] |
| 27 | Nitrogen-doped carbon dots | Jackfruit peel (JFP) | 6.4 | Dalton's lymphoma ascites cells (DLA) | 50–200 µg/mL | Low cytotoxicity only in low concentrations, i.e., below 50 µg/mL (at 200 µg/mL concentration for JFP-carbon dots 100%, cell death was observed, but in the case of TP-carbon dots, only 60% cell death happened) | [71] |
| | | Tamarind peel (TP) | 5.3 | | | | |
| 28 | Carbon dots | Arginine, chitosan, citric acid | 6–11 | AGS cells | 30:1–70:1 (carrier/DNA) | Negligible toxicity (at highest weight, cell viability decreased to 90%) | [72] |
| 29 | Folic acid-functionalized carbon dots | Red Korean ginseng | 70 | MCF-7 cells | 10–50 µg/mL | High toxicity after 48 h of incubation | [73] |

**Table 10.** *Cont.*

| Sl No. | Material | Source of Green Carbon Dot Synthesis | | Average Size (nm) | Toxicity Assay in Cell Line | Concentration | Remark | Reference |
|---|---|---|---|---|---|---|---|---|
| 30 | Carbon dots | Tea leaves | | 200 | HepG2 cells | 0–160 µg/mL | Low cytotoxicity (more 90% cell viability after 24 h of incubation at all concentrations) | [74] |
| 31 | Carbon dots | *Simarouba glauca* leaf | | 2.6 | Human breast cancer cell line (MCF-7) | 0–100 µg/mL | High toxicity with increasing concentration | [35] |
| 32 | Carbon dots | Walnut oil | | 12.3 | PC3, MCF-7, and HT-29 cells | 0–10 µg/mL | High cytotoxicity after 24 h of incubation | [75] |
| 33 | Carbon dots (CDs) | E-CDs | Citric acid and EDA | 10.0 | HepG2 and A549 cells | 0.1–3.2 mg/mL | High cellular toxicity with increasing concentration | [48] |
|  |  | U-CDs | Urea and citric acid | 5.0 |  |  |  |  |

Four types of natural carbon dots were prepared by Vasimalai and his team using four different spices (Table 10). Toxicity assay was performed after 24 h of incubation in human glioblastoma cells (LN-229 cancer cell line) with various concentrations (Table 10). A significant decrease was observed in cell viability for all the concentrations of carbon dots (Figure 3). At 2.0 mg/mL concentration, 35%, 50% and 75% reduction in cell viability was observed for cinnamon, red chili and turmeric, and black pepper-derived carbon dots, respectively. In the case of black pepper-derived carbon dots, cell viability reduced to 100% at the highest concentration (Table 10) [19]. Tejwan et al. studied the anticancer property of rutin drug loaded with folic acid-functionalized carbon dots (FA-CDs-RUT) extracted from ginseng root in MCF-7 cells. After 48 h of incubation in different concentrations (Table 10) of FA-CDs-RUT, CDs and RUT were compared with each other for anticancer property. Moreover, results showed that in the presence of rutin drug in the carbon dots (FA-CDs-RUT), anticancer property was found to be more in MCF-7 cells than in CDs and RUT drug alone. Furthermore, they observed that intracellular ROS generation was increased with subsequent cellular apoptosis when the cells were treated with FA-CDs-RUT nanohybrid [73].

Mathew et al. [35] studied the anticancer activity of carbon dots prepared from naturally occurring *Simarouba glauca* (*S. glauca*) leaf. Cell viability of prepared GCDs was tested with different concentrations (Table 10), and cellular morphologies were also checked in human breast cancer cell line (MCF-7). They observed that cell viability decreased with an increasing concentration of the GCDs. After 24 h of incubation at the highest concentration (Table 10), complete cell death was observed. Arkan et al. [75] studied the anticancer property of prepared GCDs (Table 10) in three different cancerous cells (PC3, MCF-7 and HT-29). After 24 h of incubation in various concentrations (Table 6), 50% of cell death was observed for PC3 and MCF-7 at 1.25 µg/mL and 5 µg/mL concentration, respectively, indicating that walnut oil-based carbon dots have anticancer property. However, there was no cell death observed in the case of HT-29 cell lines. At the highest concentration, cell viability was also observed to be more than 70% for HT-29 cell lines.

Arvapalli et al. [48] prepared two different types of GCDs (E-CDs and U-CDs) loaded with curcumin and then compared the GCDs with curcumin-loaded carbon dots (Curc-GCDs) to test the cell viability in HepG2 and A549 cells. Incubation was performed for 24 h in different concentrations (Table 10) of GCDs and Curc-GCDs. They found that, compared to synthesized GCDs, Curc-GCDs showed more toxicity to the cells. In HepG2 cells, cell viability was decreased to 40% and 30% at the highest concentration (Table 10) when the cells were treated with Curc-E-GCDs and Curc-U-GCDs, respectively. In A549 cell, cell viability was decreased even more than HepG2 at the same concentration to 38% and 18% when treated with Curc-E-GCDs and Curc-U-GCDs, respectively. In both cases,

Curc-U-GCDs showed higher toxicity to the cells. Furthermore, when the cell viability test was extended up to 48 h and 72 h for A549 cells at the same concentration, they observed that, in the presence of Curc-E-GCDs, cell viability decreased to 30% and 25%, and in the presence of Curc-U-GCDs, cell viability decreased to 5% and 2%, respectively [48].

From the above discussion, we can propose that GCDs are mostly biocompatible toward most of the cancer cell lines and display only negligible cytotoxicity. Besides, the size-dependent toxicity measurement was also not reported so far for both cancerous and non-cancerous cells. However, when the GCDs are conjugated with several anticancer drug molecules, they could increase the drug stability in biological media and its concentration in the target organ, thereby increasing the pharmacological action of the drug. Therefore, more research should be conducted to establish GCDs as an efficient drug delivery vehicle.

## 5. Conclusions

Carbon dots are one of the fascinating nanomaterials with a broad range of applications. Taking environmental importance into account, GCDs are considered as promising nanomaterials since the past decade and are very crucial for a sustainable future. As we discussed in this review, due to their unique biological and physico-chemical properties, carbon dots can be used in several applications, including electro-chemical sensors, bioimaging, nanomedicine, in drug delivery and in cancer therapy. In this review, we specifically focused on GCDs and their applications. Although several reports on GCDs are already documented, the research with GQDs is still in the initial phase. However, the following points need to be addressed to further explore the sensing and anticancer applications of GCDs:

1. In order to bring GCDs into commercial applications, more study should be conducted on the search for high-quality natural precursors for their synthesis.
2. In vivo toxicity studies are limited, hence, more in vivo studies involving GCDs should be considered for the implementation of GCDs in biological/clinical applications.
3. In addition, toxicity of GCDs prepared from different natural sources under different synthesis conditions should be investigated to obtain comprehensive details on GCDs toxicity.
4. Future studies should focus on evaluating the anticancer activity of GCDs using in vivo models.
5. More research should be conducted to explore the possibility of using GCDs in photodynamic and photothermal therapy.
6. GCDs-based sensors should also be used for the sensitive and selective detection of cancer biomarkers, such as miRNA and antigens, to explore the application of GCDs in clinical cancer diagnosis.
7. It is also important to check the effect of GCDs' size and surface modifications on anticancer as well as electrochemical sensing abilities.

**Author Contributions:** Conceptualization, A.P., J.D. and S.B.; writing—original draft preparation, M.H. and P.P.; writing—review and editing, M.H., J.G., S.B. and K.G.; supervision, A.P., S.B. and J.D. All authors have read and agreed to the published version of the manuscript.

**Funding:** This research received no external funding.

**Data Availability Statement:** Not applicable.

**Acknowledgments:** M.H. would like to thank DST, INDIA for DST-INSPIRE fellowship (Award number: IF210147). A.P. would like to thank DST, INDIA for the research grant in DST-INSPIRE Faculty Scheme (Award number: IFA20-CH-333). S.B. would like to acknowledge SERB, DST, Govt. of India for Start-up Research Grant (Grant number: SRG/2022/002245). P.P. thanks SERB, DST, Govt. of India for her research fellowship.

**Conflicts of Interest:** The authors declare no conflict of interest.

## References

1. Nazri, N.A.A.; Azeman, N.H.; Luo, Y.; Bakar, A.A.A. Carbon quantum dots for optical sensor applications: A review. *Opt. Laser Technol.* **2021**, *139*, 106928. [CrossRef]
2. Wang, B.; Wang, M.; Liu, F.; Zhang, Q.; Yao, S.; Liu, X.; Huang, F. Ti$_3$C$_2$: An Ideal Co-catalyst? *Angew. Chem. Int. Ed.* **2020**, *59*, 1914–1918. [CrossRef] [PubMed]
3. Chen, Z.-L.; Wang, D.; Wang, X.-Y.; Yang, J.-H. Enhanced formaldehyde sensitivity of two-dimensional mesoporous SnO$_2$ by nitrogen-doped graphene quantum dots. *Rare Met.* **2021**, *40*, 1561–1570. [CrossRef]
4. Yan, J.; Ye, F.; Dai, Q.; Ma, X.; Fang, Z.; Dai, L.; Hu, C. Recent progress in carbon-based electrochemical catalysts: From structure design to potential applications. *Nano Res. Energy* **2023**, *2*, e9120047. [CrossRef]
5. De Oliveira, B.P.; Da Silva Abreu, F.O.M. Carbon quantum dots synthesis from waste and by-products: Perspectives and challenges. *Mater. Lett.* **2021**, *282*, 128764. [CrossRef]
6. Borna, S.; Sabzi, R.E.; Pirsa, S. Synthesis of carbon quantum dots from apple juice and graphite: Investigation of fluorescence and structural properties and use as an electrochemical sensor for measuring letrozole. *J. Mater. Sci. Mater. Electron.* **2021**, *32*, 10866–10879. [CrossRef]
7. Zhu, L.; Shen, D.; Wang, Q.; Luo, K.H. Green synthesis of tunable fluorescent carbon quantum dots from lignin and their application in anti-counterfeit printing. *ACS Appl. Mater. Interfaces* **2021**, *13*, 56465–56475. [CrossRef] [PubMed]
8. Desmond, L.J.; Phan, A.N.; Gentile, P. Critical overview on the green synthesis of carbon quantum dots and their application for cancer therapy. *Environ. Sci. Nano.* **2021**, *8*, 848. [CrossRef]
9. Huo, X.; He, Y.; Ma, S.; Jia, Y.; Yu, J.; Li, Y.; Cheng, Q. Green synthesis of carbon dots from grapefruit and its fluorescence enhancement. *J. Nanomater.* **2020**, *2020*, 8601307. [CrossRef]
10. Zhang, M.; Zhao, Y.; Cheng, J.; Liu, X.; Wang, Y.; Yan, X.; Zhang, Y.; Lu, F.; Wang, Q.; Qu, H.; et al. Novel carbon dots derived from schizonepetae herba carbonisata and investigation of their haemostatic efficacy. *Artif. Cells Nanomed. Biotechnol.* **2017**, *46*, 1562–1571. [CrossRef]
11. Dias, C.; Vasimalai, N.; Sarria, M.P.; Pinheiro, I.; Vilas-Boas, V.; Peixoto, J.; Espina, B. Biocompatibility and bioimaging potential of fruit-based carbon dots. *Nanomaterials* **2019**, *9*, 199. [CrossRef] [PubMed]
12. Malishev, R.; Arad, E.; Bhunia, S.K.; Shaham-Niv, S.; Kolusheva, S.; Gazit, E.; Jelinek, R. Chiral modulation of amyloid beta fibrillation and cytotoxicity by enantiomeric carbon dots. *Chem. Commun.* **2018**, *54*, 7762–7765. [CrossRef] [PubMed]
13. Su, W.; Wu, H.; Xu, H.; Zhang, Y.; Li, Y.; Li, X.; Fan, L. Carbon dots: A booming material for biomedical applications. *Mater. Chem. Front.* **2020**, *4*, 821–836. [CrossRef]
14. Chahal, S.; Macairan, J.-R.; Yousefi, N.; Tufenkji, N.; Naccache, R. Green synthesis of carbon dots and their applications. *RSC Adv.* **2021**, *11*, 25354–25363. [CrossRef]
15. Lin, X.; Xiong, M.; Zhang, J.; He, C.; Ma, X.; Zhang, H.; Kuang, Y.; Yang, M.; Huang, Q. Carbon dots based on natural resources: Synthesis and applications in sensors. *Microchem. J.* **2021**, *160*, 105604. [CrossRef]
16. Hassanvand, Z.; Jalali, F.; Nazari, M.; Parnianchi, F.; Santoro, C. Carbon nanodots in electrochemical sensors and biosensors: A review. *ChemElectroChem* **2021**, *8*, 15–35. [CrossRef]
17. Liu, X.; Wang, Y.; Yan, X.; Zhang, M.; Zhang, Y.; Cheng, J.; Lu, F.; Qu, H.; Wang, Q.; Zhao, Y.; et al. Novel phellodendri cortex (Huang Bo)-derived carbon dots and their hemostatic effect. *Nanomedicine* **2018**, *13*, 391–405. [CrossRef]
18. Moradi, S.; Sadrjavadi, K.; Farhadian, N.; Hosseinzadeh, L.; Shahlaei, M. Easy synthesis, characterization and cell cytotoxicity of green nano carbon dots using hydrothermal carbonization of gum tragacanth and chitosan bio-polymers for bioimaging. *J. Mol. Liq.* **2018**, *259*, 284–290. [CrossRef]
19. Vasimalai, N.; Vilas-Boas, V.; Gallo, J.; de Fátima Cerqueira, M.; Menendez-Miranda, M.; Costa-Fernandez, J.M.; Dieguez, L.; Espina, B.; Fernandez-Arguelles, M.T. Green synthesis of fluorescent carbon dots from spices for in vitro imaging and tumour cell growth inhibition. *Beilstein J. Nanotechnol.* **2018**, *9*, 530–544. [CrossRef]
20. Chan, M.-H.; Chen, B.-G.; Ngo, L.T.; Huang, W.-T.; Li, C.-H.; Liu, R.-S.; Hsiao, M. Natural carbon nanodots: Toxicity assessment and theranostic biological application. *Pharmaceutics* **2021**, *13*, 1874. [CrossRef]
21. Wang, B.; Cai, H.; Waterhouse, G.I.N.; Qu, X.; Yang, B.; Lu, S. Carbon dots in bioimaging, biosensing and therapeutics: A comprehensive review. *Small Sci.* **2022**, *2*, 2200012. [CrossRef]
22. Atchudan, R.; Edison, T.N.J.I.; Perumal, S.; Vinodh, R.; Lee, Y.R. Multicolor-emitting carbon dots from malus floribunda and their interaction with caenorhabditis elegans. *Mater. Lett.* **2020**, *261*, 127153. [CrossRef]
23. Yao, H.; Li, J.; Song, Y.; Zhao, H.; Wei, Z.; Li, X.; Jin, Y.; Yang, B.; Jiang, J. Synthesis of ginsenoside Re-based carbon dots applied for bioimaging and effective inhibition of cancer cells. *Int. J. Nanomed.* **2018**, *13*, 6249–6264. [CrossRef]
24. Tejwan, N.; Saha, S.K.; Das, J. Multifaceted applications of green carbon dots synthesized from renewable sources. *Adv. Colloid Interface Sci.* **2020**, *275*, 102046. [CrossRef] [PubMed]
25. Tejwan, N.; Saini, A.K.; Sharma, A.; Singh, T.A.; Kumar, N.; Das, J. Metal-doped and hybrid carbon dots: A comprehensive review on their synthesis and biomedical applications. *J. Control. Release* **2021**, *330*, 132–150. [CrossRef]

26. Karimian, N.; Fakhri, H.; Amidi, S.; Hajian, A.; Arduini, F.; Bagheri, H. A novel sensing layer based on metal–organic framework UiO-66 modified with $TiO_2$-graphene oxide: Application to rapid, sensitive and simultaneous determination of paraoxon and chlorpyrifos. *New J. Chem.* **2019**, *43*, 2600–2609. [CrossRef]
27. Hashemi, P.; Karimian, N.; Khoshsafar, H.; Arduini, F.; Mesri, M.; Afkhami, A.; Bagheri, H. Reduced graphene oxide decorated on Cu/CuO-Ag nanocomposite as a high-performance material for the construction of a non-enzymatic sensor: Application to the determination of carbaryl and fenamiphos pesticides. *Mat. Sci. Eng. C* **2019**, *102*, 764–772. [CrossRef] [PubMed]
28. Wang, Q.; Pang, H.; Dong, Y.; Chi, Y.; Fu, F. Colorimetric determination of glutathione by using a nanohybrid composed of manganese dioxide and carbon dots. *Microchim. Acta* **2018**, *185*, 291. [CrossRef] [PubMed]
29. Hashemi, P.; Bagheri, H.; Afkhami, A.; Amidi, S.; Madrakian, T. Graphene nanoribbon/FePt bimetallic nanoparticles/uric acid as a novel magnetic sensing layer of screen printed electrode for sensitive determination of ampyra. *Talanta* **2018**, *176*, 350–359. [CrossRef]
30. Wang, X.; Feng, Y.; Dong, P.; Huang, J. A mini review on carbon quantum dots: Preparation, properties, and electrocatalytic application. *Front. Chem.* **2019**, *7*, 671. [CrossRef]
31. Hoan, B.T.; Tam, P.D.; Pham, V.-H. Green synthesis of highly luminescent carbon quantum dots from lemon Juice. *J. Nanotechnol.* **2019**, *2019*, 2852816. [CrossRef]
32. Anwar, S.; Ding, H.; Xu, M.; Hu, X.; Li, Z.; Wang, J.; Liu, L.; Jiang, L.; Wang, D.; Dong, C.; et al. Recent advances in synthesis, optical properties, and biomedical applications of carbon dots. *ACS Appl. Bio. Mater.* **2019**, *2*, 2317–2338. [CrossRef]
33. Ahmadian-Fard-Fini, S.; Salavati-Niasari, M.; Ghanbari, D. Hydrothermal green synthesis of magnetic $Fe_3O_4$-carbon dots by lemon and grape fruit extracts and as a photoluminescence sensor for detecting of *E. coli* bacteria. *Spectrochim. Acta-A Mol. Biomol. Spetrosc.* **2018**, *203*, 481–493. [CrossRef] [PubMed]
34. Zheng, J.; Xie, Y.; Wei, Y.; Yang, Y.; Liu, X.; Chen, Y.; Xu, B. An efficient synthesis and photoelectric properties of green carbon quantum dots with high fluorescent quantum yield. *Nanomaterials* **2020**, *10*, 82. [CrossRef] [PubMed]
35. Mathew, S.; Thara, C.R.; John, N.; Mathew, B. Carbon dots from green sources as efficient sensor and as anticancer agent. *J. Photochem. Photobiol. A Chem.* **2023**, *434*, 114237. [CrossRef]
36. Asghar, K.; Qasim, M.; Das, D. One-pot green synthesis of carbon quantum dot for biological application. In *AIP Conference Proceedings*; AIP Publishing LLC: Melville, NY, USA, 2017; Volume 1832, p. 050117.
37. Zhao, P.; Zhang, Q.; Cao, J.; Qian, C.; Ye, J.; Xu, S.; Zhang, Y.; Li, Y. Facile and green synthesis of highly fluorescent carbon quantum dots from water hyacinth for the detection of ferric iron and cellular imaging. *Nanomaterials* **2022**, *12*, 1528. [CrossRef] [PubMed]
38. Visheratina, A.; Hesami, L.; Wilson, A.K.; Baalbaki, N.; Noginova, N.; Noginov, M.A.; Kotov, N.A. Hydrothermal synthesis of chiral carbon dots. *Chirality* **2022**, *34*, 1503–1514. [CrossRef]
39. Yen, Y.-C.; Lin, C.-C.; Chen, P.-Y.; Ko, W.-Y.; Tien, T.-R.; Lin, K.-J. Green synthesis of carbon quantum dots embedded onto titanium dioxide nanowires for enhancing photocurrent. *R. Soc. Open Sci.* **2017**, *4*, 161051. [CrossRef] [PubMed]
40. Zhou, X.; Qu, Q.; Wang, L.; Li, L.; Li, S.; Xia, K. Nitrogen dozen carbon quantum dots as one dual function sensing platform for electrochemical and fluorescent detecting ascorbic acid. *J. Nanopart. Res.* **2020**, *22*, 20. [CrossRef]
41. Ran, X.; Qu, Q.; Qian, X.; Xie, W.; Li, S.; Li, L.; Yang, L. Water-soluble pillar [6]arene functionalized nitrogen-doped carbon quantum dots with excellent supramolecular recognition capability and superior electrochemical sensing performance towards TNT. *Sens. Actuators B Chem.* **2018**, *257*, 362–371. [CrossRef]
42. Ensafi, A.A.; Sefat, S.H.; Kazemifard, N.; Rezaei, B.; Moradi, F. A novel one-step and green synthesis of highly fluorescent carbon dots from saffron for cell imaging and sensing of prilocaine. *Sens. Actuators B Chem.* **2017**, *253*, 451–460. [CrossRef]
43. Singh, A.K.; Singh, V.K.; Singh, M.; Singh, P.; Khadim, S.R.; Singh, U.; Koch, B.; Hasan, S.H.; Asthana, R.K. One pot hydrothermal synthesis of fluorescent NP-carbon dots derived from dunaliella salina biomass and its application in on-off sensing of Hg (II), Cr (VI) and live cell imaging. *J. Photochem. Photobiol. A Chem.* **2019**, *376*, 63–72. [CrossRef]
44. Wang, X.; Yang, P.; Feng, Q.; Meng, T.; Wei, J.; Xu, C.; Han, J. Green preparation of fluorescent carbon quantum dots from cyanobacteria for biological imaging. *Polymers* **2019**, *11*, 616. [CrossRef] [PubMed]
45. Atchudan, R.; Edison, T.N.J.I.; Shanmugam, M.; Perumal, S.; Somanathan, T.; Lee, Y.R. Sustainable synthesis of carbon quantum dots from banana peel waste using hydrothermal process for in vivo bioimaging. *Phys. E Low-Dimens. Syst. Nanostruct.* **2020**, *126*, 114417. [CrossRef]
46. Liu, L.; Zhang, S.; Zheng, X.; Li, H.; Chen, Q.; Qin, K.; Ding, Y.; Wei, Y. Carbon dots derived from *fusobacterium nucleatum* for intracellular determination of $Fe^{3+}$ and bioimaging both in vitro and in vivo. *Anal. Methods* **2021**, *13*, 1121–1131. [CrossRef] [PubMed]
47. Raveendran, V.; Kizhakayil, R.N. Fluorescent carbon dots as biosensor, green reductant, and biomarker. *ACS Omega* **2021**, *6*, 23475–23484. [CrossRef] [PubMed]
48. Arvapalli, D.M.; Sheardy, A.T.; Allado, K.; Chevva, H.; Yin, Z.; Wei, J. Design of curcumin loaded carbon nanodots delivery system: Enhanced bioavailability, release kinetics, and anticancer activity. *ACS Appl. Bio Mater.* **2020**, *3*, 8776–8785. [CrossRef] [PubMed]

49. Chatzimitakos, T.; Kasouni, A.; Sygellou, L.; Avgeropoulos, A.; Troganis, A.; Stalikas, C. Two of a kind but different: Luminescent carbon quantum dots from citrus peels for iron and tartrazine sensing and cell imaging. *Talanta* **2017**, *175*, 305–312. [CrossRef]
50. Hu, Y.; Zhang, L.; Li, X.; Liu, R.; Lin, L.; Zhao, S. Green preparation of S and N co-doped carbon dots from water chestnut and onion as well as their use as an off-on fluorescent probe for the quantification and imaging of coenzyme A. *ACS Sustain. Chem. Eng.* **2017**, *5*, 4992–5000. [CrossRef]
51. Li, L.; Wang, X.; Fu, Z.; Cui, F. One-step hydrothermal synthesis of nitrogen- and sulfur-co-doped carbon dots from ginkgo leaves and application in biology. *Mater. Lett.* **2017**, *196*, 300–303. [CrossRef]
52. Li, L.; Zhang, R.; Lu, C.; Sun, J.; Wang, L.; Qu, B.; Li, T.; Liu, Y.; Li, S. In situ synthesis of NIR-light emitting carbon dots derived from spinach for bio-imaging applications. *J. Mater. Chem. B* **2017**, *5*, 7328–7334. [CrossRef] [PubMed]
53. Zhang, M.; Chi, C.; Yuan, P.; Su, Y.; Shao, M.; Zhou, N. A hydrothermal route to multicolor luminescent carbon dots from adenosine disodium triphosphate for bioimaging. *Mater. Sci. Eng. C* **2017**, *76*, 1146–1153. [CrossRef] [PubMed]
54. Liu, X.; Liu, J.; Zheng, B.; Yan, L.; Dai, J.; Zhuang, Z.; Du, J.; Guo, Y.; Xiao, D. N-doped carbon dots: Green and efficient synthesis on a large-scale and their application in fluorescent pH sensing. *New J. Chem.* **2017**, *41*, 10607–10612. [CrossRef]
55. Liu, X.; Yang, C.; Zheng, B.; Dai, J.; Yan, L.; Zhuang, Z.; Du, J.; Guo, Y.; Xiao, D. Green anhydrous synthesis of hydrophilic carbon dots on large-scale and their application for broad fluorescent pH sensing. *Sens. Actuators B Chem.* **2017**, *255*, 572–579. [CrossRef]
56. Amin, N.; Afkhami, A.; Hosseinzadeh, L.; Madrakian, T. Green and cost-effective synthesis of carbon dots from date kernel and their application as a novel switchable fluorescence probe for sensitive assay of zoledronic acid drug in human serum and cellular imaging. *Anal. Chim. Acta* **2018**, *1030*, 183–193. [CrossRef] [PubMed]
57. Ramezani, Z.; Qorbanpour, M.; Rabhar, N. Green synthesis of carbon quantum dots using quince fruit (*Cydonia oblonga*) powder as carbon precursor: Application in cell imaging and $As^{3+}$ determination. *Colloids Surf. A Physicochem. Eng. Asp.* **2018**, *549*, 58–66. [CrossRef]
58. Ma, H.; Sun, C.; Xue, G.; Wu, G.; Zhang, X.; Han, X.; Qi, X.; Lv, X.; Sun, H.; Zhang, J.; et al. Facile synthesis of fluorescent carbon dots from *prunus cerasifera* fruits for fluorescent ink, $Fe^{3+}$ ion detection and cell imaging. *Spectrochim. Acta-A Mol. Biomol. Spetrosc.* **2019**, *213*, 281–287. [CrossRef]
59. Qu, Y.; Yu, L.; Zhu, B.; Chai, F.; Su, Z. Green synthesis of carbon dots by celery leaves for use as fluorescent paper sensors for the detection of nitrophenols. *New J. Chem.* **2019**, *44*, 1500–1507. [CrossRef]
60. Sahoo, N.K.; Jana, G.C.; Aktara, M.N.; Das, S.; Nayim, S.; Patra, A.; Bhattacharjee, P.; Bhadra, K.; Hossain, M. Carbon dots derived from lychee waste: Application for $Fe^{3+}$ ions sensing in real water and multicolor cell imaging of skin melanoma cells. *Mater. Sci. Eng. C* **2019**, *108*, 110429. [CrossRef]
61. Tadesse, A.; Hagos, M.; RamaDevi, D.; Basavaiah, K.; Belachew, N. Fluorescent-nitrogen-doped carbon quantum dots derived from citrus lemon juice: Green synthesis, mercury(II) ion sensing, and live cell imaging. *ACS Omega* **2020**, *5*, 3889–3898. [CrossRef]
62. Wang, H.; Xie, Y.; Na, X.; Bi, J.; Liu, S.; Zhang, L.; Tan, M. Fluorescent carbon dots in baked lamb: Formation, cytotoxicity and scavenging capability to free radicals. *Food Chem.* **2019**, *286*, 405–412. [CrossRef]
63. Wang, M.; Wan, Y.; Zhang, K.; Fu, Q.; Wang, L.; Zeng, J.; Xia, Z.; Gao, D. Green synthesis of carbon dots using the flowers of *osmanthus fragrans* (Thunb.) lour. as precursors: Application in $Fe^{3+}$ and ascorbic acid determination and cell imaging. *Anal. Bioanal. Chem.* **2019**, *411*, 2715–2727. [CrossRef]
64. Liu, S.; Liu, Z.; Li, Q.; Xia, H.; Yang, W.; Wang, R.; Li, Y.; Zhao, H.; Tian, B. Facile synthesis of carbon dots from wheat straw for colorimetric and fluorescent detection of fluoride and cellular imaging. *Spectrochim. Acta-A Mol. Biomol. Spetrosc.* **2020**, *246*, 118964. [CrossRef] [PubMed]
65. Li, C.; Sun, X.; Li, Y.; Liu, H.; Long, B.; Xie, D.; Chen, J.; Wang, K. Rapid and green fabrication of carbon dots for cellular imaging and anti-counterfeiting applications. *ACS Omega* **2021**, *6*, 3232–3237. [CrossRef] [PubMed]
66. Liu, Y.-Y.; Yu, N.-Y.; Fang, W.-D.; Tan, Q.-G.; Ji, R.; Yang, L.-Y.; Wei, S.; Zhang, X.-W.; Miao, A.-J. Photodegradation of carbon dots cause cytotoxicity. *Nat. Commun.* **2021**, *12*, 812. [CrossRef]
67. Arul, V.; Chandrasekaran, P.; Sivaraman, G.; Sethuraman, M.G. Efficient green synthesis of N,B co-doped bright fluorescent carbon nanodots and their electrocatalytic and bio-imaging applications. *Diam. Relat. Mater.* **2021**, *116*, 108437. [CrossRef]
68. Choppadandi, M.; Guduru, A.T.; Gondaliya, P.; Arya, N.; Kalia, K.; Kumar, H.; Kapusetti, G. Structural features regulated photoluminescence intensity and cell internalization of carbon and graphene quantum dots for bioimaging. *Mater. Sci. Eng. C* **2021**, *129*, 112366. [CrossRef]
69. Emami, E.; Mousazadeh, M.H. pH-responsive zwitterionic carbon dots for detection of rituximab antibody. *Luminescence* **2021**, *36*, 1198–1208. [CrossRef]
70. Li, Z.; Wang, Q.; Zhou, Z.; Zhao, S.; Zhong, S.; Xu, L.; Gao, Y.; Cui, X. Green synthesis of carbon quantum dots from corn stalk shell by hydrothermal approach in near-critical water and applications in detecting and bioimaging. *Microchem. J.* **2021**, *166*, 106250. [CrossRef]
71. Paul, A.; Kurian, M. Facile synthesis of nitrogen doped carbon dots from waste biomass: Potential optical and biomedical applications. *Clean. Eng. Technol.* **2021**, *3*, 100103. [CrossRef]
72. Rezaei, A.; Hashemi, E. A pseudohomogeneous nanocarrier based on carbon quantum dots decorated with arginine as an efficient gene delivery vehicle. *Sci. Rep.* **2021**, *11*, 13790. [CrossRef] [PubMed]

73. Tejwan, N.; Sadhukhan, P.; Sharma, A.; Singh, T.A.; Hatimutia, M.; Pabbathi, A.; Das, J.; Sil, P.C. pH-responsive and targeted delivery of rutin for breast cancer therapy via folic acid-functionalized carbon dots. *Diam. Relat. Mater.* **2022**, *129*, 109346. [CrossRef]
74. He, Z.; Cheng, J.; Yan, W.; Long, W.; Ouyang, H.; Hu, X.; Liu, M.; Zhou, N.; Zhang, X.; Wei, Y.; et al. One-step preparation of green tea ash derived and polymer functionalized carbon quantum dots via the thiol-ene click chemistry. *Inorg. Chem. Commun.* **2021**, *130*, 108743. [CrossRef]
75. Arkan, E.; Barati, A.; Rahmanpanah, M.; Hosseinzadeh, L.; Moradi, S.; Hajialyani, M. Green synthesis of carbon dots derived from walnut oil and an investigation of their cytotoxic and apoptogenic activities toward cancer cell. *Adv. Pharm. Bull.* **2018**, *8*, 149–155. [CrossRef] [PubMed]

**Disclaimer/Publisher's Note:** The statements, opinions and data contained in all publications are solely those of the individual author(s) and contributor(s) and not of MDPI and/or the editor(s). MDPI and/or the editor(s) disclaim responsibility for any injury to people or property resulting from any ideas, methods, instructions or products referred to in the content.

MDPI
St. Alban-Anlage 66
4052 Basel
Switzerland
www.mdpi.com

*Catalysts* Editorial Office
E-mail: catalysts@mdpi.com
www.mdpi.com/journal/catalysts

Disclaimer/Publisher's Note: The statements, opinions and data contained in all publications are solely those of the individual author(s) and contributor(s) and not of MDPI and/or the editor(s). MDPI and/or the editor(s) disclaim responsibility for any injury to people or property resulting from any ideas, methods, instructions or products referred to in the content.

www.ingramcontent.com/pod-product-compliance
Lightning Source LLC
LaVergne TN
LVHW070429100526
838202LV00014B/1557